To Olin Yoder
From Bee Scheffer
— In memory of
Bob —

4/97

THE NATURE OF DISEASE IN PLANTS

This book is about how disease develops in plants, from the origins and evolution of parasites to how the great plant epidemics developed. The basic premise of the book is that the conditions favoring disease are inherent in agriculture and that diseases became destructive because of human activities. It also deals with how people have dealt with plant diseases in history – from demons to DNA.

Included in the book are the natural histories of some of the most damaging plant diseases, worldwide, with discussions of why each became destructive. Diseases are grouped according to the most significant factors in the development of epidemics; in every case, this is due to a human factor. Discussion of each model disease proceeds from observable facts to more complex concepts; thus, the reader with little knowledge of plant pathology should have no trouble with the text. Special terminology and jargon are avoided as much as possible (a glossary of special terms is included).

This is not an encyclopedia of plant disease; instead, the purpose is to give a broad general understanding, especially of the effects of human activities on plant disease. The book will have a broad audience of students and professionals in biology and agriculture.

THE NATURE OF DISEASE
IN PLANTS

ROBERT P. SCHEFFER

Michigan State University

CAMBRIDGE
UNIVERSITY PRESS

PUBLISHED BY THE PRESS SYNDICATE OF THE UNIVERSITY OF CAMBRIDGE
The Pitt Building, Trumpington Street, Cambridge CB2 1RP, United Kingdom

CAMBRIDGE UNIVERSITY PRESS
The Edinburgh Building, Cambridge CB2 2RU, United Kingdom
40 West 20th Street, New York, NY 10011–4211, USA
10 Stamford Road, Oakleigh, Melbourne 3166, Australia

First published 1997

Printed in the United States of America

Typeset in Times Roman

Library of Congress Cataloging-in-Publication Data
Scheffer, Robert P.
The nature of disease in plants / Robert P. Scheffer.
p.   cm.
Includes bibliographical references (p.   ) and index.
ISBN 0-521-48247-X (hardcover)
1. Plant diseases.   I. Title.
SB601.S28   1997
632′.3 – dc20   96-5231
                                                            CIP

*A catalogue record for this book is available from
the British Library.*

ISBN 0 521 48247 X

Contents

To the memory of
John Charles Walker and Armin C. Braun,
two giants of plant science who shaped my thinking,
and to my students who are reaching beyond

Acknowledgments

I thank the following people for critical review of certain chapters and, in some cases, for providing illustrations: Gerald Adams, Dennis Fulbright, John Hart, Patrick Hart, Alan Jones, John Lockwood, Donald Ramsdell, and Jonathan Walton. I also thank Marlene Cameron for drawings (including maps), Kurt Stepnitz for photographic services, and Jonathan Walton for preparing chemical structure figures. I am thankful to Martha Casaday for typing and other help in preparation of the manuscript. Finally, thanks to others in the Department of Botany and Plant Pathology, Michigan State University, for friendly interest and useful suggestions.

Publisher's note

Sadly, Professor Scheffer died shortly after the typescript for this book was delivered to us. We would like to thank Professor Dennis Fulbright for kindly dealing with the author's proofs and for his help in bringing this book to print.

1

Perspective

Archaeologists generally agree that agriculture began 8,000 to 10,000 years ago. Agriculture is still the basic industry, as important today as at the dawn of recorded history.

An appreciation of the origins and development of agriculture is needed for a full understanding of plant diseases. Historical and ecological viewpoints lead to an awareness of several basic premises: (1) The natural ecology is disrupted by agriculture; (2) conditions that favor plant diseases are inherent in agriculture; (3) when agriculture became more complex, pathogens followed suit; (4) diseases became destructive because of human activities. These theses and supporting evidences are core items in this book. An evolutionary point of view is taken.

Agriculture was created in prehistoric times with domestication of plants and animals. Domestication provided conditions that led to destructive epidemics, an integrating theme of this book. Most ecologists accept that disease in undisturbed plant populations has ebbs and flows, sometimes damaging but locally restricted. Natural controls of disease were removed by agriculture, a dynamic enterprise. As human populations increased and agriculture expanded, disease problems multiplied and became greater threats.

The ecology of disease in plants has many parallels with that in human populations. When people became urbanized, with locally high population densities, the stage was set for deadly epidemics, and they came. About 5,000 years ago, centers of urbanization appeared in Asia, Africa, and Europe; many new diseases and great epidemics soon appeared in these centers. In time, urban populations became adapted by elimination of the most susceptible individuals and by acquired resistance. When urban diseases were taken to isolated or essentially rural indigenous populations, such as those in the Americas, the results were devastating. Transporting plants and pathogens had comparable consequences; monoculture of plants is equiva-

1

lent to urbanization of people. Plant disease epidemics became common. Epidemics were the costs of crowding, for both animals and plants.

Plant diseases were vexing problems from the earliest days of agriculture. People in neolithic villages (6000–2000 B.C.) apparently attributed disease and other natural phenomena to powerful spirits that had to be appeased by various rituals, traditions, and symbols. There were spirits for sacred trees, deities for earth and crops, and gods for sun and moon; such were conceived from the desire to prevent famine and to increase crop yields. Plant health problems are documented in the earliest writings, beginning about 1700 B.C. in the Euphrates valley and included in treatises from Egypt, Palestine, China, and India. The Bible has many references to blasts (blights), and mildews of plants, and especially to insect pests; the Talmud, a postbiblical commentary, has many more. Trial and error gradually led from folklore to the use of chemicals and other management practices for control of plant disease (Orlob 1973).

Intellectuals of ancient Greece developed more rational explanations of disease, as indicated in an early treatise, "De Plantis." Aristotle was aware of plant disease, as was his renowned student Theophrastus, an early recorder who attempted to replace superstition with rationality. Theophrastus wrote "Historia Plantarum" and "De Causis Plantorum," texts that include discussion of disease. However, the ancient Greeks were not practical minded; they were more interested in abstract deduction of ultimate causes than in experimentation.

Romans, although good farmers, added little to knowledge of plant disease. Instead, they continued the tradition of appeasement of the spirits, as exemplified by deference to Robigus, the Roman god associated with wheat rust. Various other gods protected crops; Minerva, for example, was beseeched to protect the olive (Orlob 1973).

Nonintellectual traditions prevailed through the Middle Ages of Europe. In regard to plants, intellectual stirrings began in the eleventh and twelfth centuries, with the writings of Hildegard von Bingen, a Benedictine abbess, and Albertus Magnus, a Dominican bishop; however, cosmic origins of disease still predominated. Other notable European observers and writers on plants during the Middle Ages included Petrus Cresentius and Gottfried von Franken. The Arabs at the time were more advanced, as is evident in the writings of Ibn al Awan, a Spanish Muslim. In the Americas, some of the plant knowledge of the ancient Aztecs, Incas, and Mayas was preserved in "picture writing" and in recorded observations of the Spanish invaders (Orlob 1971).

A sound scientific basis for plant pathology was laid in Europe by the early nineteenth century. New knowledge was based on prior development of the microscope, printing, European herbals, and formation of intellectual societies. Experimental studies were spurred by the great Irish famine of 1846. An important early figure was Julius Kuhn of Germany, who wrote the first textbook on plant diseases. But a special place must be given to Anton deBary, a German biologist working at the University of Strasbourg. The works of deBary and others will be mentioned in later sections of this book. The history of plant pathology in the nineteenth and early twentieth centuries is discussed elsewhere (Large 1940; Walker 1959, 1969; Ainsworth 1981).

Several plant diseases have changed the course of human history. Two major examples are the late blight of potato that caused the Irish famine of 1846–50 and the great famine of Bengal in 1943; each led to the death of more than a million people. Chestnut blight, an ecological disaster of great magnitude early in the twentieth century, caused economic hardship and lowered the quality of life by decimating a major tree in the forests of eastern America and Europe. Likewise, Dutch elm disease, another ecological disaster, destroyed millions of valuable trees in America and Europe. A blight disease of maize caused at least a billion-dollar loss in the United States in 1970. Many other plant diseases affect our daily lives and well-being, yet few people are aware of this, and even fewer have an understanding of the situation. The social and economic impacts of plant diseases are told very well in several older books for general readers (Large 1940; Christensen 1951; Carefoot and Sprott 1967) and in a recent book for college-level use (Schumann 1991).

The development of this book was influenced by the notion that integration of academic fields is helpful; specifically, that more exchange between plant pathology, molecular biology, and ecology would be of benefit to each. Specialization is necessary, but it tends to mold our thinking. Molecular biology has the potential to contribute, in a revolutionary way, to an understanding of plant disease. On the other hand, molecular biologists working with plant diseases need reliable background information and guidance from plant disease specialists. Ecology supplies integrating concepts.

With these ideas in mind, this book was constructed with two parts. Part I is general background, designed both as an aid for the nonspecialist and as a brief but critical summary for the specialist. Some sections in Part I were included to give a broad ecological view of plant diseases. Part II includes natural histories (origins, evolution, and analyses) of specific diseases, each representative of a group.

Human activities were involved in the development of every serious epidemic in plants. Foreign pathogens were brought to virgin areas, where the aliens encountered plant populations with no tolerance (Chapter 7). Foreign plants were brought into new areas, where they were exposed to virulent pathogens that previously existed in tolerant endemic plants (Chapter 8). Some crop plants were moved to new areas without their endemic pathogens; over time, cultivation and selection in the absence of such pathogens created a uniformly susceptible crop. Eventually, pathogens overtook their hosts (Chapter 9), and the results were devastating. Changes in agriculture and forest management created new epidemic situations (Chapter 14). Annual reintroduction of some diseases, by human activities, to an area has had serious consequences (Chapter 15). Some abiotic diseases resulted from direct effects of human activities (Chapter 16).

Monoculture has had great impacts on disease in crops. Spread of plant pathogens was favored when agriculture adopted monoculture; the trend was accelerated with modern use of crops with genetic uniformity (Chapters 10, 11, 12, and 13). Pathogens are adaptable, and we have, in effect, bred pathogens as we developed agriculture (Chapters 11, 12, and 13). Monoculture undoubtedly is a factor in many serious plant disease problems.

Major discussions in Part II are limited to disease examples that are easily classified in the several categories. They include many of our very serious disease problems. Many other plant diseases lack sufficient documentation for easy classification or have complex causal circumstances. However, the disease categories are inclusive for plant diseases in general. Several representative diseases in each category are discussed because each differs from the others in natural history. Some diseases could be placed in several categories, as indicated in the discussions; in those cases, each is classified according to the factor that is perceived to be essential, without which there would be no epidemic or serious damage.

Every specialized field of knowledge has its jargon and terminology, and plant pathology is no exception. A blizzard of unfamiliar terms confronts anyone new to this specialized area. I have attempted to avoid jargon as much as possible, for the benefit of a wider audience. However, some special terms are necessary for precision and brevity. These are defined on first encounter, and special terms used more than once are listed in a brief glossary.

First, what is disease? A precise definition is difficult, although plant diseases are easily recognized as such. Simply stated, disease is any condition that interferes with normal functions or structures, caused by a constant irritation. The latter qualifier is to exclude mechanical injuries and the effects of plant-eating insects, which generally are left to entomologists. The definition begs the question of normal, but this usually is not a practical difficulty. Disease can be infectious (caused by microbes) or noninfectious (usually caused by factors such as nutrient-element deficiencies). The term *disease* often is confused with *pathogen:* The former is a condition of the host; the latter is the causal agent. Epidemiology requires use of several corollary terms; there are *epidemic, endemic,* and *pandemic diseases.* An epidemic disease refers to one with a widespread increase in its incidence. A pandemic disease is an epidemic that extends over several seasons or years and involves large geographical areas. An endemic disease is always found, either mildly or severely, in a defined area; *endemic* also is used in reference to a place of origin.

A final reminder: This book is not an encyclopedia of plant diseases, although a number of important examples are discussed. This book does not follow traditional patterns of many plant pathology books, with diseases classified by causal agents and treated as production problems for agriculture and forestry. Instead, this book is a natural history of some representative and damaging diseases. The text includes discussion of origins, analyses of how each disease became damaging, and discussions of anthropogenic factors involved. Finally, modern concepts of ecology and molecular biology are incorporated.

This book is for people with a serious interest in plants. The most obvious are biologists and agricultural specialists: students, researchers, teachers, and practitioners. Text discussion proceeds from observable phenomena to more complex scientific concepts; thus, the nonspecialist reader should have no trouble with much of the discussion. This book contains background information that should be useful for molecular biologists who work with plant disease problems.

.

Part I

Biology and Control of Plant Diseases

The purpose of Part I is to provide a brief, analytical discussion of the basic biology of plant disease. Included, among other topics, are discussions of our present understanding of how pathogens attack plants, how plants resist infection, and the origins, ecology, and evolution of pathogens and disease. Many readers may find that Part II is understandable without the information provided in Part I. However, the background provided in Part I should be especially helpful for the nonspecialist to better understand and appreciate the ideas and themes that are developed later.

2
Causes and Spread
of Plant Disease

Disease in plants is caused by the same types of agents that cause disease in animals; for both, there are biotic and abiotic causes. Biotic causes include fungi, viruses, bacteria, and nematodes, listed in order of their frequency of occurrence on plants. Parasitic angiosperms (for example, mistletoe) and algae also infect plants; an alga is a common cause of leaf necrosis of many plant species in the tropics (Joubert and Rijkenburg 1971). Abiotic causes of disease in plants include extremes of temperature, water, oxygen, and soil pH, plus nutrient-element deficiencies and inbalances, excess heavy metals, and air pollution.

More than two hundred species of bacteria infect plants. Most of these species are in the following genera: *Agrobacterium, Clavibacter, Erwinia, Pseudomonas, Streptomyces,* and *Xanthomonas.* Of these, only *Clavibacter,* cause of some wilt and leafspot diseases, and *Streptomyces,* cause of scab diseases, are gram positive (retain color when exposed to Gram's stain). Almost half of bacterial plant pathogens are in the genus *Pseudomonas,* with species that cause rots, wilts, blights, and cankers. *Xanthomonas* species cause some serious leaf blights and leafspots. *Erwinia* species cause blights, wilts, and soft rots. *Agrobacterium* species are best known as the cause of crown gall (Chapter 3, section 3) in many plant species. A species of *Streptomyces* causes potato scab. Phytoplasmas (mycoplasmalike organisms that lack firm cell walls) cause some important plant diseases (for example, aster yellows) (Maramorosch and Raychauduri 1988). There also are fastidious vascular bacteria that are difficult to culture (for example, the cause of Pierce's disease of grape; see Chapter 8, section 1). Characteristics of plant pathogenic bacteria and the diseases they cause are described elsewhere (Sigee 1993). Several serious bacterial diseases are discussed in later chapters.

Viruses are second only to fungi in frequency of occurrence as plant disease agents; most species of higher plants are hosts to one or more viruses.

9

The virus is a submicroscopic entity that consists of a protein coat (capsid) enclosing a nucleic acid (RNA or DNA) core. Morphologically, viruses are either rod-shaped, bacilliform, or polyhedral structures, with either a single- or double-stranded nucleic acid core. Some have well-defined membranes. Infectious nucleic acids that lack protein are known as viroids, the cause of several plant diseases including the infamous cadang-cadang (sick-sick) disease of coconut trees (Semancik et al. 1987). There are excellent sources of information on plant viruses and on virus diseases of plants (Matthews 1992). Viruses involved in major epidemics will be discussed in later chapters.

Tobacco mosaic virus was the first virus to be isolated and characterized, in 1936. This accomplishment won a Nobel prize for W. M. Stanley in 1946; however, he did not recognize the nucleoprotein nature of the particle. This oversight was clarified by N. W. Pirie and F. C. Bawden of Britain. Many other advances in virology that later were applied to animal viruses were first discovered in work with plant viruses as models (Kelman 1995).

Some species of nematodes (class Nematoda, phylum Nemathelminthes) cause important plant disease; examples are root knot, root lesion, and the pine wilt disease. Nematodes are very small, eel-shaped, unsegmented animals that can be endoparasitic (live within host tissues) or ectoparasitic (live on the surface, feeding by stylets that pierce host cells). Some nematodes transmit so-called soil-borne virus diseases; others increase the chances of root invasion by soil fungi and bacteria. There are nematode predators on plant-infecting forms (Dropkin 1989).

Some seed plants are parasitic and cause disease. Included are the mistletoes (Loranthaceae: genera *Arceuthobrium, Viscum,* and *Phoradendron*), (Kuijt 1969; Calder and Bernhardt 1983); dodder (*Cuscuta* spp.); broomrape (Orobanchaceae); witchweed (*Stiga* spp.); and members of the families Scrophulariaceae and Santalaceae (Musselman 1980). Parasitic angiosperms obtain sustenance via haustoria that penetrate host tissue (Kuijt 1977). Dwarf mistletoe causes a serious disease of conifers in western North America (Hawksworth and Wiens 1972). There is one report of a conifer parasitic on other conifers (DeLaubenfels 1959).

Fungi are the most common parasites of plants. All higher plant species are susceptible to one or more fungal diseases, and most crop plants have many. There are many taxonomic groups of fungi, and all major groups contain plant pathogens. Their taxonomy, based on morphology, is complex and often changed. No attempt will be made here to cover the whole system; however, a few essential ideas will be helpful in later discussions. There are excellent modern sources for further information on fungi (Webster 1974; Alexopoulos, Mims, and Blackwell 1996).

Fungi are classified in two kingdoms and several phyla. Primitive fungi are in the kingdom Protocista, as are algae; these have flagellated swimming spores. The Protocista contains three fungal phyla, but only one (Oomycota, order Peronosporales) contains appreciable numbers of pathogens on higher plants. Important diseases caused by fungi in the order Peronosporales include late blight of potato (Chapter 9, section 1) and the downy mildews of grape and tobacco (Chapter 15, section 2).

True fungi (kingdom Fungi) are classified in two phyla: Zygomycota and Dikaryomycota. Zygomycota (conjugative fungi) are further divided into several orders; only two orders have members with direct effects on higher plants: Mucorales (for example, *Rhizopus nigricans*) and Endogonales (vesicular-arbuscular mycorrhizae). Plant pathologists are more familiar with the Dikaryomycota, which has two subphyla: Ascomycotina (= Ascomycetes) and Basidiomycota (= Basidiomycetes). Each of these two subphyla has a vast number of plant pathogens. The Ascomycetes produce sexual spores in an ascus (a microsac). Basidiomycetes form sexual spores on a basidium, a clublike structure; included in this group are rusts, smuts, and many of the mushrooms, including some that are pathogenic. Sexual stages of many fungal species have never been found; these are classified as imperfects or Fungi Imperfecti. From time to time, perfect, or sexual, stages are found; usually they are Ascomycetes. Pathogens in these different groups will be identified and discussed further in chapters on specific diseases.

The rust fungi are complex and very important as plant pathogens. A basic knowledge of their life cycles is essential for understanding wheat rust, white-pine blister rust, coffee rust, and others that will be discussed later.

Complete, or long-cycle (macrocyclic), rusts have several spore stages; that is, they are pleomorphic. Some species require an alternate host for a complete life cycle; the term for this is *heteroecious*. Wheat rust is an example of a macrocyclic rust that is heteroecious (Figure 2.1; see Chapter 11, section 1). The teliospore, found in all rust fungi, is produced by the fungus on wheat as the crop matures. This spore has a thick wall and is resistant to desiccation; it can remain dormant for some time. The teliospore germinates under the proper conditions, forming a basidium on which basidiospores (usually four) are produced. Basidiospores, which are carried by air currents, are capable of infecting barberry (*Berberis vulgare*) but not wheat. The fungus on infected barberry produces pycniospores, which act as spermatia in a sexual process. After sexual union, the mycelium forms an ae-

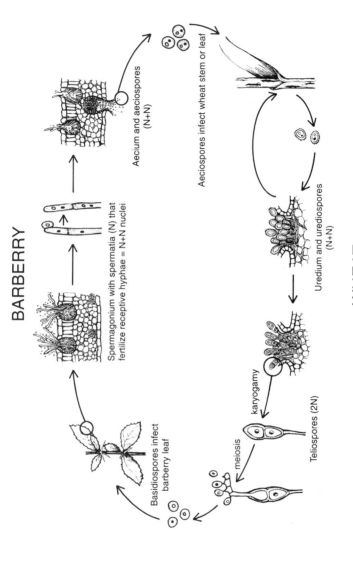

BARBERRY

Spermagonium with spermatia (N) that
fertilize receptive hyphae = N+N nuclei

Aecium and aeciospores
(N+N)

Aeciospores infect wheat stem or leaf

Basidiospores infect
barberry leaf

meiosis

karyogamy

Teliospores (2N)

Uredium and urediospores
(N+N)

WHEAT

Figure 2.1. Disease cycle of black stem rust of wheat (*Puccinia graminis* f. sp.
tritici). (Drawing by Marlene Cameron)

cium (spore-bearing pustule, or sorus) with aeciospores that are carried by wind, infecting wheat but not barberry. The fungus in infected wheat then forms sori with urediospores that cause further infections on wheat. Finally, the black teliospores are formed, completing the cycle.

Other rust species differ in their life cycles; some lack one or more spore stages. Short-cycle, or microcyclic, species lack pycnia, and some have only one type of spore, the microtelium. Some rusts are macrocyclic and auto-ecious; that is, they have all spore stages on one host (for example, asparagus rust). The rust genera *Puccinia, Gymnosporangium, Melampsora,* and *Cronartium* will be discussed in later chapters (Chapter 7, section 3; Chapter 9, section 2; Chapter 11, section 1; and Chapter 14, section 3).

The spores of many fungal pathogens are well adapted for dissemination by wind, even over great distances; examples are discussed in later chapters on wheat rust, downy mildews (Chapter 15, section 2), coffee rust (Chapter 9, section 2), and potato late blight (Chapter 9, section 1). Dry spores produced on diseased plants are carried locally by air currents; some are lifted by updrafts and carried for many kilometers, even over ocean barriers, by prevailing winds. Other fungi exude sticky masses of spores that are adapted to dissemination by splashing raindrops and, somewhat farther, by driving rain; anthracnose of maize is an example (Chapter 14, section 4). Spores of some fungi adhere to the bodies of insect vectors and are carried to noninfected host plants, as is the case with the Dutch elm disease fungus (Chapter 7, section 2). Pathogenic soil fungi can be carried to new areas with tools, drainage water, and infected planting stocks, as occurs with Fusarium wilt of banana (Chapter 14, section 2). Other fungal diseases are carried with infected seed from diseased plants. Root grafts to adjacent trees allow the spread of systemic infections, as with Dutch elm disease.

Many plant pathogenic bacteria are disseminated by rain splash and driving rain. Some species are carried by insects (Purcell 1982); for example, bees can carry the fire blight bacterium that affects pome fruit (Chapter 8, section 2), and leafhoppers can carry the phytoplasma that causes Pierce's disease of grape (Chapter 8, section 1). Some bacterial species are disseminated by contaminated tools, particularly pruning tools, and by drainage water.

Many viruses are spread by piercing, sucking insects that take in the virus while feeding on a diseased plant, then inject the virus, with their digestive juices, into healthy plants. In some cases, the virus multiplies in the insect vector, which then becomes a persistent vector, able to transmit the virus for the rest of the insect's life. Aphids and leafhoppers are the most common insect vectors, but thrips, mealybugs, beetles, whiteflies, and others are vectors of some virus species. Other vectors include mites, nematodes, dodder, and fungi (Grogan and Campbell 1966; Costa 1974; Harris and Maramorosch 1980; Harris 1981; Fulton, Gergerich, and Scott 1987). Some viruses, such as tobacco mosaic virus, are spread mechanically by tools and workers' hands or on the bodies of insects. A few viruses and other systemic diseases can spread by root grafts to adjacent trees. Pollen transmission is known in a few cases.

3

How Pathogens Attack Plants

3.1. Toxins

One obvious means of attack is by secretion of toxic compounds by the pathogen. Toxins, which are now known to be involved in a number of plant diseases, are perhaps the best known of mediating mechanisms (Gross 1991). Three essential criteria must be met before a compound is recognized as a toxin in plant pathology: (1) It is a product of a microbial pathogen; (2) it causes obvious damage to plant cells and tissues; and (3) the compound is known with confidence to be involved in disease development (Scheffer 1983). The last feature is included because bacteria and fungi in culture produce many compounds that are toxic to plant cells but have no known roles in the diseased plant. Most of the plant disease toxins known to date fail to act as antigens in serological studies because they are small molecules; this is in contrast to toxins involved in animal diseases.

Some plant disease determinants are known as host-specific toxins because they affect only hosts of the producing microorganism; there is little or no effect on nonhosts and resistant genotypes at physiological concentrations. The difference in host sensitivity can be striking; in one case, the resistant genotype tolerated a millionfold higher concentration of toxin than did the suscept, with only a one-gene difference between the two host genotypes (Scheffer 1983, 1991). Most specific toxins are considered to be pathogenicity factors because they determine hosts and because they are required for a specific type of pathogenicity. At least eighteen different host-specific toxins are now known from at least seven fungal genera: *Cochliobolus, Alternaria, Corynespora, Fusarium, Periconia, Phyllosticta,* and *Pyrenophora* (Scheffer 1983; Kohmoto and Otani 1991). Selective toxins have been reported from several other genera, but the reports have not been confirmed. To date, no low molecular weight host-specific toxins are known from bacteria. The diseases and the causal fungi appear to be typical of plant diseases, in most respects (Scheffer and Briggs 1981). Host-specific toxins will be discussed in detail for the *Cochliobolus* (Chapter 12), *Alternaria* (Chapter 13), and *Periconia* diseases (Chapter 8, section 3).

There are several credible reports of proteins that are selectively toxic to hosts of the producing microorganism. The best example is with the fire blight disease of rosaceous plants, caused by *Erwinia amylovora*. A toxic protein (harpin) associated with the bacterial cell surface was characterized; it appears to be required for pathogenicity (Wei et al. 1992, 1993) (see Chapter 8, section 2). A second example is a protein that is selective for wheat genotypes that are hosts of the producing fungus *Pyrenophora triticirepentis* (Ballance, Lamari, and Bernier 1989). *Fusarium oxysporum* f. sp. *lycopersici* was reported to produce in cultures a protein that is selectively toxic to genotypes of tomato that are susceptible to the disease (Sutherland and Pegg 1992). These cases deserve much more study. Finally, *Ceratocystis ulmi,* the cause of Dutch elm disease, produces a small protein with selective toxicity at the host species (not genotype) level (Takai, Richards, and Stevenson 1983; Takai 1989).

A number of nonspecific toxins have now been isolated, characterized, and shown to be involved in disease development (Mitchell 1984). The purified toxins are active against both host and nonhost species; thus, host selectivity is determined by other (unknown) factors. Mutants of the pathogen that have lost ability to produce toxin can still be pathogenic; therefore, we think of nonspecific toxins known to date as "virulence factors," or perhaps as "secondary determinants of disease." Tabtoxin, or wildfire toxin, will be discussed as an example.

Tabtoxin was the first toxin known with confidence to be involved in the development of a plant disease. The toxin is produced by *Pseudomonas syringae* pv. *tabaci,* a tobacco pathogen that causes small necrotic spots on the leaf. Each spot becomes surrounded by a chlorotic zone, or "halo," caused by toxin diffusing from bacteria in the necrotic zone; no bacteria are found in the halo. The bacterium also produces the toxin when grown in culture.

We think of tabtoxin as a virulence factor because mutants that lose ability to produce it can still be pathogenic but have decreased virulence. The mutants induce necrotic lesions without haloes. Toxin-minus mutants are very similar to a bacterium (*P. syringae* pv. *angulata*) that causes a disease known as angular leafspot. There are suggestions that *P. syringae* pv. *tabaci* originated as a toxin-producing mutant of *P. syringae* pv. *angulata* (Fulton 1980).

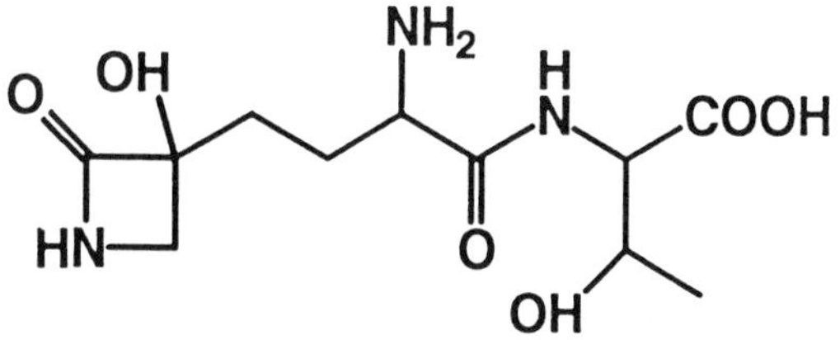

Figure 3.1. Chemical structure of tabtoxin from *Pseudomonas syringae* pv. *tabaci.*

Tabtoxin was the first authentic plant disease determinant to be isolated and characterized (in 1955); the first proposed structure was later modified (Figure 3.1) (Stewart 1971). Tabtoxin also was first of the authentic toxins to be examined for mode of action (Braun 1955); it was said to interfere with methionine utilization in protein synthesis. There are similarities in the effects of tabtoxin to those of methionine sulfoximine, a known antagonist of methionine utilization. Later work showed that highly purified toxin did not affect methionine uptake; instead, the toxin is converted in the plant to tabtoxin-B-lactam, which inhibits glutamine synthetase, thus indirectly interfering with methionine metabolism. Glutamine synthetase is as important to the bacterium as to the plant. Then why does the bacterium not poison itself? *P. syringae* pv. *tabaci* has a Zn-containing amino peptidase that hydrolyzes tabtoxin (Durbin 1991).

A number of other nonspecific toxins are now accepted as factors in plant disease development. Most of these are listed: fusicoccin from *Fusicoccum amygdali,* cercosporin from *Cercospora* sp., tentoxin from *Alternaria alternata* f. sp. *tenuis,* coronatine from *Pseudomonas syringae* pv. *atropurpurea,* phaseolotoxin from *P. syringae* pv. *phaseoli,* rhizobitoxinine from *P. andropogonis* and from certain strains of *Rhizobium japonicum,* syringotoxin from *P. syringae* pv. *syringae,* tagetitoxin from *P. syringae* pv. *tagetis* (Scheffer 1983; Mitchell 1984, 1991; Durbin 1991), and oxalic acid from *Sclerotinia sclerotiorum* (Godoy et al. 1990). Mechanisms of action are known for some toxins. The ones that are most studied are phaseolotoxin, which inhibits ornithine carbamoyltransferase, and tagetitoxin, which inhibits chloroplast RNA polymerase (Durbin 1991).

A vast number of toxic compounds from microorganisms have been isolated and characterized, but few of these have been shown conclusively to be involved in disease development. We must assume that these compounds are innocent until proved guilty. Some proposed toxins were shown later to have no role in disease (Scheffer 1983).

3.2. Pathogen-secreted Enzymes

The possible role of enzymes in plant disease development has been studied for years, beginning about 1860 with the work of deBary, long before enzymes were fully understood. Early leaders in enzyme studies in disease include H. M. Ward, L. R. Jones, and W. Brown. The studies peaked in popularity around 1970. At that time, two related hypotheses stimulated much research: (1) All pathogenicity to plants was postulated to be based on ability of a fungus or bacterium to degrade cell wall constituents; and (2) enzymes from the pathogen were thought to degrade the unique polysaccharides in host plants but not the polysaccharides in nonhost plants (Albersheim, Jones, and English 1969; Dimond 1971). These ideas were modified over the years, to accommodate enzyme inhibitors in plant tissue, as well as other factors. There is a very large literature (Bateman and Basham 1976), but only the pertinent general ideas will be discussed here.

Many different enzymes are released by plant pathogens and by saprophytes. These enzymes include the pectin-degrading enzymes, cellulase, various esterases, protease, cutinase, phosphatidases, amylase, galactosidases, B-glucosidase, and hemicellulase. The pectic enzymes (polygalacturonases and transeliminases, or lyases) were given most attention, because some of them in purified form can degrade intact plant tissue (Bateman and Basham 1976; Collmer and Keen 1986). Cellulase degrades wood but has no proven role in pathogenicity in metabolically active tissues. Cutinase was given serious attention because it should aid in penetration (Dickman and Patil 1986); there are contraindications for some pathogens (Bonnen and Hammerschmidt 1989; Valent and Chumley 1991; Stahl and Schäfer 1992). Early studies did not establish roles of extracellular enzymes as significant factors in pathogenicity. Molecular genetics methods recently have been used, and it is hoped that the questions will be clarified (Boucher, Gough, and Arlat 1992; Barras, van Gijsegem, and Chatterjee 1994).

Clearly, cell-wall degrading enzymes are involved in tissue maceration. The enzymes must be major factors in soft rot of fruit (Howell 1975; Barras, van Gijsegem, and Chatterjee 1994) and in damping-off of seedlings, both caused by opportunistic pectolytic pathogens affecting senile, juvenile, or stressed tissues. However, even here tissue maceration may be a late event in pathogenesis, not involved in the critical early stages. Soft rot could be biologically gratuitous in the disease syndrome (Collmer and Keen 1986). It is now clear that these enzymes have little or nothing to do with pathogenicity or virulence of many specialized pathogens. In recent years, the significance of tissue-macerating enzymes as key or general factors in pathogenicity and virulence has been deemphasized for lack of conclusive evidence and because of negative evidence with some diseases.

Several considerations must be remembered in evaluating the roles of extracellular enzymes. First, attempts to prove production of cell-wall and tissue-degrading enzymes by several important pathogens have not succeeded. These include *Erwinia amylovora,* which causes fire blight (Seemüller and Beer 1976); *E. stewartii,* which affects maize (Braun 1990); *Cladosporium fulvum,* which affects tomato (VanDijkman 1972); and the bacterium that causes Pierce's disease of grape (Hopkins 1985). Next, many saprophytes have cell-wall-degrading enzymes in abundance; why are they not able to infect? It is possible that the enzymes are more important in saprophytism than in parasitism.

Many pathogenic fungi will penetrate cell walls of both resistant and susceptible plants (for an example, see Valsangiacomo and Gessler 1988); generally, the difference is expressed only after contact with host plasmalemma or protoplasm. In most cases, the cuticle is no barrier to penetration by virulent pathogens (Bonnen and Hammerschmidt 1989), thus raising questions about cutinase as a significant factor. Finally, there are no confirmed examples that show a relationship between host specificity and the action of extracellular enzymes, even after many attempts to find correlations.

The first doubts regarding roles of wall-degrading enzymes in pathogenicity were raised by experiments with mutants that lacked specific enzymes. Several examples: Mutants of *Verticullium albo-atrum* that lacked endopolygalacturonase were still virulent (Howell 1976); a mutant of *Cladosporium cucumerianum* that lacked protease was pathogenic (Robertsen 1984); both virulent and hypovirulent strains of *Endothia* (= *Cryphonectria*) *parasitica* produced polygalacturonase, cellulase, and protease (McCarroll and Thor 1985); mutants of *C. lagenarium* that lacked cutinase were pathogenic (Bonnen and Hammerschmidt 1989); and mutants of *Erwinia carotovora* without polygalacturonases were virulent (Willis, Engwall, and Chatterjee 1987). There are contradictory data for some of these (Durands and Cooper 1988), but the reports are enough to raise doubts.

Molecular genetic studies, in which genes for specific enzymes are deleted or inserted, should be more conclusive. However, such studies to date have not provided conclusive evidence for a general role of extracellular enzymes in pathogenicity of specialized and virulent pathogens. Several cases were interpreted as support for a role in virulence by opportunistic pathogens, but there are conceptual and methodological problems with some of this work. How is pathogenicity separated from relative virulence? What constitutes infection? Were the proper inoculation procedures followed and the proper plant tissues used for testing?

In several cases, deletion of genes for specific enzymes has given isolates that were as pathogenic as the wild-type parental forms (Willis, Engwall, and Chatterjee 1987; Boccara et al. 1988; Brooks, Collmer, and Hutcheson 1989; Dow et al. 1989; Valent and Chumley 1991). Endopolygalacturonase production by *Cochliobolus carbonum* is not required for pathogenicity (Scott-Craig et al. 1990), whereas the host-specific toxin (HC-toxin) is necessary (Chapter 12). Other reports indicate that specific enzymes contribute to virulence (Howell 1975; Dickman and Patil 1986; Schell, Roberts, and Denny 1988) or that pectic enzymes are necessary for soft rot (Liao, Hung, and Chatterjee 1988; Barras, van Gijsegem, and Chatterjee 1994), a part of some disease syndromes. Simultaneous production of several wall-degrading enzymes may be necessary for virulence (Reid and Collmer 1988).

To summarize the findings to date: Extracellular, tissue-macerating enzymes appear to have no more than a minor role in pathogenicity of many plant-infecting bacteria and fungi. There is a somewhat more convincing case for virulence of certain soft-rotting microbes that are opportunistic pathogens on senile or storage tissues. However, there is room for skepticism. It is hoped that molecular methods eventually will clarify the uncertainties.

3.3. Genetic Colonization: The Crown Gall Story

Transfer of genes from pathogen to host (genetic colonization) is known for a few plant diseases. The best-known example is crown gall (Figure 3.2), a disease that is now among the best understood of all plant diseases; its study has led to some significant advances in biology. The importance of this research justifies more than usual attention. Crown gall is the first known example of both genetic colonization and "genetic engineering" in nature (Hooykaas and Schilperoort 1992); there are evolutionary and ecological implications. However, it is unlikely that genetic colonization is involved in most plant diseases.

Crown gall, a true tumor or "plant cancer," is initiated by a bacterium (*Agrobacterium tumefaciens*) that infects many plant species. The disease is a problem in the plant nursery industry, causing damage to grapes and berries, fruit trees (Figure 3.2), roses, and many ornamental plants. True plant tumors differ from ordinary galls in that tumors are not self-limiting in

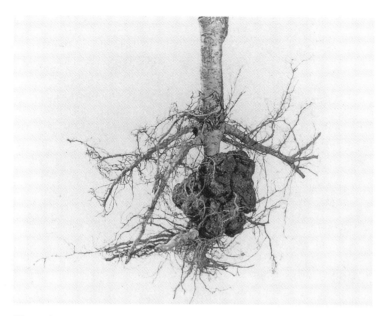

Figure 3.2. Crown gall on roots of young peach tree. (Picture from A. L. Jones)

growth. In normal plants and animals, the individual has mechanisms that control growth of its parts. The basic question regarding all tumors and cancer in animals or plants: How do cells of a multicellular organism escape this overall control (that is, become autonomous)?

Our basic understanding of crown gall was gained via a series of significant discoveries over the years. Each new finding was based on previous knowledge, and each will be summarized in turn; this chronology illustrates how science advances. First, a brief background is needed for perspective of the disease in nature.

A. tumefaciens is a soil bacterium that infects plants at wounds, often those from insect bites on roots and crown (transition zone, root to shoot). The resulting galls can become very large, up to 45 kg or more. Many dicotyledonous species are affected; moncotyledonous plants, including grasses, seldom are infected. Some isolates of the bacterium have wide host ranges, inducing galls on many species; other isolates have limited hosts, even to a single species. Isolates vary from highly virulent to intermediate to avirulent on a given host. Some isolates induce fast-growing galls; others

Figure 3.3. Armin C. Braun (1911–85). (Courtesy of American Phytopathological Society)

induce slow-growing galls because they do not cause complete transformation of all metabolic steps necessary for rapid growth (Braun 1982).

Morphology of galls induced by a single bacterial isolate can differ, depending on the site of infection. If a plant is severed and inoculated at both cut surfaces, then a fully disorganized gall will develop at the base of the upper portion, and a teratoma often forms at the top of the lower part (Braun 1953). A teratoma is a gall that contains buds and partially differentiated tissues; it appears to grow from cells with incomplete transformations.

Soon after the significance of DNA was recognized, Armin Braun (Figure 3.3) suggested that the tumor-inducing principle for crown gall is a DNA. He did not succeed in proving this, as the techniques were not yet available. The idea was "proved" and disproved several times over the years, with each claim failing, in turn, for various reasons. Finally, research since 1975 established that the tumor-inducing principle really is a DNA.

Each significant step in the crown gall story will now be identified and discussed briefly (steps 1–10). These basic facts were established by 1980; more recent research has added many important details (Ream 1989; Hookyas and Beijersbergen 1994).

1. *The discovery of crown gall and its cause.* Erwin F. Smith (see Chapter 8, section 2) first described crown gall and its causal bacterium in 1907. Smith, a leading biologist of his day, was aware of the tumorous nature of the disease, and his work was widely acclaimed; at the time, no animal tumor had been induced experimentally. Smith stressed similarities of crown gall and cancer but was not aware of the autonomous nature of both. A leading cancer expert postulated in 1910 that crown gall must be autonomous, but Smith was never convinced; he believed instead that all cancers are caused by parasites (Braun 1982).

2. *The autonomous nature of crown gall.* P. White and Armin Braun reported in 1941 that "once formed, crown gall tumors persist and grow without the inducing bacteria, and can be maintained indefinitely in culture without loss of the tumorous character" (see Braun 1982). All later progress was predicated on this discovery. Autonomy is the normal condition for single-cell organisms; in multicellular organisms, is the change to autonomy a reversion to the primitive?

3. *The role of wounds in tumor induction.* It soon became clear that both wounds and *A. tumefaciens* are required for tumor induction (reviewed by Braun 1982). Wounds trigger a few divisions by cells adjacent to the site, but division soon stops in normal tissues. When the bacterium is present in the wound, adjacent cells continue to divide and grow. The cells have been transformed to tumor cells; they continue to grow without regulation.

   Lippincott, Whatley, and Lippincott (1977) proposed bacterial attachment sites in cells adjacent to the wound. Avirulent bacteria can occupy these sites and prevent attachment of virulent cells, thus preventing tumor formation. Avirulent bacteria have been used as a biocontrol. However, avirulent isolates of *Agrobacterium* also produce an antibiotic known as Agrocin 84, which is another element in biocontrol (Kerr 1980). There are reports of lipopolysaccharides from *A. tumefaciens* that occupy attachment sites and protect tissues from virulent bacteria.

   Another factor involved in binding of bacteria to sites is the nature of the host surface; cellulose fibers from bacteria are thought to be involved (Matthysse, Holmes, and Gurlitz 1981).

4. *Temperature relationships in tumor induction.* Knowledge of temperature relationships made way for further progress on the crown gall problem. When temperature is raised to 46°C for three days, all virulent bacteria in the tissue are killed; tumor development, in plant species that tolerate this temperature, is not affected. When bacteria are killed 36 to 48 hours after inoculation, the resulting tumors are slow growing. When bacteria are killed four or more days after inoculation, the resulting tumors are fast growing (reviewed by Braun 1982).

   Further studies showed that a temperature of 32°C prevents the conversion of normal to tumorous condition but does not affect wound healing or bacterial growth. Conversion occurs at temperatures up to 29°C; once conversion occurs, 32°C (or higher) does not prevent the tumor from developing. No conversion takes place during the first 24 hours after wounding, and none occurs if inoculation is delayed until the wound starts to heal at five days. During the critical time period (24 to 120 hours after wounding), the normal cell is transformed to a tumor cell, regularly and with precision, by exposure to bacteria for 10 hours. A 10-hour exposure during the sensitive period results in a slow-growing tumor; a 20-hour exposure results in a fast-growing tumor (Braun 1947). The slow- and fast-growing tumors maintained these characteristics in culture, indefinitely; even years after the transformation event, cultures maintained their relative growth rates.

5. *The transformation from normal to tumorous condition occurs stepwise.* The discovery of transformation steps was a final distillation of much work on temperature relations and other aspects of transformation. Several different degrees of transformation were achieved experimentally; resulting tumors on minimal media varied from slow to very fast growing (Braun 1978, 1982). The explanation came with later work.

6. *The tumorous condition is reversible, and each host cell has a complete set of genes* (that is, is genetically totipotent). Braun found that the crown gall condition is reversible, as are other tumors (Braun 1951, 1980). Next, he showed that each vegetative cell is totipotent (contains all the genetic information for the species) (Braun 1959); this was demonstrated by starting with a single tumor cell and recovering from it a phenotypically normal plant. Later, F. C. Stewart proved that a single normal cell can be developed into a normal plant. These were significant intellectual achievements. The tumorous condition usually survives

mitosis, but not meiosis (Turgeon, Wood, and Braun 1976). Development of a whole plant from a single tumor cell of the teratoma type was accomplished by forcing the cells in culture into rapid division and then grafting the resulting budlike tissue to the tip of a healthy stock plant (Braun 1959). The plant that developed from this bud carried some crown gall genes, but it set seeds; progeny plants usually had a normal genotype (Turgeon, Wood, and Braun 1976; Braun 1982).

7. *Transformation involves activation of certain biosynthetic systems.* To understand how this knowledge was gained, we must know the requirements of normal cells to grow in culture as callus tissue. Requirements vary with plant species; *Vinca rosea* (periwinkle) and tobacco were used in the crown gall experiments. When normal pith cells of these species are placed on a minimal medium, containing only the essential elements with a carbon source, they do not grow. The addition of auxin (indole acetic acid), a plant growth hormone, causes cells to enlarge but not divide. The addition of auxin plus cytokinin (another plant growth hormone) causes cells to enlarge and divide slowly. Certain other additives, notably inositol, will speed growth. These normal cells stop growing when they are removed from the supplemented medium (Braun 1958); thus, at least two growth hormones are essential for growth. The biosynthetic systems for producing the growth factors are blocked in pith cells; that is, genes for hormone production are "turned off." In the normal healthy plant, growth hormones are produced in meristems, the growing parts; mature cells, such as those in pith, make no further growth unless growth factors are supplied from other tissues.

In contrast, fully altered tumor cells grow rapidly on minimal media. Tumor cells synthesize abundant amounts of auxin and cytokinin, allowing cells to enlarge and divide; adding these growth factors to the medium does not increase the rate of growth (Braun 1958). These findings led to the conclusion that transformation to a tumor cell is achieved by unblocking, or "turning on," the genes for synthesis of growth-controlling compounds. This activation can be achieved in various ways, including exposure to *A. tumefaciens*. Fast-growing tumors result from activation of several biosynthetic systems in addition to those for auxin and cytokinin.

Increased concentrations of certain ions in the culture medium also can overcome the need of normal cells for supplemental growth factors; only the need for cytokinin is not eliminated. Excesses of KCl, $NaNO_3$, and NaH_2PO_4 (10- to 100-fold) in the medium, along with cytokinin, allowed normal cells to grow as rapidly as do crown gall cells. Ion-activated callus tissues contain excess levels of the necessary growth factors, as do tumor cells on the minimal medium (Wood and Braun 1961; Braun and Wood 1962). Normal cells stop synthesizing growth factors and stop growing when they are removed from the high-ion medium. The data show that the normal cell has all the genes needed for unregulated tumorous growth. However, later work has shown that bacterial genes are moved into the host genome, activating hormone production.

8. *Crown gall tissues contain novel amino acids (opines) that are not normally found in plants.* French scientists reported that crown gall tissues contain some novel compounds known as opines; the reports were confirmed elsewhere (Menage and Morel 1964; Petit et al. 1970; Montoya et al. 1977; Nester et al. 1977). There are two major opines: octopine and nopaline. There also are minor opines such as agropine. Opines are found in detectable concentrations only in crown galls; normal plant cells have no more than traces. Octopine was first discovered in the octopus.

    The type of opine present in a gall is determined by the strain of *A. tumefaciens* that induced the gall, not by the plant species. *A. tumefaciens* does not contain opines, but it induces their production in the plant. Some bacterial strains induce octopine; others induce nopaline. The octopine inducers can utilize octopine, but not nopaline, as a carbon source, and nopaline inducers utilize only nopaline (Petit et al. 1970). Very few other microorganisms can utilize opines. Thus, *A. tumefaciens* creates a special food base and an ecological niche for itself.

    These discoveries suggested that bacterial genes for opine synthesis are transferred to the plant cell (Petit et al. 1970); this suggestion was later confirmed. The findings also tell much about *A. tumefaciens* as a parasite. The bacterium alters its host in such a way that the environment is improved for bacterial multiplication and survival. The host plant is not killed under ordinary circumstances; the parasite continues to multiply in and on the tumor. *A. tumefaciens* spreads readily to new sites, even though it cannot multiply in the soil. This is an elegant arrangement for parasitism; no other examples among plants have been described in such a definitive way.

9. *A bacterial plasmid carries the genes for induction of crown gall.* Plasmids are nonchromosomal DNA fragments that carry genetic information. A large plasmid (about 200 kb of DNA) in *A. tumefaciens* encodes for tumor-inducing ability, opine production, and production of several plant growth hormones. This landmark discovery was made indepen-

dently in two different laboratories, one at the University of Washington in Seattle by E. Nester and associates, and the other in Belgium by J. Schell and associates. The plasmid, now known as the Ti plasmid, is present in all pathogenic isolates of *A. tumefaciens* and is absent in all avirulent forms (Zaenen et al. 1974; Watson et al. 1975; Montagu et al. 1980; Kahl and Schell 1982). The Ti plasmid can be transferred to avirulent isolates of *A. tumefaciens,* making them tumor inducers. The plasmid also can be transferred to *Rhizobium* and other related bacteria, which then induce tumors.

The work of Hamilton and Fall (1971) provided the key to the later discovery of the Ti plasmid. Their work showed that certain strains of *A. tumefaciens* lose tumor-inducing ability when grown in culture at 36°C. The Seattle and the Belgian groups found that bacteria in such cultures had lost a large plasmid.

The Ti plasmid encodes functions other than the ability to induce tumors. It gives a bacterial isolate the ability to use a specific opine and the ability to induce plant cells to produce the same opine (Petit et al. 1970). Certain mutant bacteria induce tumors but not opines, showing that opine metabolism is genetically independent of tumor-inducing ability (Montoya et al. 1977). Host range also appears to be determined by the Ti plasmid. Some isolates have narrow host ranges; others cause tumors on many species, suggesting a family of Ti plasmids. Petunia protoplasts in culture were transformed to the tumorous condition with a Ti plasmid; previously, transformation had been accomplished only in intact tissue (Davey et al. 1980). Other anamolies of crown gall are associated with the Ti plasmid.

10. *The tumor-inducing principle is a fragment of the Ti plasmid.* This very significant discovery was announced by Chilton et al. (1977) from the University of Washington. The fragment (T-DNA) is integrated into host DNA in the cell nucleus, giving new characteristics to the host plant (Chilton 1982). T-DNA has proved to be a good vector for introducing other genes into plants (Ream 1989). More recently, many more details of the Ti plasmid and T-DNA have become known (Hooykas and Beijersbergen 1994), resulting in better understanding of tumors, in progress for molecular biology, and in the use of T-DNA as a gene vector to introduce other foreign genes into plants. The Ti plasmid has now been mapped thoroughly.

The ten significant discoveries just summarized led to our present understanding of genetic colonization as a disease mechanism. More recently, many details of how this occurs have been elucidated (Hookyas and Bei-jersbergen 1994). Host specificity is determined by *vir* genes carried on the Ti plasmid; however, control of host range still is not fully understood. Genes that control various steps in the colonization process have been iso-lated; included are those that control binding of bacteria to the host cell, insertion of T-DNA into the cell, and binding of T-DNA to the host DNA. Some genes that control tumor induction are homologous with genes that control vital cellular functions such as DNA transport across membranes and cellular import/export of proteins. Also, transfer of T-DNA into plant cells is related to mechanisms of bacterial conjugation. These are all very signifi-cant discoveries but are beyond the scope and purposes of this book.

At least three other true tumor diseases of plants have had careful study: (1) the wound tumor disease, (2) genetic or Kostoff tumors, and (3) habitua-tion tumors. Similar physiological and biochemical factors appear to be in-volved in development of all the tumors. There are differences in triggering agents, which can include carcinogenic compounds, radiation, other inju-ries, bacteria, viruses, and certain genetic combinations. Habituation is a spontaneous change to a tumorous condition.

The wound tumor disease is a disorder induced by a virus that affects many plant species; in some species, the tumor is similar in appearance to crown gall. The virus is transmitted from plant to plant by leafhoppers, in which the virus can multiply as it does in the plant. Strains of the virus differ in their capacity to induce galls. Species and genotypes of plants differ in their ability to respond with gall formation; sweet clover has been the favored experimental host. Wounds are required for tumors to start; some plants carry the virus systemically, but no tumors appear until there is a wound.

The tumor virus can disappear from plant tissues in culture, but the tu-morous condition persists, as with crown gall. Virus-induced tumors synthe-size excess growth-promoting compounds, as do crown galls. The virus can be manipulated to eliminate it from galls in cultures, but the tissues still grow on a minimal medium that does not support growth of normal callus, as in the case with crown gall. The tumors are capable of indefinite, non–self-limiting growth, on the host and in culture. However, opines are not present (Black 1982).

The same transformations appear to be involved in virus tumors and crown galls, but to date there is no evidence that viral genes are incorporated into the plant genome.

Genetic tumors occur spontaneously on certain plants, without the in-volvement of infectious agents. This is the usual case with certain interspecific hybrids; at least eighty different hybrid combinations are known to throw tumors spontaneously. Most of the studies have been with certain interspecific hybrids of *Nicotiana* (tobacco). The hybrid plants appear to be normal until wounds are made; then, they always develop tumors. Fully mature plants usually have tumors on all parts, especially at sites of abscissed leaves. These are known as Kostoff tumors, after their discoverer.

Kostoff tumors grow on minimal media; parental lines do not grow on such media. Kostoff tumors are indistinguishable from crown gall, except that no opines are present (Bayer 1982). It is of interest that certain strains of mice will form tumors spontaneously.

Habituation sometimes occurs in tissue cultures. This is a spontaneous change to a state that is very similar to that of crown gall; it appears to result from progressive changes at the cellullar level. Habituated cells produce an excess of growth-promoting compounds, as do crown galls, but they do not contain opines. Habituated cultures can be grafted to normal plants; the tumor usually remains autonomous. As with crown gall, habituation can be complete, producing fast-growing tumors, or it can be partial, producing slow-growing tumors (Meins 1982, 1990).

Habituated cells are totipotent and can be made to revert to normal at will. Complete plants have been recovered from single habituated cells; recovered plants are phenotypically normal but have the capacity to revert to the tumorous condition, as is the case with plants grown from crown gall cells. However, no foreign genes are involved in habituation. When recovered plants set seed, the next generation is fully normal; apparently, the habituated condition does not pass through meiosis. Habituation is reversibly lost when tissues of some cell lines are grown at 16°C rather than 26°C. Also, reversions often occur in a high-salts medium (Meins 1982). Habituation can be induced in cell cultures of some species by temperature shock; the change possibly could occur in nature as a result of irradiation and other injuries.

Habituation may involve heritable expression of genes that normally are silent (turned off) in cultured cells. Habituation also appears to be epigenetic, because rates of appearance and reversion are one hundred to one thousand times greater than expected from somatic mutations in the same species (Meins 1982).

The parallels between habituation and crown gall are striking, although no foreign DNA is involved in habituation. Crown gall T-DNA may activate certain biosynthetic systems via a "master switch"; a comparable switch activation may occur spontaneously in habituation. We still have much to learn.

The hairy root disease, caused by a bacterium (*Agrobacterium rhizogenes*), is not a typical tumor, but it does involve genetic colonization

(White et al. 1982). Infection results in abundant proliferation of roots at wound sites. There are other hairy root diseases; for example, hairy root of beets (rhizomania) is caused by a virus. Bacteria that cause hairy root infect species that are hosts of the crown gall bacterium.

With the exception of symptoms, there are striking similarities between hairy root and crown gall (Holsters et al. 1982; White et al. 1982). The causal bacteria are closely related. *A. rhizogenes* has a large plasmid (the Ri plasmid) that is correlated with the ability to induce root proliferation (Chilton et al. 1982). The Ri plasmid can be transferred to plasmid-free isolates, which then induce hairy root. The plasmid also can be transferred to *A. tumefaciens*, which then induces hairy root. The Ti and Ri plasmids are compatible but show only a limited base-sequence homology. Hairy roots contain agropine, and *A. rhizogenes* can utilize agropine (White and Nester 1980b). Whole plants were regenerated from cultured hairy root tissues; although the plants were normal in appearance, they spontaneously erupted into rooty forms and were tumor prone (White and Nester 1980a, 1980b; Sinkar et al. 1988; Ream 1989).

There is a gall disease of brambles known as cane gall, caused by *A. rubi*. Cane galls do not contain opines, but the bacteria do contain Ti plasmids; otherwise, cane gall and crown gall are similar.

Summary

Genes from a bacterium (*A. tumefaciens*) are transferred in nature to plant cells, where the bacterial genes are replicated and expressed. The transferred genes include those for tumor induction, for the synthesis of opines, and for the synthesis of growth-promoting compounds. Data on crown gall explain in definitive terms an example of parasitism, give insights into genetic ecology, define a possible mechanism for certain evolutionary changes, and are useful in genetic engineering. Gene transfer between bacterial species now is well known. Other examples of gene movements from bacteria to multicellular organisms have been reported.

In addition to crown gall, three other neoplastic (tumorous) plant diseases have been studied: (1) wound tumor disease, incited by an RNA virus; (2) Kostoff tumors, involving chromosomal inbalance; and (3) habituation, involving spontaneously activated synthesis of growth-controlling compounds. There also is a neoplastic condition known as hairy root, which has many features in common with crown gall. The end result for each case is similar, suggesting basic underlying mechanisms in common. All the known neoplastic tissues synthesize growth-promoting compounds in excess. Normal cells do not do this; their growth is regulated by hormones from other parts of the plant. Production of the growth-regulating compounds depends on either new enzyme systems or the activation of existing systems.

3.4. Pathogen-induced Hormonal Changes

Many plant pathogens and saprophytes produce plant hormones or hormonelike compounds in culture. Infected plant tissues often contain excess growth hormones, and correlations of levels with host responses sometimes are evident. The obvious hypothesis is that pathogen-produced hormone is a determinant of disease. However, the situation is often complicated; in some diseases, the excess hormone is of host-cell origin (see Chapter 3, section 3). All known hormones that control growth in plants (auxin, cytokininus, gibberellins, and ethylene), plus a growth-hormone inhibitor (abscisic acid), have been studied for possible roles in various diseases.

Growth responses to infection include stunting, excessive growth, and changes in morphology; all could be related to changes in growth hormones. However, there are many cases in which it is difficult or impossible to reproduce such changes by application of known hormones.

There are a few reports that decreased growth after infection is correlated with excess abscisic acid in tissues (Wiese and DeVay 1970). In general, however, increases in abscisic acid in diseased tissue are thought to reflect changes in host metabolism, not production by the pathogen (Stedman and Sequeira 1970). Only one pathogen is known to produce abscisic acid (see Scheffer and Briggs 1981). No role of pathogen-produced abscisic acid in disease can be assumed at this time. There are, of course, many possible causes of stunting, such as defoliation.

Several examples of overgrowth in diseased plants have been attributed to gibberellins produced by the pathogen. Best known is the bakanae disease of rice, in which infected plants are greatly elongated. The causal fungus (*Gibberella fujikuroi*) produces gibberellins in culture, and application of the compound to plants causes the characteristic overgrowth. Another example is a rust disease of thistle, which changes normally erect plants to a vinelike condition. Infected vines have high levels of gibberellic acid, unfortunately identified only by paper chromatography (reviewed by Scheffer and Briggs 1981). In both cases, the excess gibberellin was assumed but not definitely proved to be from the fungus. In some other diseases, dwarfing has been attributed to decreases in gibberellin (Pegg 1976b).

The major significance of the bakanae disease was that it led to the discovery of gibberellins, a landmark event in plant physiology. Gibberellins, which are present in all normal plants, are significant in control of growth and development (Pegg 1976b). Synthetic "antigibberellins" are used commercially to grow short, dark green plants in the florist industry and to produce short, stout cereal plants that resist lodging in the field.

Cytokinins are involved in development of several diseases, but in some cases the excess cytokinin is from the altered host cell (see Chapter 3, section 3). Excess cytokinins have been suggested as causal factors in many gall diseases, but the evidence usually is minimal or lacking. In addition, there is always the problem of source: Does excess cytokinin come from the pathogen or from activated host cells?

There are significant studies on cytokinins from *Rhodococcus* (= *Corynebacterium*) *fascians,* cause of a disease of pea and many other species. Infected plants produce a mass of buds and undifferentiated tissue (fasciations) at a node; a similar condition results from application of cytokinin. The causal bacteria produce a cytokinin (cis-zeatin) in culture; the same compound was extracted from infected tissue (Thimann and Sachs 1966; Murai et al. 1980). The logical conclusion was that cytokinin from the bacterium is the disease determinant. However, there are still problems. Relatively few bacterial cells induce the effect in the host; do they produce enough (Sequeira 1973)? Symptoms of the disease are similar to those of crown gall teratomas. The possibility remains that the bacteria activate host cells for cytokinin production.

Auxin and related compounds were the first-known plant growth hormones. Auxin is a common product of bacteria and fungi; thus, it is tempting to give auxin from pathogens a general role in disease. However, the microbial source has not been established, with a few exceptions. Auxin is high in crown gall and in habituated (crown gall–like) tissues, but the source is activated host cells (see Chapter 3, section 3). Auxin also is high in other types of galls and in tissues with many other diseases, but usually the source and the significance are not known (Sequeira 1973).

There are diseases that suggest lack of auxin. One such is the coffee disease caused by *Omphalia flavida* (Sequeira 1973). Data indicate that auxin is inactivated or not produced in infected leaves, resulting in leaf drop. The fungus in culture destroys indole acetic acid, and culture filtrates do the same; the destruction appears to be enzymatic. The coffee disease deserves more study, now that effects of ethylene and abscissic acid are known.

Studies on the knot disease of oleander and olive gave convincing evidence that auxin is a determinant of this disease. The causal bacterium, *Pseudomonas syringae* pv. *savastanoi,* produces abundant indole acetic acid. Mutant isolates that do not produce auxin do not induce galls, even when bacterial populations in tissues are high; thus, auxin is a virulence factor in this disease. Virulence in oleander isolates of the pathogen is controlled by plasmid genes that direct synthesis of indole acetic acid. Insertions into the plasmid locus cause loss of auxin production, attenuation of virulence, and loss of enzyme activities involved in auxin production (Smidt and Kosuge 1978; Comai and Kosuge 1980; Comai, Surico, and Kosuge 1982). These findings still hold, although more recent work indicates many complications in the system (Yamada 1993). Indole acetic acid is produced by different metabolic pathways in the plant and in the bacterium (Sequeira 1973; Yamada 1993); use of appropriate precursors has shown that auxin in diseased oleander tissue is from the pathogen, although genetic colonization (see Chapter 3, section 3) was not completely ruled out. Changes in cytokinin levels also are involved (Surico et al. 1985).

Tissues of plants with other diseases, including wilts, rusts, and mildews, can have high auxin levels. Sources of high levels can be host cells (Sequeira 1973) or pathogen, or could result from inhibition of auxin movement and degradation by plant enzymes (Deverall and Daly 1964).

Diseased and injured plant tissues release ethylene, as do some microorganisms (Sequeira 1973; Pegg 1976a). There have been many studies, but, to date, the possible role of ethylene in disease development and disease resistance is not clear.

Summary

It is clear that the levels of growth hormones (auxins, cytokinins, gibberellic acids) in plants often are raised or lowered, depending on the type of disease. Hormonal changes can lead to stunting, overgrowth, loss of control (galls), or morphological changes. Excess hormones can come from the pathogen (virulence factors), from host colonization by pathogen genes for hormone production, or from activation of hormone production in infected plant cells. Excess auxin in the olive knot disease arises from production by bacterial cells. In the fasciation disease caused by Rhodococcus (= *Corynebacterium*), excess cytokinin is thought to be a bacterial product that stimulates abnormal growth. Overgrowth in the bakanae disease of rice usually is conceded to result from gibberellin produced by the pathogenic fungus. Growth hormones from pathogens or from activated host cells are evident in many other diseases, but the general significance is not clear at this time.

3.5. Models Using *hrp* (Virulence/Avirulence) Genes

Specific genes in microbes control pathogenicity and virulence to plants. These genes can be used as a route to understanding how pathogens attack and how hosts defend; there is further discussion in Chapter 4, section 4, on host defense. In regard to mechanisms of attack, the most advanced models to date are fire blight of apple and pear and leaf mold disease of tomato. Significant work has been done with several bacterial diseases (see Chapter 4, section 4).

Work with fire blight (Chapter 8, section 2), caused by *Erwinia amylovora,* has provided solid data regarding the nature of susceptibility, resistance, and virulence. The bacterium has a gene cluster (*hrp*) that governs pathogenicity to susceptible plants and induction of hypersensitivity (rapid localized killing of plant cells) in resistant plants. The *hrp* genes were cloned as a single cosmid and transferred into nonpathogenic mutants of *E. amylovora;* transformed mutants were then pathogenic to susceptible hosts and induced hypersensitivity in resistant hosts. Also, moving *hrp* to other bacterial species gave them the ability to induce hypersensitivity. Wild-type *E. amylovora* and transposed isolates produce harpin, a protein associated with the bacterial surface (Wei et al. 1992, 1993). Purified harpin causes cell collapse (hypersensitivity) in nonhost but not in host cells. Thus, resistance or susceptibility of a plant to *E. amylovora* is determined by reaction to a protein on the pathogen's surface (Bauer and Beer 1991; Laby and Beer 1992; Wei et al. 1992; Wei and Beer 1993). Interacting compounds in the host are still unknown.

Cladosporium fulvum, cause of leaf mold of tomato, may have a pathogenesis mechanism similar to that of *E. amylovora.* There are indications that factors promoting compatibility (susceptibility) are present in intercellular fluids of infected tissue (DeWit 1992). This promising lead has not to date been developed; such a factor (or factors) could be comparable in some respects to harpin from *E. amylovora* or to the host-specific toxins. Molecular work with *C. fulvum* is more definitive regarding host resistance; this will be discussed in Chapter 4, section 4.

4
How Plants Defend against Pathogens

Mendelian genetics of resistance to disease in plants is well known. Resistance often is controlled by a single gene, either dominant, semidominant, or recessive; multigene control also is well known. However, the biochemical mechanisms or gene products involved in resistance are known in only a few model cases; summaries of these will be given. The popular notion for years has been that phytoalexins are the basis of resistance; this will be considered, with no attempt at a complete review.

4.1. Preformed Inhibitory Compounds

The most obvious possibility regarding resistance is that resistant plants contain substances, lacking in susceptible plants, that inhibit microbes. This idea gained strength from early work by J. C. Walker (reviewed 1955) (Figure 4.1) on the onion smudge disease, caused by *Colletotrichum circinans.* Resistant onion bulbs have pigmented outer scales that contain inhibitory levels of catechol and protocatechuric acid; susceptible bulbs lack this. However, onion smudge is a unique case. The outer scales are dead; when the scales are removed, the underlying tissues are susceptible. Similarly, decay resistance is evident in wood (dead tissue) of certain trees that is high in tannins. The onion work stimulated much research, but to date there is no conclusive evidence that disease resistance in general is based on presence of constitutive compounds in living tissues. Researchers turned to a search for inhibitory substances synthesized around points of entry by microbes.

There is now some evidence that resistance to another disease is based on a preformed inhibitor (Bowyer et al. 1995). Many plant species contain saponins that are general inhibitors of fungi; saponin in oat roots (avenacin A-1) apparently forms a barrier to infection by some microbes. *Gaeumannomyces graminis,* cause of the "take-all" root disease in many cereals, has

Figure 4.1. John Charles Walker (1893–1994). (Courtesy of American Phytopathological Society)

an enzyme that detoxifies avenacin, and some strains of the fungus are
pathogenic to oats. Presence of avenacin can give resistance to this pathogen. The findings raise several questions. Various fungi invade oat roots: Do
such pathogens also have avenacin-detoxifying enzymes? Many plants contain saponins: Do all their pathogens have such enzymes? Are some pathogens insensitive to avenacin?

The nature of the take-all disease must be considered in relation to the
avenacin findings. *G. graminis* is an opportunistic soil fungus that needs
undecomposed crop residues for survival and parasitic activity; competing
microbes and nitrogen levels are involved. Oat is affected only by a strain
found in Britain; most other isolates do not affect oat, which is resistant. In

areas other than Britain, oat is recommended in rotations to control the disease. Genetics of resistance to *G. graminis* is not understood, but many genes are involved.

4.2. Phytoalexins

Phytoalexins are produced in disturbed, stressed, or wounded plant tissues; such compounds often accumulate after invasion by potential pathogens. Phytoalexins inhibit growth of bacteria and fungi, *in vivo* and *in vitro*. The idea that such inhibitors are responsible for disease resistance was first stated by K. O. Muller, a German phytopathologist, based on experiments with potato and *Phytophthora infestans*. The hypothesis was developed by I.A.M. Cruickshank of Australia, working with fresh pea and bean pods. Many researchers have joined the bandwagon, and now there is a very large literature (Bailey and Deverall 1983; Van Etten, Matthews, and Matthews 1989). Plant pathologists have generally accepted the phytoalexin hypothesis.

Phytoalexins have been isolated and characterized from many plant species, notably legumes, solanaceous plants, various grasses, orchids, sweet potato, carrot, cotton, and beet. The original idea was elaborated to consider phytoalexin inducers (elicitors) and suppressors. Phytoalexins are not found in inhibitory concentrations in normal healthy tissues but are present in infection droplets on plant surfaces, in intercellular spaces and cell walls around infection sites, and in dead or dying cells. Concentrations are high in the vicinity of aborted penetrations in resistant tissues. Pathogens usually are more tolerant of phytoalexins than are nonpathogens.

Chemical analyses gradually revealed several different types of compounds acting as phytoalexins. Among the first discovered were pterocarpans from legumes (examples: pisatin from pea; phaseollin from bean; medicarpan from alfalfa) and sesquiterpenes from solanaceous plants (rishitin from potato and tomato; capsidiol from *Capsicum* spp.). The list of characterized phytoalexins now includes cryptophenols, isocoumarins, isoflavenoids, and others. Fragments of some cell-wall components will elicit phytoalexin production; other elicitors include certain polysaccharides, glycoproteins, chitosans, glucans, arachidonic acid, elaidic acid (Stoessl 1983), many nonorganic materials, and ultraviolet light. Heavy metal ions are among the best elicitors. Possibly, any stress or injury will lead to phytoalexin production by plant tissues.

From the first, a few skeptics raised questions regarding the phytoalexin hypothesis. In some cases, inhibition by the compounds *in vivo* and *in vitro* was poorly correlated. High levels sometimes were present in infected susceptible tissues, and low levels sometimes were in resistant tissues. Single-gene control of resistance is difficult to fit into the hypothesis, because a whole set of genes is required for phytoalexin production; thus, it is difficult to rationalize how phytoalexins account for host specificity. In some cases, resistance is expressed immediately following penetration. Are phytoalexins produced early enough and in sufficient concentration to account for such resistance? Phytoalexins have not been found in some plant species, and resistance in a few cases has been found to reside elsewhere; thus, no general role in resistance can be assumed. Such questions usually were rationalized or overlooked.

Most doubts were erased by reports from Van Etten et al. (1989), with convincing data showing that phytoalexins can account for resistance. Their experiments were with *Nectria haematococca* (anamorph = *Fusarium solani*), a fungus that causes root and stem rot of pea, bean, and other plants. Results indicated that pathogenicity depends on ability of the fungus to detoxify phytoalexins via an enzyme, pisatin demethylase. Strains lacking the enzyme caused much smaller lesions than did strains with the enzyme. Recombinant DNA techniques were used to transfer genes for pisatin demethylase to a fungus that lacked both the enzyme and the ability to colonize pea stems; the receiver then caused a small but nonspreading lesion. The conclusions were that pisatin demethylase is a virulence factor and that phytoalexin is significant in resistance, at least for an opportunistic fungus that needs wounds for entry and a stressed plant for significant colonization. There are other examples of fungal inactivation of phytoalexins, but none have been examined this carefully.

Later work, however, did not support the conclusions. Mutants that lacked pisatin demethylase were created by disrupting the gene cluster in control of demethylase; some of the mutants were still pathogenic. Thus, pisatin demethylase is not essential for pathogenicity or virulence. How can the conflicting data be resolved? The gene for pisatin demethylase and gene(s) for pathogenicity in this fungus are located on a B-chromosome that is unstable and not necessary for life functions (Miao et al. 1991); the whole B-chromosome seems to segregate intact under most conditions. Pisatin demethylase and pathogenicity genes probably were separated in the disruption process (Yoder and Turgeon 1994).

The phytoalexin hypothesis was not supported by research with *Arabidopsis thaliana,* a species that is resistant to many pathovars of *Pseudomonas syringae.* Resistance to the pathovars was formerly thought to depend on accumulation of a phytoalexin known as camalexin. However, no correlations were found between camalexin production by the host and virulence/avirulence of the bacterium in that host. Three mutants of *A. thaliana* deficient in camalexin potential were tested; the mutant plants were found to be as resistant to *P. syringae* as were the wild-type plants. Thus, phytoalexin is not needed for resistance in this case. Virulent strains of *P. syringae* made somewhat more growth in plants of two of the three mutants than the same strains did in wild-type plants, suggesting that phytoalexins may have a minor role in the restriction of virulent pathotypes but apparently no role in the more dramatic restriction of avirulent pathotypes (Glazebrook and Ausubel 1994).

There are still no conclusive evidences that phytoalexins are key factors in plant disease resistance, although they may contribute in secondary ways. The research on phytoalexins nevertheless has been of value because it has added to our understanding of wound healing. Phytoalexins may be higher around infection sites than around ordinary wounds because a continuous irritation is present.

4.3. Toxin Inactivation

We now have a much better understanding of the biochemistry of resistance in a few model diseases. The first were the diseases mediated by pathogen-produced host-specific toxins (Scheffer and Livingston 1984; see Chapters 12 and 13). Resistant genotypes are insensitive to these toxins; the basis of insensitivity has been determined for at least two of the toxins. Resistance in maize to *Cochliobolus carbonum* race 1 is controlled by a single dominant gene (*Hm*). The gene encodes a reductase enzyme that converts the fungal-produced toxin (HC-toxin) to a nontoxic derivative (Meeley and Walton 1991). This reductase activity is present only in resistant, not in susceptible, maize (Meeley et al. 1992). Clearly, the biochemical basis of resistance in maize to *C. carbonum* race 1 is enzymatic detoxification of HC-toxin from the pathogen. The gene for resistance (*Hm*) was cloned by transposon tagging; its product is a dinucleotide-dependent carbonyl reductase (HC-toxin reductase) (Johal and Briggs 1992).

Resistance to *C. heterostrophus* race T and to its toxin (T-toxin) depends on a different mechanism. This toxin has a mitochondrial site of action; mitochondria in susceptible (Tms) maize have a unique 13-kd protein (T-URF13) not found in resistant plants. Genetic control of this protein is encoded in a mitochondrial gene, an open reading frame (*T-urf13*) presumably originating from spontaneous rearrangement of the mitochondrial genome (Dewey, Timothy, and Levings 1987; Wise et al. 1987; Dewey et al. 1988).

When the *T-urf13* gene was transferred to a bacterium, to a yeast, and to tobacco, these species became sensitive to T-toxin, making them the only known suscepts other than Tms-cytoplasm maize (Dewey et al. 1988; Huang et al. 1990; Vonallmen et al. 1991). T-toxin is bound reversibly to the T-URF13 protein (Braun, Siedow, and Levings 1990). Thus resistance to *C. heterostrophus* and to its toxin depends on lack of a protein receptor for toxin.

4.4. Avirulence/Virulence Genes in Pathogen

Erwinia amylovora and several other pathogenic bacteria have *hrp* genes that control ability to cause a hypersensitive (resistant) reaction. Mutants of several bacterial species that lack *hrp* do not cause hypersensitivity; thus, *hrp* genes in the pathogen are essential for expression of resistance in the host. The *hrp* genes in *Pseudomonas solanacearum, P. syringae,* and other species are known to control both hypersensitivity in nonhost genotypes and susceptibility in hosts. The *hrp* genes in *P. solanacearum* are on a large plasmid; most mutants lacking *hrp* were shown to be nonpathogenic and to lack ability to induce hypersensitivity in resistant hosts. The products of *hrp* genes are unknown, except for *E. amylovora* (see Chapter 3, section 5, and Chapter 8, section 2), but they probably are needed for an early step in colonization of susceptible tissues (Panopoulas and Peet 1985; Boucher, Gough, and Arlat 1992).

The leaf mold disease of tomato, caused by *Cladosporium fulvum,* has become a major model for molecular studies. There are many races of the fungus and several host genes for resistance; a gene-for-gene relationship, for avirulence in the pathogen and resistance in the host, is now assumed. Fungal proteins (elicitors) were extracted from infected tissues but were not detected in cultures. The elicitors caused hypersensitivity in tissues of resistant tomatoes, as did the fungus; the elicitor did no damage to susceptible tomato plants. Hypersensitivity is thought to be a major factor in some types of disease resistance.

One elicitor from *C. fulvum* is the presumed product of an avirulence (*avr9*) gene. The elicitor was not detected in plant tissues infected with a fungal isolate that lacked the *avr9* gene. The elicitor was said to interact with the presumed product of a host resistance gene (*Cf9*). The elicitor gene was cloned and shown to form an mRNA that, in turn, encoded a precursor protein that was sequenced for sixty-three amino acid residues. The data show that fungal races carrying the recessive allele for *avr9* are virulent on hosts with the *Cf9* gene (DeWit 1992).

A race of *C. fulvum* with virulence to tomato plants that carry the *Cf9* gene was transformed with the *avr9* gene. The transformed race was avirulent, proving that *avr9* is responsible for incompatibility (resistance) in tomato plants with the *Cf9* gene. There are similar analyses of another virulence gene (*avr 4*) from *C. fulvum* (Van Kan et al. 1994). Further discussion of the *C. fulvum* model is in Chapter 3, section 4.

Several avirulence (*avr*) genes from bacteria have been cloned, but the primary products of such genes are unknown; the products should be involved in induction of hypersensitivity. However, *Pseudomonas syringae* pv. *tomato* with an avirulence (*avrD*) gene produces a secondary product, a low-mol-weight, lipidlike elicitor of phytoalexin (DeWit 1992).

4.5. Acquired Resistance

Some susceptible plants become systemically resistant in response to localized infections, a phenomenon known as acquired resistance. This is best known in cucurbits and tobacco. When a lower leaf is infected, the whole plant becomes resistant to the same and to other pathogens and remains so for weeks. Plants with acquired resistance have high levels of pathogenesis-related proteins, salicylic acid, peroxidase, and other factors. Obviously, there is a signaling mechanism that carries information to distant parts of the plant, but the nature of the signal is unknown (Hammerschmidt and Kuc 1995). There are significant theoretical and practical implications.

5

Ecological Considerations

5.1. Plant Disease in Natural and Managed Systems

Plant disease is present in both natural and managed ecosystems, but the situation differs in the two environments. In undisturbed natural systems, most experts agree that diseases usually are not a serious threat to the host, although many pathogens exist there. Past disease threats presumably have been restrained, resulting in a balance between pathogen and host. In managed and mismanaged systems of agriculture and forestry, diseases often become serious factors, requiring still further management.

Anthropogenic factors usually are involved in plant disease epidemics. This conclusion is based largely on observations, but it is a general one, as can be shown by a few quotations.

> Strong epidemics, as far as is known, have always followed some identifiable human activity. Epidemics are largely man-induced (Zadoks and Schein 1979).

> Epidemic diseases . . . are not seen in natural vegetation – except after some major disturbance (Harper 1977).

> Man is the direct or indirect cause of most (possibly all) of the epidemic inbalances we know about. The high virulence and extreme susceptibility of an epidemic situation is an unnatural inbalance usually brought about by human disturbance (Harlan 1976).

The problem is that we have few, if any, undisturbed situations for comparison.

The term *natural balance* is popular; in general, it is a valid but overused concept. The extreme is the notion that disease is absent in plants that are grown "organically." In fact, local or patchy epidemics are common in natural ecosystems; there is an ebb and flow, depending on local conditions and variations in both host and pathogen populations. In nature, there are genetically diverse populations of hosts and pathogens, all intermixed. Host populations may be overcome locally by pathogens, but resistant genotypes soon fill the void. Next, virulent genotypes of the pathogen may move in (Burdon 1993), accounting for the ebb and flow. Consider that even a "climax" flora adjusts to changes in conditions; there is no firm and final climax. Nevertheless, the undisturbed ecosystem clearly is well buffered; epidemics that affect large geographical areas are rare or nonexistent without a major disturbance such as introduction of a virulent alien pathogen, as occurred with chestnut blight. The concept of "natural balance" is speculative because there are few, if any, undisturbed areas left for observation and comparison.

Definitive evidences for anthropogenic factors in disease are available in data and observations on recently domesticated native plants. Here, some microorganisms that were mild or opportunistic pathogens in the natural system suddenly became serious problems, causing destructive epidemics in plantings. There are many examples in forest trees. When *Populus* species were planted in monoculture in the northern United States, *Septoria* canker and rust became limiting factors (see Chapter 10, section 2); the pathogens were present but of no consequence in the natural forest. When pines were planted in blocks, especially outside their native range, *Dothistroma, Diplodia, Tympanis,* and other previously innocuous fungi became destructive (see Chapter 8, section 4, and Chapter 10, section 2). When wild rice was domesticated in Minnesota, fungi of the genus *Cochliobolus* soon limited cultivation (Chapter 10, section 3). Wild rubber trees in Amazonia had a trivial blight disease; plantation trees in the area were soon destroyed by the same disease (Chapter 10, section 1). There are many other examples.

Why do these problems occur in managed systems? First, managed systems usually are monocultures, with no barriers to dissemination of inoculum. In contrast, natural systems usually are composed of a mixture of species, and this inhibits dissemination of host-specific pathogens. Also, each host species can contain diverse genotypes that can vary from susceptible to resistant, again a deterrent to inoculum and disease buildup. Plantations usually are started from individuals selected for high yield or fast growth; genetic diversity, including genes for resistance, is lost. The environment in a plantation differs greatly from that in a natural stand; there usually are more extremes in moisture, temperature, and other environmental factors, and there are different weed competitors. Populations of vectors such as leafhoppers and aphids are more likely to reach epidemic levels in monocultures, increasing the possibilities of the spread of certain diseases.

All this leads us to think of the serious diseases as "crowd diseases"; they became destructive because genetically uniform crops were grown in monoculture, typical of modern agriculture (Browning 1974).

Endemic pathogens tend to reach equilibrium with hosts under undisturbed conditions, much as a climax flora achieves with its environment. When a virulent pathogen first reaches an area with susceptible hosts, the ensuing disease tends to be severe. With time, the host population tends to become tolerant, and the pathogen tends to become less virulent. Milder strains usually survive better because the host is not eliminated. This tendency is well known for certain animal diseases, such as myxomatosis of rabbits in Australia (Harlan 1976, 1977); historical records show that it holds as well for human diseases such as smallpox, measles, and others. Maize rust in Africa (Chapter 9, section 3) is a clear example with plants; the disease was severe when first introduced because the crop was very susceptible, and great losses occurred. Over time, the fungus became less virulent, the surviving maize was more tolerant, and the severe epidemic subsided. Maize had been taken to Africa without rust, and the crop had lost the tolerance possessed by its ancestral forms in America (Borlaug 1972). Clearly, pathogen and host tend to become "coadapted" by genetic adjustments. Even the notorious chestnut blight has eased in Europe.

Disease resistance involves a genetic load or cost, and yield is likely to be lower than yield of a susceptible crop without its pathogen (Harlan 1976, 1977). Genes for resistance give fitness in the presence of disease, but this seems to be at the expense of other kinds of fitness.

The situation, in some cases, is more complicated than that implied previously. The general impression is that natural populations became resistant to endemic pathogens because the most susceptible genotypes are eliminated over the years, but this is not always the case. Damaging epidemics generally do not occur in wild grain in Israel and adjacent areas, the endemic home of wheat and barley. Yet susceptible individuals were not eliminated by coevolution with endemic pathogens; suscepts survived along with tolerant genotypes. The present population is well buffered; it contains plants with general nonhost immunity, plants with polygenic resistance, and plants with race-specific resistance (Browning 1974). There are successful natural monocultures in some areas, as with conifers in the far north, possibly because of genetic diversity within the species (Dinoor and Eshed 1984).

Resistance in a population that is no longer exposed to a pathogen is likely to be lost, whereas it usually is maintained in a population that is constantly exposed. However, there are examples of resistance genes being maintained without the pathogen, because of close linkage with other genes.

In some cases, such linkages eventually have been broken (Mather 1973). Also, genes for virulence in the pathogen may survive even without a susceptible host, possibly because of linkage to other genes important for survival (Leonard 1977). This seems to be the case with *Cochliobolus victoriae,* which was found in areas where susceptible oat was never grown (Chapter 12) (Scheffer and Nelson 1967).

Tropical rain forests are extremely diverse, with many species growing together. This creates a strong buffer against spread of disease, and epidemics are unlikely without some significant disturbance. Nevertheless, endemic diseases are common. Primitive cultivators seem to follow this cue, although they may not comprehend the reasons, and plant many species together in their gardens (Harlan 1976). Their cultivars, or "land races," are selected for survival; this maintains resistance to local diseases, but yields are not spectacular, never reaching levels attained in modern agriculture, where tolerance to disease often is lost. However, stability may be more important than yield in subsistence agriculture. The tolerance in native land races often is multigenically controlled; it is similar in this respect to that in natural ecosystems. Plant breeders often turn to land races for a source of resistance.

5.2. Origins and Evolution of Pathogens and Disease

Diseases of plants and animals probably were present almost from the beginnings of such life forms, but because fossil records are scarce, we can only postulate on origins. There are few undisputed facts, and many of the hypotheses will never be verified; nevertheless, we can make interesting and rational suggestions, based on the nature of primitive organisms that exist today.

A possible beginning of disease is illustrated very well by some simple observations of amoebae (Burnet and White 1972). Amoebae envelop and digest bacteria and microscopic algae, extracting usable compounds; unusable remains are simply pushed out. Evidently, the amoeba can recognize "non-self"; otherwise, why do its digestive enzymes not destroy its own structure? Sometimes an amoeba ingests a microorganism that is not recognized as non-self, and the prey is excreted unharmed. Sooner or later, strains of the unrecognized prey could have adapted and multiplied at the expense of the predator. These or comparable circumstances could be involved in the origin of disease in animals. Disease in plants may have started with somewhat different circumstances.

Self-recognition is involved in disease, and the sponge is a useful model.

The sponge, a primitive animal, comes in several colors. Sponges can be forced through sieves, which break the sponges into single cells and small groups of cells that are still alive. Red sponge cells can then be mixed with white sponge cells in seawater and left undisturbed. Red cells aggregate to form a new red sponge, and white cells form a new white sponge (Burnet and White 1972). Obviously, there is recognition of self or non-self. Recognition systems are the rule throughout the animal kingdom, an indication that the mechanisms developed early in evolution. Recognition and defense systems in animals have become very complex.

Plants also recognize non-self. For example, almost any apple cultivar can be grafted successfully to another apple cultivar, but apple and oak cannot be grafted; if such grafting is attempted, the adjoining cells die and the graft fails. Pollen provides another example of recognition of non-self; if placed on the stigma of its own species, pollen will germinate, grow down the style, and fertilize the egg. Pollen placed on the stigma of another species will germinate, penetrate, and die, often followed by the death of a few pistil cells. Comparable reactions occur when pathogenic microorganisms contact host or nonhost plants. For example, rust spores placed on the leaf of a susceptible plant will germinate, penetrate the surface, and colonize the tissue. When placed on the leaf of a nonhost, the spore germinates, penetrates, and dies. Plant cells adjacent to the invader often die as well; this is known in plant pathology as the hypersensitive reaction. Susceptible host cells appear not to recognize the rust fungus as non-self, whereas the nonhost does so. The biochemistry of recognition and nonrecognition in plants is poorly understood; it may have to do with protein structure.

A wide range of pathogenic microorganisms interact with plants, following similar patterns of recognition or nonrecognition. Both obligate pathogens (for example, rust fungi) and nonobligate pathogens (for example, bean anthracnose fungus) first have a compatible relationship with the host and cause hypersensitive reactions in nonhosts. In fact, a basic requirement of a specialized and successful pathogen is that the host not recognize it. Careful studies show that the distinctions between so-called biotrophs (such as the obligate parasites) and necrotrophs (said to live only on killed host cells) have little meaning in the early stages of infection.

Most plant pathogens seem to have evolved from saprophytes, an opinion supported by many observations. Many fungal genera contain pathogenic, opportunistic, and saprophytic species; in some cases, some or all such variants are sexually compatible. This suggests that pathogenicity has arisen many times among various microorganisms. Other observers have speculated that parasitic fungi were derived from biotrophic parasites. However, this goes against a common assumption by evolutionists that new life forms arise from unspecialized forms with genetic plasticity, not from specialized climax forms. Pathogens and parasites, in general, occupy ecological niches that require more specialization than is possessed by saprophytes. Still other observers argue that pathogens developed from symbionts such as *Rhizobium* or endomycorrhizal fungi. This could be true in some cases. *Rhizobium*, under adverse conditions, can live entirely at the expense of its host. However, this scenario is unlikely for most pathogens, since some symbionts belong to specialized taxonomic groups with unknown affinities. More detailed discussion of the evolution of plant parasites was provided by Heath (1987). No hypothesis can be verified at this time, and some are unreasonable.

It is impossible to know precisely how each plant pathogen evolved to its present relationship with its host. Nevertheless, there are rational speculations that fit many cases. Many bacterial and fungal species live as true saprophytes, utilizing plant and other organic debris. Some races of saprophytes developed the ability to live on the surfaces of living plants, without harming living cells. Further adaptations allowed surface feeders to colonize senile or stressed tissues and, finally, to exploit metabolically active tissue.

New and destructive diseases of plants are on record and continue to appear. How does this happen? One explanation of new diseases is genetic adaptations (microevolution) that occur in both opportunistic and specialized pathogens. If the genes of a pathogen are altered in a way that allow it to exploit the prevalent environment, including host tissue, then the pathogen has at least a temporary advantage: It may multiply explosively and may spread into new areas. This is well known with several *Cochliobolus* and *Alternaria* species (Chapters 12 and 13). Newly acquired ability to excrete certain metabolites has transformed previously weak or opportunistic pathogens to virulent and host-selective forms (Scheffer 1991). In other cases, benign or "well-balanced" pathogens became destructive when they encountered new hosts that lacked resistance; this happened with chestnut blight (Chapter 7, section 1). Pathogens readily evolve to exploit monocultures of genetically uniform plants, a trademark of modern agriculture. These and other ideas will be developed in chapters to follow.

Fungal genera that inhabit leaves, with species that are either saprophytes or parasites, include *Alternaria, Ascochyta, Botrytis, Cladosporium, Cochliobolus, Fusarium,* and many others too numerous to mention. There appear to be adaptive or evolutionary lines among such fungi; the perceived progressions go from true saprophytes to opportunistic forms that invade only senescent tissue and, finally, to host-selective forms that colonize metabolically active tissue. Such lines are present in root-infecting fungi as well.

A classification of fungal–plant relationships (Scheffer 1991) will help to understand the significance of such a spectrum within a given genus; the classification reflects possible evolutionary developments. The most simple fungus–plant relationship is saprophytic growth on plant surfaces (category A). The microorganism never invades or damages normal living cells; it subsists on nutrients leaked from the plant. There is a more intimate relationship with fungi that can invade tissues as opportunistic pathogens. Fungi in this category (B) are low in virulence, require wounds for entry, and need senescent or stressed host tissue for colonization and disease expression. The pathogen appears to have the ability to prevent recognition as "non-self." A slightly more specialized group of opportunists has the capacity to penetrate living tissues via special structures but fails to colonize and produce disease except in senescent or stressed tissue. Some of these fungi (category C) affect certain host species, not others; that is, they are host-specific.

A still more evolved and specialized group (category D) includes fungal species or genotypes that are host-selective and virulent on metabolically active tissues. These fungi can exist as saprophytes in culture and under special conditions in nature; however, they primarily are parasites, not competitive with aggressive saprophytes in soil or debris. Members of this group cause some of our best-known and most destructive diseases. The pathogens have evolved mechanisms that bypass, suppress, or ignore the normal plant responses to disturbance. Generally, the physiological basis of this is unknown. One known explanation is the excretion of minute amounts of very active compounds, as occurs with some species of *Alternaria* and *Cochliobolus* (Scheffer and Livingston 1984; Scheffer 1992). Resistance in the host plant also is specialized, affecting only the specific pathogen; often, resistance is controlled by a single gene, either dominant, semidominant, or recessive. Such resistance is superimposed on the normal plant response to disturbance, which includes production of inhibitory compounds known as phytoalexins (Baily and Mansfield 1982). In general, all this applies to both bacterial and fungal pathogens. Isolates of a given species that fit each of these categories can be obtained without difficulty.

There are two more categories of microbe–plant relationships that are recognizably different from those mentioned previously. One category includes the so-called obligate parasites that are highly evolved in host-parasite interactions and that often follow a "gene-for-gene" pattern of virulence–host resistance (Ellingboe 1981). Rust fungi, downy mildews, and powdery mildews are in this category. Such pathogens are difficult to grow in culture and do not complete their life cycles in nature without the host. Viruses may be the ultimate obligate parasites. Finally, there are a few special fungi that are both saprophytes and highly virulent pathogens on a wide range of host species. This group includes a root pathogen (*Phymatotrichum omnivorum*), the southern blight fungus (*Sclerotium rolfsii*), the "take all" fungus (*Gaeumannomyces graminis*), and a root rotter (*Phytophthora cinnamomi*).

Many genera of fungi have species, races, or genotypes that fit each of the first four categories (A–D). In some cases, there are no morphological differences between saprophytic and pathogenic forms, and they are placed in the same species. In a few cases, sexual mating of saprophytic and pathogenic forms prove close relationships. The variable fungus *Phytophthora megasperma* is a good model; the various form species and races are closely related, as shown by chromosome counts, DNA homology, and protein profiles (Hansen 1987). Many other genera deserve such study. Four genera of fungi will be discussed as examples of leaf-inhabiting fungi that have saprophytic, opportunistic, and virulent species or genotypes: *Alternaria*, *Cochliobolus*, *Phytophthora*, and *Botrytis*.

Alternaria species include many saprophytes; *A. alternata* and *A. tenuissima* are the best known. *A. alternata* grows as a saprophyte on foodstuffs, textiles, and plant debris and on the surfaces of living plants. It is perhaps the most cosmopolitan of fungi, occurring worldwide in temperate and tropical climates. The spores are common allergens for people. The species includes a varied collection of forms, both opportunistic and specialized virulent types. The species designation is based entirely on morphology. *A. tenuissima* also includes forms that are opportunistic pathogens on many plant species (Domsch, Gams, & Anderson 1980); one form is virulent and host-selective (Nutsugah et al. 1993, 1994).

A. alternata as a leaf-surface inhabitant has special meaning in the present context. The fungus can exist on nutrients leaked from plant tissues; it has been called a "common primary saprophyte" (Hudson 1971). The leaf-surface inhabitants quickly invade plant tissues as they die; thus, they are the initial saprophytic invaders of plant residues. Adaptations by *A. alternata* have made it possible for some strains to invade living but senescent or stressed plant tissues. In some unknown way, the fungus appears to have developed at least a limited compatibility with living host cells.

Many pathogenic races or forms of *A. alternata*, some with host-selectivity, have been identified; the morphology of all pathogenic forms is the same as that of saprophytic forms. Examples are *A. alternata* f. sp. *citri* (=*A. citri*), which infects senescent leaves of many citrus species, and *A.*

alternata f. sp. *mali* (=*A. mali*), which infects senescent tissues of apples and related plants. Known genotypes of each of these are specialized further, affecting only certain species or cultivars of the host. These specialized forms affect metabolically active tissues of specific hosts. Evolution to selective virulence in *Alternaria* is based on acquisition of genes for production of toxic compounds with the same selectivity as the pathogen itself (Scheffer 1992). Ability to produce host-specific metabolites had major ecological consequences, which will be discussed in Chapters 12 and 13.

A. alternata has many other forms that are specialized to single hosts or to a group of related plants. In fact, most of the *Alternaria* species listed by Ellis (1971) are said to be host-specific, even though many are opportunistic pathogens on senescent or stressed tissues. Many of these species are now classified as form species of *A. alternata*. The biochemical basis of selectivity by most of the *Alternaria* pathogens is unknown. The highly virulent and more specialized forms of *A. alternata,* in addition to *A. citri* and *A. mali,* include those pathogens formerly known as *A. kikuchiana* (affecting Japanese pear), *A. brassicae* (on *Brassica* spp.), *A. lycopersici* (on tomato), and *A. longipes* (on tobacco) (Domsch, Gams, and Anderson 1980; Scheffer 1992). It is apparent that *A. alternata* is genetically diverse and prolific in spawning new races or genotypes. New forms appear after widespread planting of new crops or new crop genotypes.

Various species, races, and genotypes of the Helminthosporia fit into one or the other of the categories (A–D) of microbe/plant relationships previously described. The term *Helminthosporia* is used herein to designate a group of related fungi, the most important of which are those in the genus *Helminthosporium* as formerly constituted. The group includes the sexually reproducing genera *Cochliobolus, Pyrenophora,* and *Setosphaeria,* plus the asexual genera *Bipolaris, Exserohilum, Dreschlera, Curvularia,* and *Helminthosporium.* A few fungi in the group are pure saprophytes; other species, races, or genotypes clearly are opportunistic, nonspecialized pathogens. In general, the Helminthosporia are poor saprophytes that do not compete well with aggressive saprophytes (Garrett 1956).

Many members of the Helminthosporia, as well as many other fungi, excrete diverse secondary metabolites that could contribute to their survival as either pathogens or saprophytes (Scheffer 1991). Among the metabolites are the host-specific toxins that affect only hosts of the producing fungus; nonhosts are insensitive. Host-specific toxins are known for *Cochliobolus victoriae* on oats, *C. carbonum* race 1 on maize, *C. heterostrophus* race T on maize, and *Bipolaris sacchari* on sugarcane (Chapter 12). These species are specialized and virulent; their special pathogenicity depends on secretion of host-specific toxins. Other specialized species include *C. miyabeanus* (on rice), *Bipolaris gramineum* (on cereals), *Exserohilum turcicum* (on maize), *Pyrenophora bromi* (on certain grasses and cereals), and many others (Sprague 1950; Ellis 1971). There are indications but no conclusive evidence that some members of this latter group produce host-specific toxins. Unfortunately, the nomenclature of the Helminthosporia is confusing; some so-called species are based only on single gene differences in pathogenicity (Scheffer, Nelson, and Ullstrup 1967).

C. sativus (anamorph, *H. sativum* = *Bipolaris sativus*) appears to be an ancestral species for several of the specialized virulent forms in the genus. Generally, *C. sativus* is considered to be nonspecialized, with a wide host range on grasses and cereals (Garrett 1956; Domsch, Gams, and Anderson 1980). Nevertheless, the species as now recognized includes mild opportunistic strains, possibly even saprophytes, and aggressive specialized strains or genotypes (Kline and Nelson 1963; Scheffer 1989a; Scheffer 1991). These forms may have minor differences in morphology in addition to the differences in pathogenicity, and some can be mated sexually (Nelson 1960; Nelson and Kline 1963; Scheffer 1991). All the types, both opportunistic and specialized, can colonize senescent or weakened tissues of many grasses and cereals. New genotypes of *C. sativus* sometimes appear, apparently from genetic adaptations. Differences in virulence and pathogenicity appear to be under single-gene control, in some cases (Nelson 1961; Nelson and Kline 1969; Nelson 1970).

The major characteristics of *C. sativus* are shared with *C. carbonum* and *C. victoriae,* presently recognized as different species. These three species are morphologically similar, with sexual compatibility between *C. carbonum* and *C. victoriae* (and possibly with certain isolates of *C. sativus*), indicating close relationships. There also is chemical evidence that indicates a close relationship between *C. sativus* and *C. victoriae* (see Chapter 12, section 1). Each of the three can colonize senescent tissue of many grasses and cereals. The basic form of *C. carbonum* (race 2) is opportunistic, affecting many genotypes of maize and many grasses. Several new races of *C. carbonum* have appeared over the years; each is highly selective and virulent on only certain genotypes of maize (Ullstrup 1941); similar patterns of pathogenicity, virulence, and host-selectivity are evident in other genera and species of fungi (Domsch, Gams, and Anderson 1980; Scheffer 1991; Scheffer 1992).

There are gene pools in natural populations of *Cochliobolus* that control selective pathogenicity to many grass species; selectivity to a given species often is controlled by one or two gene pairs (Nelson 1961). The fungus identified as *C. carbonum* is closely related to and hardly distinguishable from *C. sativus*. Progeny of crosses involving six isolates of *C. carbonum* that differed in pathogenicity were tested against seven grass species; results for F_1 progeny indicated at least five different genes for selective pathogenicity (Nelson 1970). Some interspecific hybrids of *C. sativus* and *C. carbonum* were said to have a wider host range than those of the parents. The potential for evolution of pathogenicity is great.

The *Cochliobolus* races that produce host-specific toxins appeared following widespread planting of new crop genotypes (Scheffer 1989a, 1991). *C. victoriae* was discovered after a new gene for resistance to rust was introduced into cultivars of oats, a self-pollinated and, therefore, genetically uniform crop. Cultivars that carried the new gene were superior in yield and quality and were widely planted; they had to be discarded after the appearance of *C. victoriae*. *C. carbonum* race 1 appeared in monocultures of certain maize inbred lines and their hybrids. *C. heterostrophus* caused a major epidemic in genetically uniform maize that carried Texas male-sterile cytoplasm, used throughout the grain belt of the United States for economy in production of hybrid seed. *Bipolaris sacchari* became a serious problem in certain new clones of sugarcane, which is vegetatively propagated and, therefore, genetically uniform. New diseases of this type are not unusual; we have detailed histories of these and several others that will be covered in Part II.

Adaptations in pathogenicity and virulence are evident in the fungal genus *Phytophthora* (Hansen 1987). In general, *Phytophthora* species are soilborne and not readily dispersed over long distances in nature. Agricultural practices, especially movement of rooted plants, have moved various *Phytophthora* species to new areas, resulting in isolated populations and rapid local adaptations (microevolution) to new conditions. The new forms at first show little or no morphological change but readily develop selective pathogenicity to new hosts. Restriction to specific hosts brings further genetic isolation; morphological changes can gradually occur, to the point that new species are recognized. This is evident in *P. megasperma*, a general opportunist that causes root rot on weakened plants of many species. *P. megasperma* appears to have spawned many races or form species with virulence to certain hosts. One of these new forms is *P. megasperma* f. sp. *glycinea* *(= P. sojae)*, cause of a destructive disease of soybean (Chapter 11, section 2). Relationships between the *P. megasperma* complex and related forms were studied by chromosome counts, DNA homology, and protein profiles, powerful tools for determination of genetic relationships (Hansen 1987).

The fungal genus *Botrytis* should be useful for studies on origins and evolution of plant pathogens. *B. cinerea* is a collective form species so variable that isolates sometimes are designated as "*Botrytis* of the *cinerea* type" (Coley-Smith, Verhoeff, and Jarvis 1980). It is a ubiquitous fungus, growing as a saprophyte on plant debris in temperate areas of the world. Certain strains grow on plant surfaces without harming healthy tissues; other strains are a common cause of rot in stored fruit, vegetables, and flowers. Some strains are relatively virulent on many plants, causing blight or gray mold (Domsch, Gams, and Anderson 1980) on more than two hundred species; the fungus is a major problem on grapes, grapevines, and nursery plants. Variants of the species differ in virulence. *Botrytis* cells are multinucleate, and much of the variability may be traced to heterokaryosis.

The genus *Botrytis* has species other than *cinerea* that are more specialized. Examples are *B. allii* on onion, *convovula* on iris, *gladioli* on gladiolus, *narcissicola* on narcissus, and *tulipae* on tulips and lilies. Several of these, as well as *B. cinerea,* have been examined serologically, and all appear to be closely related. Overall, these species fall into the same categories of host–pathogen relationships described for *Alternaria* and *Cochliobolus.* The sexual stage *Botryotinia* (a cup Ascomycete) is known for some of the species, including *B. cinerea* (Coley-Smith, Verhhoeff, and Jarvis 1980), and this should help in genetic analyses. Other fungal genera, including *Sclerotinia,* are closely related to *Botrytis.* Data on DNA homology and protein profiles, correlated with host–range studies, should help to clarify relationships.

Origins and evolution of pathogens in the genera *Alternaria, Botrytis, Cochliobolus,* and *Phytophthora* are believed to be representative of many other plant pathogens and the diseases they cause. Still others, such as the rusts and mildews, appear to be more complex; they may have evolved in much the same way but went further in adopting the parasitic life-style. The possibility for saprophytic existence thereby was lost, and living hosts became necessary for completion of the life cycle. Another possibility is that rusts and mildews evolved from symbionts (Heath 1987).

The origin of viruses is still more difficult to comprehend. Viruses are the ultimate obligate parasites; they live and reproduce only in appropriate host cells, and they alter host metabolism to produce more virus. There have been several thoughtful and rational suggestions as to their origin; one is that they descended (or degenerated) from bacteria. Perhaps after long periods of a host–parasite relationship, the organism gradually lost all structures and functions not essential to parasitic life. Another possibility is that viruses originated from microorganelles of the host itself, becoming self-replicating (as are mitochondria). Viruses continue to evolve or change by alterations in their chemical structures.

Summary
Pathogens belonging to the fungal genera *Alternaria, Cochliobolus,* and many others have adapted or evolved in response to agronomic changes, especially to widespread planting of new crop cultivars. Virulent, specialized pathogens continue to appear, presumably evolving from low-grade, opportunistic pathogens that affect only senescent plants. The adaptations, in some cases, are known to involve gene mutations or recombinations. Parental forms have wide host ranges on weakened tissues; some forms evolved from them are still opportunists on many hosts but, in addition, have host-specific virulence. The opportunists and the specialized forms of some fungi have characteristics in common, including morphology, host compatibility, and sexual compatibility.

Species of the genera *Alternaria* and *Cochliobolus* are informative models. Some genotypes are opportunistic, whereas others are specialized and virulent on metabolically active tissues. Some of the latter have genes for production of selectively toxic metabolites.

Overall, it seems likely that certain saprophytes adapted to life on plant surfaces and that the surface feeders eventually "learned," in turn, to invade senescent plant tissues. Finally, opportunists developed the ability to exploit metabolically active tissue, causing destructive diseases. Eventually, some pathogens reached a more or less stable or balanced condition with hosts, due to defense mechanisms developed by plants. The balance can be destabilized by genetic changes in either pathogen or host.

5.3. Effects of Virulence and Resistance Genes on Populations

It is obvious that increased virulence and changes in pathogenicity are of short-term advantage to the pathogen, leading to increases in its population, and that this is a disadvantage to the host. Conversely, new genes for resistance in the host are to its advantage and are likely to depress pathogen populations. Of more interest is involvement of the plant breeder in these relationships. In effect, plant breeders are creating new pathogens as well as new plant cultivars. This can be seen in many crops; several examples that are clear and well documented will be highlighted here.

The population of the fungal pathogen *Exserohilum* (=*Heminthosporium*) *turcicum* soared, and the disease (northern leaf blight of maize) became severe throughout the American grain belt after hybrid maize was widely planted. Plant breeders found a gene (*Ht*) for resistance in *Tripsacum floridanum,* a related species, and incorporated it into maize (Harlan 1976); widespread planting of cultivars with the *Ht* gene led to decreased populations of *E. turcicum,* although no quantitative data are available. The *Ht* gene does not give immunity or a high degree of resistance, hence does not exert strong pressure for development of new fungal races. *E. turcicum* is still present, but overall it became a minor problem. After many years, two new races emerged (Leath and Petersen 1986).

Bipolaris (=*Helminthosporium*) *maydis,* anamorph of *Cochliobolus heterostrophus,* caused a minor disease of maize in the southeastern United States until 1970, when it created a continental epidemic. The reason was a new form (race T) of the fungus, which had been detected a few years earlier. Race T is thought to have originated as a mutant or other single Mendelian gene change. The new race became destructive on genetically uniform maize that contained a cytoplasmic factor for male sterility, used throughout the United States and elsewhere for economy in hybrid seed production. This genetic uniformity led to greatly increased populations of *B. maydis* and to an expansion of its geographical area, from the southeastern United States to the whole area of maize production in North America and elsewhere. The disease abated when T-cytoplasm maize was abandoned. At present, the fungus has largely disappeared in the northern United States. Further discussion is in Chapter 12.

The oat crop in North America was greatly improved by the use of genes from a cultivar (Victoria) from Argentina; the resulting new cultivars were popular and soon comprised more then 50 percent of the crop. Within a few years after the new cultivars were introduced, a new disease appeared, with devastating results. A single gene (*Vb*) was shown to account for susceptibility in oat, and a single Mendelian gene in the causal fungus accounts for pathogenicity to oat. The causal fungus *Cochliobolus victoriae* is closely related to widespread, opportunistic species, *C. sativus* and *C. carbonum; C. victoriae* may be only a race or genotype. Apparently, the fungal race affecting oat occurred widely (Scheffer and Nelson 1967) but was never obvious until susceptible oat cultivars were widely planted; then, populations of *C. victoriae* soared. Oat with the *Vb* gene was abandoned, and *C. victoriae* is again a rarity. Further discussion is in Chapter 12.

There are similar situations with a number of diseases caused by host-specific forms of *Alternaria alternata.* When certain new host cultivars were widely planted, new *Alternaria* diseases appeared and devastated the crops.

This has happened with apple, pear, strawberry, tomato, and tobacco in Japan; with apple, citrus, and tomato in the United States, and with citrus in Australia. Planting highly susceptible cultivars led to great increases in pathogen populations and to adverse effects on host populations. Further discussion is in Chapter 13.

Familiar examples of striking decreases in natural populations of hosts, with concomitant increases in pathogen populations, are discussed in Chapter 7, which covers chestnut blight (section 1), Dutch elm disease (section 2), and white pine blister rust (section 3). There are many other examples.

5.4. Environmental and Climatic Effects

Weather, climate, and other environmental factors have decisive effects on the incidence and severity of plant diseases. There are direct effects on pathogen, host, and host–pathogen interaction; there also are indirect or predisposing effects on host plants. All stages of disease inception and development are influenced by environmental conditions: Wind, driving rains, or vectors are required for dissemination of many pathogens; high humidity or free water on plant surfaces is necessary for germination of most fungal spores; and disease inception and development require certain temperatures. Severity of soil-borne disease is affected by fertility, soil type, moisture, oxygen, carbon dioxide, and pH of the soil. Poor light and suboptimal nutrient levels can enhance susceptibility to some diseases but decrease susceptibility to others.

The exact conditions that favor every important plant disease have been determined; for example, cool, wet weather favors some (late blight of potato), whereas warm, humid weather favors others (brown rot of stone fruit and fire blight of apple trees). Several textbooks contain good general discussions of environmental effects on disease (Stakman and Harrar 1957; Walker 1969; Agrios 1988). The APS compendia are good sources for a number of crops; an example is W. J. Hooker's *Compendium of Potato Diseases* (1981).

Environmental factors affecting disease development have been rated on numerical scales and collectively factored into mathematical analyses of the rise and fall of epidemics. Such epidemiological analyses are useful for predicting the incidence of disease (Fry 1982).

Climate determines the distribution of many plant diseases. Some diseases are prevalent in the humid climates of Europe and eastern North America but are missing in dry areas such as the western United States. Apple scab and bean anthracnose, which are rare in dry areas, are good examples. Seeds of annual crops often are produced in dry areas and sold elsewhere to eliminate seed-borne diseases. Other diseases may be favored in dry areas; examples are certain powdery mildews and soil-borne diseases. Low winter temperatures restrict many diseases by precluding overwintering of pathogens, by restricting buildup and dissemination of inoculum, by destroying vectors, or by restricting infection. Some examples of pathogens that are eliminated by cold in the northern United States and Canada include the bacterium of Pierce's disease of grape; the bacterium that causes wilt of tobacco, tomato, and other crops (*Pseudomonas solanacearum*); and a fungus (*Phymatotrichum omnivorum*) that causes root rot of many plants.

The effect of weather is illustrated dramatically with plant diseases that were major factors in several great famines. The Irish famine of 1845–7 threatened the social and economic fabric of Europe and North America; it was created in part by social conditions in Ireland but more directly by a fungus (*Phytophthora infestans*) that causes late blight of potato, the staple food crop of Ireland at the time. Historical records show that late-blight years in Ireland and throughout northwestern Europe were unusually wet and cool, conditions that favored the disease. The 1845 season started well, and the potato crop was of unusual vigor until July, when the weather changed to wet, dark, and cool. By September, late blight was rampant, the situation obviously was serious, and the potato crop failed. During the following years, an estimated 1.5 million Irish people died of starvation and disease, and another million emigrated to North America. This is discussed further in Chapter 9, section 1, on potato blight, and there is an excellent account in *The Advance of the Fungi* by Large (1940).

Plant disease was involved in other famines, notably the one in Bengal, India, in 1943. As in most famines, there were, of course, other contributing factors; administrative and transport failures usually are involved. But the immediate problem in Bengal was a leaf-blight disease of rice, the staple crop, caused by *Bipolaris* (=*Helminthosporium*) *oryzae*. This disease ordinarily caused some decreases in rice yields, but weather conditions in 1942 and early 1943 led to a serious epidemic, with catastrophic losses. There was exceptionally heavy rainfall late in 1942; 40 to 60 cm fell in September, compared with the average of 9 to 33 cm. Above average rainfall, cloudiness, and high temperatures continued through December. Rice is most susceptible to blight at the time of flowering; in 1942, the susceptible period coincided with conditions favorable for sporulation and infection by *B. oryzae*. Heavy rains also leached nutrients from the soil, which predisposed rice plants to infection. All these factors favored disease development. Harvests varied locally, but were 40 to 90 percent below those of previous years. An estimated 2 million people died of starvation (Padmanabhan 1973).

6
Disease Controls
and Their Limitations

6.1. Host Resistance and Its Manipulation

Overall, the best strategy for plant disease control is to use resistant cultivars selected or developed by plant breeders for this purpose. Resistance is especially important for major crops such as the cereals, sugarcane, potato, and soybean. There will be further discussion in later sections dealing with individual diseases.

The limitation in use of disease resistance in modern agriculture is adaptability by pathogens; new races overcome resistance. Highly effective resistance often is controlled by a single gene, implying that resistance depends on a single metabolic site. In the pathogen as well, virulence often is under single-gene control; thus, a mutation or other simple genetic change in the pathogen can allow it to overcome resistance in the host. Multigene control of host resistance also is well known; this indicates multiple sites for resistance, with multiple changes required for the pathogen to overcome, which is statistically less likely. However, multigene resistance usually is intermediate or less effective than is monogene resistance, and it is more difficult to manipulate by breeders. Specific examples will be covered in later sections.

Breeding for resistance has a long history (Stakman and Harrar 1957; Walker 1959), no doubt starting with primitive farmers who selected the best plants to save for seed. Later, farmers recognized cultivar differences in disease susceptibility, possibly as soon as varietal differences in crops were recognized; in Greece, Theophrastus recorded such differences several centuries B.C. By 1850, British farmers were aware of cultivar differences in resistance of wheat to rust. A potato cultivar (Magnum Bonum) resistant to *Phytophthora infestans* was in use in Britain by 1876; the Magnum Bonum potato was the result of efforts to improve the crop after the great Irish famine in the 1840s. Unfortunately, new races overcame this resistance by the end of the century. Resistance to downy mildew in American grapes (*Vitis labrusca*) was soon evident; in 1880, Professor Millardet in France hybridized the European grape (*V. vinifera*) with American grapes and thus obtained resistance in a suitable wine type.

Organized breeding of wheat for rust resistance began with Farrer (Australia), Cobb (United States), and Freeman (United States). A breeding program was soon started and is still active in the U.S. Department of Agriculture. A notable first occurred in Britain in 1905, when Biffen showed clearly that resistance to rust in cereals is inherited according to Mendel's genetic principles (Walker 1969). There was much skepticism at the time; difficulty in repeating the work was later traced to the fact that rusts have many races.

Early work in the United States on resistance to Fusarium wilt diseases, which proved to be of economic significance, was started by W. A. Orton about 1900. Resistant cultivars of cotton, watermelon, cowpea, flax, tomato, and cabbage were developed empirically, by selecting plants that survived in heavily infested soil. Next, Orton hybridized watermelon (susceptible) with stock melon or citron (resistant), and backcrossed repeatedly to the watermelon parent until a desirable melon with Fusarium wilt resistance was obtained. The "backcross" was Orton's pioneer contribution; it has been a standard procedure ever since. Walker (Figure 4.1) showed in 1920 that resistance to the Fusarium wilts is under either single-gene or multigene control. Monogenic resistance often is high, is easily manipulated, and is much used. Again, Walker's report first met with skepticism. After 1920, breeding for resistance became common, resulting in resistance in many crops to various diseases caused by all types of microorganisms (Walker 1953; Stakman and Harrar 1957; Walker 1969).

At present, standard Mendelian genetics is still used to breed for resistance to plant diseases. In addition, resistance is manipulated by molecular methods, opening much larger sources for resistance genes. Molecular genetics techniques hasten the process of combining resistance with other desirable characteristics and make possible the use of novel approaches, such as use of resistance to viruses obtained by incorporating genes for viral coat proteins into the host genome (Beachey, Loesch-Fries, and Tumer 1990).

There are similarities between the effects of plant disease resistance of the two Mendelian types. Monogenic resistance can have either large (that is, giving extreme resistance) or small effects that are either differential (resistance to one or few races of the pathogen) or nondifferential (affecting many races). Polygenic resistance also can have small or large effects that are differential or nondifferential (Fry 1982). Only two of these eight possibilities were considered by Van der Plank (1963, 1984), who coined the terms *vertical resistance* (monogenic, differential) and *horizontal resistance* (polygenic, small differential effects). Van der Plank's terminology is confusing; Nelson (1978) has recommended against its use.

In epidemiological terms, host resistance can be considered either rate reducing or differential (Fry 1982). When resistance is low to moderate, the pathogen colonizes tissue, but plants are not severely affected, and the rate of disease increase (that is, epidemic development) is reduced. Rate-reducing resistance often is controlled by several genes, each giving a minor but additive effect. Such resistance usually, but not always, is durable. Single genes for resistance that give minor effects also can be rate reducing. The term *field resistance* sometimes is used for the rate-reducing type.

Differential resistance generally is monogenic, often very high, but it is effective only against certain races of the pathogen. For example, type A resistance in tomato will protect against race 1 of the *Fusarium* wilt pathogen, but not against race 2. This type of differential resistance has been called selective, specific, discriminatory, major gene, vertical, or hypersensitive by various researchers. In some cases, resistance is both differential and rate limiting, adding further complication. The terms are not mutually exclusive because they are used in an epidemiological context (Fry 1982). It may help to think of the susceptible host as compatible because the pathogen grows readily in it without a drastic initial reaction; the resistant host is incompatible because the pathogen fails to grow and the invaded host cell often is killed quickly (hypersensitive).

A high level of resistance that is not differential is ideal but rare. It should be stable over time because several genes usually are involved. However, even single-gene resistance sometimes is stable. If nondifferential resistance is low, additional control measures are required when conditions favor disease.

Regarding the pathogen, single genes can control pathogenicity, or the ability to infect specific host species or cultivars (Scheffer, Nelson, and Ullstrup 1967). New races that differ in pathogenicity (hosts affected) can appear, as can new and more virulent genotypes that are unchanged in host specificity. Relative virulence of pathogens seems to be under multigene control in the few cases that have been examined (Nelson, Scheffer, and Pringle 1963).

Some crop species with a long history of cultivation have many genes for resistance, each specific against a given race of the pathogen. Each host genotype or cultivar differs from others in the resistance genes that it carries.

The pathogen species also can have many genes for pathogenicity, each allowing it to overcome or avoid the effects of a single host gene for resistance. This is the "gene-for-gene" situation, first observed by Flor (1947) from work on flax rust and elucidated further by others (Ellingboe 1976). A gene-for-gene situation has been demonstrated or postulated for many plant diseases caused by all types of microorganisms. The hosts in all these cases have one to many genes giving differential resistance, which usually, but not always, is dominant. The pathogen gene that can overcome each gene for resistance can be recessive, but this has seldom been demonstrated because many fungi are haploid. Gene-for-gene systems are no doubt the result of coevolution of host and pathogen (Fry 1982). The theory has been a guide in breeding for resistance.

6.2. Chemical Control by Fungicides and Bactericides

Chemical control is practical for certain diseases of high-value crops. Fungicides are important with fewer than fifteen crop species, including apple, banana, grape, citrus, coffee, cotton, tobacco, peanut, and potato; these crops occupy relatively small areas as compared with those occupied by grains. The estimated total amount of fungicides used worldwide in 1974 was approximately 6 million kg. This amount was only 7 percent of the total for all pesticides used in agriculture; insecticides and herbicides accounted for the rest (Fry 1982). These relative values probably are still valid.

There are significant limitations in the use of pesticides. Unintended effects include human and animal health hazards, environmental hazards, kill of nontarget species, and development of resistance by target species. Dangers of fungicides to people and animals have been minor as compared with the dangers of insecticides. Carcinogenic effects are always a potential, but they may take years to detect and are difficult to evaluate. To date, the only acute danger from fungicides has been from the use of methyl bromide for fumigation, from mercury compounds used as seed treatment, and from some obsolete products.

The environmental hazards of fungicides also have been minor in comparison to those of insecticides and herbicides. Benzimidazole fungicides are known to destroy earthworms that decompose litter in orchards and to inhibit predatory mites, resulting in increased populations of herbivorous forms. Some of our modern fungicides affect only one metabolic site in target microbes; among these are the benzimidazoles, which give excellent control of some fungi (Ascomycetes) but can lead to increased populations of others

(Oomycetes and Basidiomycetes) (Fry 1982). Most of the older fungicides affect many metabolic sites in the pathogen, making development of resistance less likely.

Development of tolerance by target pathogens has become a major limitation for some highly effective chemicals (Dekker 1976). This has generally been the case with antibiotics (produced by microorganisms) and systemic, narrow-spectrum fungicides that affect a specific metabolic site. Resistance can result from genetic changes such as single-gene mutations; also, resistance can result from explosive multiplication of a few resistant pathogen genotypes that had very low populations prior to first exposure to the toxicant. Resistance to several antibiotics has developed in areas of intensive use (Cooksey 1990). For example, streptomycin had wide use for control of fire blight, a bacterial disease (see Chapter 8, section 2); resistance was first noticed in the northwestern United States and is now common elsewhere. Streptomycin is no longer effective in these areas. Antibiotics also have been used against some fungi; Kasugamycin effectively controlled rice blast, but the antibiotic became useless in parts of Japan only three years after its initial use. Polyoxins gave good control of Alternaria blights of apple and pear in Japan, but polyoxin use was soon limited by the development of resistance, first detected in 1973 (Nishimura, Kohmoto, and Udagawa 1976; Fry 1982). These are examples of many similar occurrences.

Problems with fungicide tolerance can be minimized or eliminated by concomitant use of toxicants that have different sites of action (Staub 1991). Several fungicides can be applied together, or alternate applications can be used (Delp 1980). Crop failure because of fungicidal tolerance is still relatively rare, in part because the broad-spectrum compounds are still the most used. Fry (1982) has discussed these and other aspects of the use of fungicides.

Fungicides have a long history. The first to be used was sulfur; its beneficial properties were discovered empirically at least two thousand years ago in Greece. Sulfur as a fungicide was then forgotten, to be rediscovered during the 1800s, when sulfur dusts were applied to grapevines to control powdery mildew. Effectiveness was improved by mixing with lime. Sulfur is still used in great quantities to control powdery mildews and rose black spot. Copper sulfate was used as early as 1807 as a fungicidal seed treatment, but phytotoxicity prevented its use on foliage. A mixture of copper sulfate and lime was used on grapes in France to prevent pilfering; Professor Pierre Millardet, in 1882, noticed that it prevented downy mildew as well, and the famous Bordeaux mixture was born (Millardet 1885). This was a very effective fungicide, used on grapes, potatoes, and other crops until it finally was displaced during World War II by organic fungicides. Other copper-based formulations, such as the "fixed coppers," have been widely used; the copper compounds are effective against both bacteria and fungi. Inorganic sulfur and copper formulations were the only effective protectants until the 1930s (Fry 1982).

Most fungicides, until recently, were protective rather than therapeutic, implying that complete coverage is necessary so that toxicant is present when spores germinate. There is little or no effect on the pathogen once it is internal. A protective fungicide must have several properties: It must be toxic to the spores; it must spread over the whole surface to give adequate protection; it must have low solubility in water to remain on the plant during rains; it must be compatible with insecticides; and it must be safe for the applicator. Various adjuvants often are used to enhance these properties.

The use of organic fungicides started in the 1930s, with discovery of the useful properties of the dithiocarbamates. A product known as Thiram was much used as a seed treatment. Metal salts of dimethyl dithiocarbamate became important; the ferric salt known as ferbam is still in wide use. Ethylenbisdithiocarbamates were discovered by Dimond, Huberger, and Horsfall in 1943; the zinc salt is known as zineb, and the manganese salt is maneb. These compounds have excellent fungicidal properties and are still used as protectants. There are and have been many other useful fungicides. The organic mercuries were once widely used for seed and soil treatments; they are now prohibited in the United States and elsewhere because of toxicity to animals. Chlorothalonil was formerly used widely, as was hexachlorobenzene and pentachlorobenzene. Biphenyl is used to protect stored fruit. Captan, a heterocyclic nitrogen compound with many desirable properties, was a widely used broad-spectrum fungicide. There are many others, mostly with limited use (Fry 1982).

The next development was systemic fungicides, starting with the carboxamides (Carboxin and Oxycarboxin) (von Schmeling and Kulka 1966). The benzimidazoles, especially Benomyl, are the most effective and widely used systemics. Other benzimidazoles have trade names such as Nimrod, Dexon, Bayleton, and Baytan. These compounds are taken up by plant roots and transported via the xylem to all plant parts, giving protection against specific microorganisms (Fry 1982). A few systemics are moved in the phloem and can give systemic protection to roots after application to foliage. Benomyl is effective against internal infections such as the Fusaria that cause wilts, which previously had no effective chemotherapy.

The limitation of systemics and of antibiotics is based on single metabolic sites of action; thus, resistance in target pathogens has developed quickly.

The systemics were initially very effective, and they became popular, with the expectation that they would displace protectants. Unfortunately, the appearance of resistance is limiting their use (Davidse 1986; Delp 1988). The protective, broad-spectrum fungicides are still important.

We now recognize many dangers in the use of pesticides. Various steps have been taken to decrease usage, including integration with other control measures, determination of exactly when and how many applications are needed, increased use of disease resistance genes in crops, and utilization of biological controls. Usage per unit crop area has been decreased, but there probably will be continued need for fungicides in some crops because of limitations in other control measures. Dreams of a world free of pesticides probably are unrealistic in modern agriculture, which is, by its very nature, an ecological disturbance.

6.3. Exclusion, Eradication, and Certification

Some of our most destructive plant diseases are caused by alien pathogens that were moved into new areas by human activity. The aliens arrive at an ever-increasing pace because of increases in commerce. Still, there are many more pathogens that can invade new areas, and at least some of these could be catastrophic. There are six hundred or more diseases, now localized by natural barriers, that are potential threats if moved (Thurston 1973; Kingsolver, Melching, and Bromfield 1983); in addition, there are many insignificant diseases that could prove disastrous when moved to a new environment. These are the reasons behind laws for exclusion or quarantine as a strategy to prevent disease.

Plant pathogens often arrive with imported plants, seeds, and other propagative parts, with produce for the market and with insect vectors, often via aircraft. Thus, most nations now restrict or prohibit movement of plants, unprocessed plant products, and animals across borders. The restrictions vary in scope and in enforcement; thus, they often are ineffective, and new diseases and pests continue to arrive. Plant hobbyists and even scientists often flaunt the exclusion laws; enforcement is difficult. The usefulness of quarantines often is questioned, sometimes without justification.

We have striking examples of destruction by introduced pathogens. Dutch elm disease is one that should have been prevented or at least delayed by existing law (the U.S. Plant Quarantine Act of 1912). White pine blister rust could have been prevented if the 1912 law had been in effect at the time of introduction. Pierce's disease of grape in California could have been prevented if present laws had been in effect. However, chestnut blight and other diseases may not have been intercepted because there was inadequate knowledge at the time of their arrival. Estimates over a twenty-five-year period indicate that three new pathogens become established in the United States each year. This number might be much higher without exclusion laws.

In general, exclusion succeeds only when it has a sound scientific base and efficient administration; these requirements sometimes are overlooked. Natural barriers to spread are necessary if exclusion is to succeed; barriers include those of climate, oceans, mountain ranges, and deserts. Some diseases can move across such barriers, if there is the potential for long-distance dispersal by wind (example, coffee rust). Political barriers without natural barriers are meaningless, and exclusion efforts can be wasted.

A serious problem is that the public has lost respect for exclusion laws because, in some cases, the laws were created with no scientific basis. Other factors that can lead to distrust by the public include quarantines established for political purposes, excessive costs compared to benefits, and inability to sustain long-term exclusions. Large segments of the traveling public now lack understanding of the laws and tend to ignore them.

The pros and cons of exclusion legislation have been discussed at length (Stakman and Harrar 1957; Mathys and Baker 1980; Waterworth and White 1982; Kahn 1991). The major arguments against such laws are that they are ineffective over time, are too costly, and frequently are abused. Arguments in favor of such laws are that they clearly are of economic value, and their costs are minor compared with costs of controlling new diseases; they generally are effective, completely so in some cases, and at least delay the introduction of pests; and abuse is not a valid reason for elimination. Most thoughtful authorities believe that elimination of exclusion laws would be irresponsible, but neither is it wise to have laws that are ineffective and without consideration of scientific and economic factors.

Another major problem is lack of a data base that can prove or disprove conclusively that exclusion laws are effective; such data are difficult to obtain. One possibility is to compare the number of new pests introduced immediately before and after a new exclusion program is introduced. Forty-seven known pathogens were introduced during the twenty-five years prior to the U.S. Plant Quarantine Act of 1912; seventy-five are known during the twenty-five years following 1912. However, the comparison has little meaning because early records were poor and there was increased traffic in plants during the latter period. Even so, effectiveness of the exclusion act can still be questioned.

Once introduced, eradication of pests has both failures and successes. A few examples are informative. Eradication of chestnut blight and Dutch elm disease did not succeed in North America and Europe. Eradication of fire blight of apple and pear trees failed in Europe. In contrast, white pine blister rust is controlled by local eradication of *Ribes* (alternate host). Local eradication of *Juniperus* sp. effectively controls cedar-apple rust in apple. The ravages of stem rust of wheat were decreased by large-scale eradication of barberry, the alternate host. Citrus canker in Florida was held in check for some time by thorough eradication of infested orchards and seedling nurseries. Eradication of peach trees with the phony peach disease in the southeastern United States has slowed the spread.

Voluntary inspection and certification programs are used to protect buyers from insect pests and diseases in shrubs, trees, bulbs, seeds, and other plant materials. The producer simply allows the crop to be inspected during and after the growing season and, if free of pests, to be certified as such. Rules governing inspections and certifications are determined and enforced either by growers' organizations or by government agencies. Such programs were first used with seed potatoes (Shepard and Chaflin 1975); major programs now involve seeds of cereals, cotton, and vegetable crops, roots of sweet potatoes used for propagation, and tomato plants started in the south for planting in more northern areas of the United States.

The quality of fresh vegetables and fruit was greatly enhanced when producers and shippers were able to certify that their products were free of pests and pesticides. Again, inspections can be made by private organizations or by government agencies. The procedures protect growers, shippers, and consumers.

Disease-free planting materials sometimes are unavailable, and their development may be necessary for certification programs. This applies especially to vegetatively propagated crops such as fruit trees, grapes, potato, sugarcane, cassava, strawberry, chrysanthemum, and carnation. Such crop plants may have systemic virus or vascular wilt infections, often without symptoms (Spiegel, Frison, and Converse 1993) and not obvious at propagation time; the pathogens are disseminated with planting materials such as tubers and cuttings. Some systemic pathogens can be eliminated by heat, provided they are more sensitive than is the host. Heat therapy has been used with cereal and cabbage seeds, strawberry plants, and sugarcane seed pieces. Certain foliar pathogens are absent in dry climates, and seed can be produced in such areas.

Chrysanthemum and carnation have special disease problems; commercial production of these plants in North America faced ruin because the plants carried vascular wilt pathogens (*Fusarium* and *Verticillium*) and systemic viruses. The problems were solved by use of "indexed plants" to establish disease-free stocks for the florist industry. Individual cuttings were tested for wilt pathogens by laboratory isolations from pieces of the stem, and for systemic viruses by serological assays or by inoculations from the cutting to sensitive indicator plants. Plants that tested negative were grown as "mother plants" under carefully controlled conditions; plants were propagated from mother plants by specialists and were sold to growers of cut flowers (Dimock 1962; Hollings 1965; Nyland and Goheen 1969). Modifications of the procedure have included use of meristem culture to obtain disease-free stock (Raychaudhuri and Verma 1977). These methods are applicable to other vegetatively propagated plants; for example, cassava cultivation in Africa could be improved and yields significantly increased by the use of such methods, provided skilled personnel were available.

6.4. Biological Control

Biological control obviously occurs in nature. This explains, in part, the lack of widespread and damaging epidemics in undisturbed ecosystems. In contrast, epidemics are common in agriculture, an unbalanced system with disruption of biocontrol. Biocontrol of plant pathogens in nature is achieved, in part, by diverse microorganisms through competition for resources, predation/parasitism, and antibiosis (Fravel 1988). However, biocontrol seldom is complete, even in undisturbed situations.

In a broad sense, many agricultural-management practices foster biocontrol because of effects on microbial pest-control biota. Decreases in the incidence or severity of some diseases can be achieved by crop rotations, intercropping, fallowing, flooding, or the addition of manure and other organic materials to the soil. These practices alter microenvironments, often encouraging (in some cases, discouraging) the microbes that compete with pathogens. Some management practices also predispose crop plants toward either tolerance or sensitivity to pathogens.

Roots of plants that grow in certain soils may escape infection, even with the pathogen present; such "suppressive soils" often are high in clay and organic materials, thus are favorable for high populations of various microbes. Suppressive soils have been utilized for growing banana, avocado, and other specialty crops.

Biocontrol sometimes develops on its own; for example, continuous cropping with a single species can result in decreases in disease incidence. This happens with the take-all disease, caused by *Gaeumannomyces graminis* var. *tritici;* the disease often increases for the first two to four years of wheat monoculture, then declines during subsequent years. Another such case is cotton root rot, caused by *Phymatotrichum omnivorum*. Biocontrol in both cases is correlated with increased populations of saprophytic microbes.

The best example of biological control by use of indigenous microbes was developed by Ko (1982) in Hawaii, for root rot of papaya. This crop in Hawaii is grown by direct seeding into a porous gravel created by blasting holes in old lava flows. Papaya orchards usually are replanted after four years. The first planting in an area bears fruit that often is infected with *Phytophthora palmivora*. Diseased fruit falls and infests the soil; replants in such areas always failed because of root rot caused by *P. palmivora*. Ko found that seedlings remained healthy when planted in a small volume (9 liters) of virgin soil that was placed in the planting hole. Virgin soil is any soil from an area where papaya had not grown; in Hawaii, this usually meant soil from a sugarcane field. Roots of seedlings become resistant once the plant grows beyond the juvenile stage. The method is inexpensive, non-hazardous, and almost completely effective.

The principle of the virgin-soil method is competition between saprophytic microbes and the pathogen. Under proper conditions, the method is effective against other opportunistic soil pathogens, but not against *Rhizoctonia* root infections. At first, the papaya growers would not believe that such a simple method could work; they had to be subsidized to even try it. Now, the method is in general use in Hawaii.

Biological control sometimes is defined as the direct use of a specific microbe (or a specific group) for control of a given disease (Cook and Baker 1983). The phenomenon is demonstrated easily in laboratory or greenhouse experiments, but few practical results have been achieved; only seven cases have reached commercial application (Lewis and Papavizas 1991). The two best examples involve control of crown gall and Heterobasidion rot of tree roots. Formulations containing *Agrobacterium radiobacter* strain 84 (a saprophyte) are marketed for control of *A. tumefaciens*, the cause of crown gall. Liquid slurries containing *A. radiobacter* are used to treat seed and transplants of peaches, roses, and other nursery-grown crops. This has been the most effective known control in some areas but, unfortunately, is not effective against all races of the pathogen, including the race most common on grape. It also can lose effectiveness with development of new races of *A. tumefaciens*, as occurred quickly in Greece. Control depends on production of an antibiotic (Agrocin 84) by *A. radiobacter* 84; most strains of *A. tumefaciens* are sensitive to Agrocin 84 (Kerr 1980).

Root rot of conifer trees caused by *Heterbasidion annosum* is controlled by use of a competing fungus. The disease can be severe in managed forests after thinning. The fungus is spread by spores that enter wounds or new stumps, moving from such infection sites to adjacent trees via root grafts (see Chapter 14, section 4). Control is achieved by inoculating the newly cut stumps with the saprophytic fungus *Peniophora gigantea*. This is the most successful control to date of *H. annosum* root rot and is used in some managed forests.

Direct biological control to date has been practical only with root diseases. The few successful cases may be, in part, a reflection of the research effort expended; time and resources given to biological control are small, compared with those expended on development of a resistant cultivar or a new fungicide. More use of biological control is a hope for the future, but there clearly are limitations. There is no shortage of books and reviews on the subject (Cook and Baker 1983; Hornby 1983; Hoitink and Fahy 1986; Campbell 1989).

Another type of biological control is known as cross-protection. For example, tomato plants can be inoculated with mild or nonexpressive strains of tobacco mosaic virus; this gives protection against virulent strains. A similar procedure has been used to control tristeza disease of citrus (Fulton 1986) (Chapter 14).

Blight of chestnut has essentially eliminated the chestnut tree in North America, and the blight was severe in Europe. Many trees in southern Europe are recovering because a virus has invaded the causal fungus, leading to loss of virulence (Chapter 7, section 1). Similar hypovirulence has appeared in America but has spread only in Michigan. Hypovirulence can be utilized to save valuable trees.

Induced resistance is still another phenomenon that can be considered a biocontrol. When lower leaves of certain plants, notably cucurbits, become infected, the whole plant becomes resistant to further infections and remains so for weeks or months (see Chapter 4, section 5). There are possibilities for practical use (Hammerschmidt and Kuc 1995).

Part II

Natural History of Some Destructive Diseases

The effects of human activities on the incidence and severity of plant diseases form an integrating thesis of Part II. Many diseases are easily categorized according to the major anthropogenic factors involved in their epidemiology. For example, pathogens were moved by people to new geographical areas, with devastating effects on native plants (see Chapter 7, section 1). Several such anthropogenic categories are used herein. Discussions of several example diseases are included for each category; each example differs from the others in its natural history. Some of the examples used could be placed in two or more categories; the placement selected may seem arbitrary, but it is meant to reflect the key factor, without which the disease would have been of little significance.

Most of the highly destructive plant diseases probably would fit into one or another of the several categories. However, it is not now clear where some would fit, possibly because too many factors are involved or because there is insufficient evidence.

7

Native Plants,
Alien Pathogens

Some of the most dramatic plant diseases are caused by pathogens brought from other continents. Imports were common before quarantine and regulatory services were organized; the disasters were justification for such agencies. Our current understanding of epidemics caused by alien pathogens will be stated in its simplest form.

Continental drift or changing ocean levels can separate populations of plant species, resulting in divergent evolution. Pathogens that arose in one isolated area gave constant selection pressure on the plants in that area, resulting, over time, in tolerant host species. This often is stated to have occurred by elimination of the most susceptible genotypes; clearly, genetic factors regarding resistance are involved. The geographically separated but related suscept line had no such exposure and developed no tolerance. Disastrous epidemics occurred when pathogens from an area of host tolerance were moved into an area occupied by a related plant that lacked tolerance. This is what happened in Europe and North America with chestnut and white pine trees; they were devastated by foreign fungi from far regions of Asia. The source of the pathogen that killed elms is less certain, but clearly it was an alien in North America. Full understanding of several model cases is needed to appreciate these simple ideas. There are many other examples.

7.1. Death of the Chestnut Forest

Chestnut blight, caused by a fungus, is the most prominent example of disaster from an alien pathogen. There was only one significant causal factor in the epidemic: introduction of the fungus *Cryphonectria parasitica* (=*Endothia parasitica*) from the Orient to North America and Europe. Other factors, such as genetic adaptability and monoculture, were scarcely involved.

Figure 7.1. Chestnut trees in the southern Appalachian Mountains around 1900. Chestnut was the most valuable forest tree in eastern America; its destruction by disease had severe ecological and economic consequences. (Courtesy of the American Chestnut Foundation)

American chestnut (*Castanea dentata*) (Figure 7.1) probably was the most valuable forest tree in the eastern United States, prior to 1900. It grew rapidly and did well, even on poor sites. The wood was of superior quality and had decay resistance; thus, it was useful for construction, fencing, telegraph poles, ship masts, and rail ties. Chestnut wood also was used to make furniture, panels, and musical instruments. Tannins in the wood and bark gave decay resistance; the tannins became the basis for a leather-tanning industry. The nuts were the best of all chestnuts and were a major food for man, wildlife, and farm animals. Many trainloads of chestnuts were shipped each year from the Appalachian Mountains to the cities; this was a source of income for the people of Appalachia. Finally, the chestnut was an excellent and beautiful tree for shade.

Chestnut was a major component of the eastern forest of the United States (Figure 7.2). Under good conditions, the tree could increase its trunk diameter by 2.5 cm per year and reach heights of 39 m, with trunks 2.5 m in diameter. Harvested trees replaced themselves by sprouts from the roots. The species accounted for one-fourth of the timber cut in its geographical range, which included mixed stands on 811,000 sq. km, with more than 3,400,000 ha of chestnut trees.

The European chestnut (*C. sativa*) was endemic around the Black Sea, in eastern Europe and northern Turkey. During classical times, Romans moved the tree westward throughout the Mediterranean region, where it was cultivated as an important source of food and timber. The European chestnut is similar to the American, but is somewhat more tolerant of the blight disease.

Chestnut blight was first observed in 1904, as a disease of shade trees in the New York Zoological Garden. The causal fungus was described as a new species. Several researchers at that time speculated that the fungus had been brought from the Orient on imported nursery stock. David Fairchild, a great plant explorer, suggested a search in China for the fungus; following Fairchild's suggestion, F. N. Meyer discovered *C. parasitica* in China in 1911, as an endemic species; later, it was found in Japan. Populations of both the Chinese (*C. mollissima*) and Japanese (*C. crenata*) chestnuts are resistant to the disease (Graves 1950), although susceptible variants are known. These facts support the theory that *Cryphonectria parasitica* is endemic in China or Japan, or both. Generally, endemic areas contain host populations that coexist with native pathogens. For more complete accounts of the disease and the fungus, see reviews by Anagnostakis (1987 & 1988) and the monograph by Roane, Griffin, and Elkins (1986).

The blight pathogen soon moved beyond New York, causing a plant disease epidemic greater than any previously known. It enveloped the entire

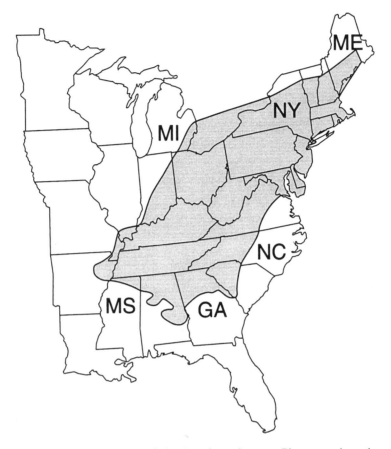

Figure 7.2. Natural range of the American chestnut. Pioneer settlers planted the species west of this area.

range of the chestnut tree, from Maine to Alabama and west to Michigan and Missouri. The spread was at the rate of about 37 km per year; in fifty years, an estimated 3.5 billion trees were killed. In its natural range, the American chestnut survives at present only as root sprouts from killed trees; the sprouts usually die before they grow trunk diameters of more than a few centimeters. There are a few large survivors outside the original range of the species. The fungus was spread by wind (ascospores), by infected nuts, by movement of infected seedlings, and possibly by birds and mammals. Rapid spread such as this often occurs when an alien pathogen encounters large populations of a very susceptible host. The Asiatic, European, and American chestnuts are interfertile, indicating a common origin in the distant past. Evolutionary changes of isolated populations through the ages account for differences, including differences in susceptibility to *C. parasitica*.

The European occurrence of chestnut blight was more recent. It was first observed in Italy in 1938, probably brought in from Japan on nursery stock that had been imported for experimental nurseries several years earlier. Blight soon spread throughout the range of the European chestnut; the epidemic was similar to that in North America, but spread was slower and total destruction much less (Pavari 1949). Damage was first seen in 1938 around Genoa (Biraghi 1950) but may have occurred in Spain at about the same time. The fungus invaded France by 1946, Switzerland by about 1951, Greece by 1964, and Turkey by 1967 (Anagnostakis 1987). Scattered stands in Britain and northern Europe are still blight-free. Slower spread in Europe than in the United States is attributed to the fact that the European chestnut is a little more tolerant of the fungus than is the American species (Heiniger and Rigling 1994); this tolerance results in a longer generation time and lower inoculum levels. Lack of continuous distribution of the host may have been a factor.

There were early attempts to breed American chestnuts for resistance to blight; these efforts made use of interspecific hybridizations to incorporate resistance from Asiatic chestnuts (Jaynes 1964), irradiation of American nuts followed by selection of seedlings for resistance (Dietz 1978), and selective breeding of large remaining American trees. Unfortunately, most of the early programs were discontinued by 1960, for lack of support and interest. In Europe, breeding for resistance was started in 1951; a number of European selections with moderate to high resistance were found (Bazzigher 1981). Chestnut breeding is long-term because tree generation times are long; many years may pass before definitive results for good tree type, good nuts, and resistance are obtained.

The most extraordinary event in the history of chestnut blight was the discovery of fungal hypovirulence. This phenomenon offers hope for the eventual control of the disease; it also has wider implications. Hypovirulence first appeared in Italy, about fifteen years after the epidemic started. Cankers on some trees appeared to be healing, leading to recovery (Biraghi 1950). The French mycologist J. Grente (1965) obtained samples from healing cankers in Italy and isolated fungal strains that had low virulence. These isolates were called hypovirulent; when they were inoculated into a lesion caused by the virulent fungus, growth of the canker was arrested. French researchers then reisolated from healing cankers and found that most of the new isolates were of the hypovirulent type. They suggested that placing hypovirulent isolates into cankers caused by the virulent fungus resulted in hyphal anastomosis and conversion of the virulent or normal type to hypovirulence. If true, there must have been transfer of genetic information; in culture, the normal isolates never segregated to hypovirulence, but hypovirulent isolates often segregated into normal and hypovirulent types.

The recovery of chestnut stands in Italy was remarkable, and hypovirulence has spread (Heiniger and Rigling 1994). Hypovirulent strains were successfully introduced into France and now have spread naturally but slowly. After exposure to hypovirulent isolates, potentially deadly cankers start to recover; heavy callus develops around the infection, and hypovirulent isolates always are recovered from the lesions. All this happened in Europe within twenty-five years after blight first appeared. This is in contrast to the situation in the United States, where only limited pockets of hypovirulence are evident after ninety years. In Italy, chestnut trees are again a source of timber and of nuts, even for export (Turchetti 1982). Hypovirulence is a form of biocontrol in Europe (Heiniger and Rigling 1994).

The important findings on hypovirulence in Europe were followed further in the United States. Workers at the Connecticut Agricultural Experiment Station obtained hypovirulent cultures from Grente in France. Experiments with French and Italian isolates showed that cankers caused by the virulent American fungus were "cured" (Anagnostakis and Jaynes 1973). Reisolations from the cured cankers gave cultures that were hypovirulent and that differed in appearance from the original virulent isolate. Next, virulent strains were identified by marker genes and exposed to the hypovirulent forms. Results proved that hypovirulence is cytoplasmically inherited and is transferred when fungal strands (mycelium) of the two types fuse (anastomosis). Double-stranded ribonucleic acid (dsRNA) was present in hypovirulent strains but not in virulent strains (Van Alfen et al. 1975).

The Connecticut research was taken to the field, after preliminary greenhouse tests to determine susceptibility of many native and exotic woody species. Only American chestnut, certain cultivars of Chinese chestnuts, and some hybrid chestnuts supported growth of the hypovirulent fungus (Jaynes, Anagnostakis, and Van Alfen 1976). Hypovirulent isolates derived from French sources were inoculated into cankers on three hundred forest trees. After one year, growth of 86 percent of the cankers was arrested or controlled (Jaynes & Elliston 1980), but there were some new cankers on the trees, apparently from wild inoculum in the area. The experiment proved that hypovirulence has the potential to control blight, but it did not demonstrate natural spread of hypovirulence.

Native hypovirulence was found in American chestnuts in Michigan in 1976. Fungal isolates taken from healing cankers contained a high-mol-weight dsRNA, indicating similarity with hypovirulent strains from Europe. However, the dsRNA bands and the colony appearance differed from those found in European isolates (Paul and Fulbright 1988). American hypovirulence was transmissible, and it cured existing cankers (Day et al. 1977; MacDonald and Fulbright 1991). Many hyporvirulent strains that differ in the dsRNA bands and in appearance of cultures have now been found in other parts of Michigan, and in Pennsylvania, Tennessee, Virginia, and elsewhere (Jaynes and Elliston 1982; Fulbright et al. 1983; Elliston 1985; Enebak, MacDonald, and Hillman 1994). Hyporvirulence in Michigan now appears to be spreading naturally to other trees within a grove, following initial inoculation of hypovirulent isolates (Garrod, Paul, and Fulbright 1985).

Most researchers now consider the dsRNA to be a virus that attenuates virulence when it moves into the virulent fungus (Newhouse, MacDonald, and Hoch 1990). Elimination of the dsRNA from a fungal isolate causes reversion to virulence (Fulbright 1984).

Recent research has shown that some isolates of *C. parasitica* contain a 12-kb dsRNA but are virulent (Enebak, MacDonald, and Hillman 1994). There appear to be many types of dsRNA, some that give hypovirulence and some that do not. Also, lesions induced by some isolates with no dsRNA are healing, and the hypovirulence is cytoplasmically transmissible; mutation of mitochondrial genes may be involved (Mahanti et al. 1993). The relative degree of resistance in the host may be a factor; European chestnut is more tolerant than the American species, and hypovirulence seems to be more effective in Europe. There is still much to learn about hypovirulence.

Hypovirulent strains of American origin were first found some 200 miles west of the natural range of chestnut; trees had been carried to the west and planted by pioneer settlers. Thus, chestnut groves were established in western Michigan and in Wisconsin and Minnesota; some of these groves escaped blight for many years. There are some surviving trees in the western areas, even surviving groves with two thousand or more trees, which are competing well with the native flora. Blight-recovering groves are found only in areas west of the chestnut range, although individual recovering cankers are observed elsewhere. The only areas in the United States where spread of hypovirulence is now known is in Michigan; there are at least thirty chestnut stands that are surviving infection (MacDonald and Fulbright 1991). In a few stands, almost all evidence of *C. parasitica* is gone. The similarity of the situations in Michigan and in Europe needs further study. Perhaps there are fungal vectors of hypovirulent strains in Michigan that are missing elsewhere in the United States. Hypovirulence in *C. parasitica* has been reviewed by MacDonald and Fulbright (1991).

Compatibility relationships in the fungus may explain spread of hypovirulence in Europe and lack of spread in the United States. A given isolate of *C. parasitica* will fuse (anastomize) on contact with another isolate with which it is genetically compatible. If the hypovirulent isolate is incompatible with a virulent isolate, then there is no merger and no transfer of hypovirulence. More than seventy compatibility groups are known from Appalachia in the United States, but relatively few have been found in Europe (Heiniger and Rigling 1994) and Michigan. Relatively few compatibility groups should simplify the spread of hypovirulence in the latter places. However, the situation in Appalachia may be more hopeful than indicated because some carrier isolates can transmit hypovirulence to many compatibility groups (Anagnostakis 1983; Kuhlman et al. 1984; MacDonald and Fulbright 1991). If natural spread does not occur, then constant additions of isolates of many compatibility groups to the system may be required for successful biocontrol (MacDonald and Fulbright 1991).

There is a wider implication of hyporvirulence because it has been found in other fungi that are pathogenic to plants. The phenomenon is under study as a possible control for several plant diseases (Nuss and Koltin 1990).

Hypovirulence is not the only hope for saving the chestnut; by 1989, breeding for blight resistance in the United States was restarted, thanks to support from the American Chestnut Foundation. Chinese and Japanese chestnuts are good sources of resistance, and these species mate readily with the American chestnut. The objective of breeding is to produce a forest tree with the desirable tree form and nut quality of the American chestnut, combined with the disease resistance of Asiatic species. Crosses of resistant and susceptible chestnuts give F_1 progeny with intermediate resistance, suggesting incomplete dominance. Possibly only two genes control resistance and susceptibility. The most resistant progeny of the crosses are being backcrossed to the American species, followed by selection of desirable progeny. The progeny should have 15/16 of the American genes by the third backcross (Mulcahy and Bernatzky 1992). There are expectations of success, but it will come slowly. Earlier breeding programs were disappointing, in part because proper use of the backcross method was neglected. Cooperative work is under way in Connecticut, Virginia, West Virginia, Michigan, Minnesota, and elsewhere in the United States. The techniques of molecular genetics also are being used, to introduce foreign genes into the American chestnut.

Summary and Outlook

Chestnut blight, a bark canker disease, is perhaps the most devastating of all plant diseases, in terms of impact on a natural ecosystem. The single epidemiological factor behind the catastrophe was accidental importation of the causal fungus *Cryphonectria parasitica* from the Orient to North America and Europe. The disease has killed several billion trees of the most important forest species in the eastern United States, with accompanying losses to forest industries. It also destroyed an invaluable source of food for wildlife, man, and domestic animals. Essentially, American chestnut survives only as root sprouts with a short life. Oriental chestnuts are tolerant or resistant.

There are two hopeful possibilities for restoration of chestnuts. First is the discovery and exploitation of hypovirulence, which depends on a virus infection in the fungus that decreases its virulence. Hypovirulence is transmitted in cultures and *in vivo* from infected fungal strains to virulent strains. Healing of active cankers occurs after exposure to a hypovirulent fungus; this biological control has been effective in France and Italy. The second hope is to transfer resistance from Asiatic to American chestnuts, via traditional hybridization and backcrossing techniques and by use of recombinant DNA.

We now have expectations of saving the chestnut as an orchard tree. The challenge is to return it to the forest.

7.2. Dutch Elm Disease

Dutch elm disease (DED) (Figure 7.3) is another classic example of the destructive effects of alien pathogens. DED has killed the trees on the streets where we live; thus, it is one of only a few plant diseases known to the general public in Europe and North America. The name was given because the problem was first discovered in the Netherlands, where excellent early research was done by a group of seven women scientists.

Elms were highly valued trees in the northern hemisphere. The American elm (*Ulmus americana*), in particular, has a graceful form, and prior to the advent of DED, it was the most-planted shade tree in North America. European elms, including *U. carpinifolia* (the European field elm) and its variant, the English elm (sometimes labeled *U. procera*), were widely planted, especially in the Netherlands, Belgium, France, and Germany. Elm foliage at one time was used for cattle food in Europe, and the inner bark as human food during famines (Richens 1983). Native Amerindians used the bark to

Figure 7.3. Dutch elm disease has destroyed trees in urban and forest areas in North America and Europe. (Picture by R. P. Scheffer)

build shelter. The wood is useful for special purposes; it is strong and flexible and holds screws well. The tree made excellent windbreaks on coastlines and prairies. Loss of elm trees in North America and Europe has been a cultural disturbance of some significance.

Early attachment to the elm tree in the United States is illustrated by the story of the liberty tree. The tree was a rallying place for rebels in Boston during the revolution of the American colonies; consequently, British soldiers cut it down. Soon, elm liberty trees appeared all over the colonies (Klein 1979).

The elm genus *Ulmus* is a complex group. The European field elm is a highly variable species, with about one hundred subspecies and cultivars. It is endemic in most of Europe, North Africa, the Middle East, and into adjacent Asia. Southern England has the stately English elm, which also has been planted in North America and Australia. *U. glabra,* the Wych or Scots elm, is found in northern Europe and in high mountains. There are hybrids of these. *U. americana* has a large natural range in the United States and southern Canada (Figure 7.4), from the Atlantic west to the Great Plains and

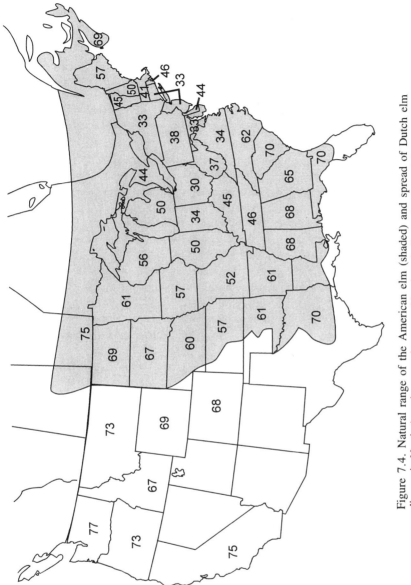

Figure 7.4. Natural range of the American elm (shaded) and spread of Dutch elm disease in North America over the years. Numbers indicate the year of first record in each area: First introduction was in Ohio, 1930; second introduction was in New York, 1933; and third in Quebec, 1946. (Source: Stipes and Campana, 1981)

south to Texas and northern Florida. It was widely planted elsewhere because of its desirable characteristics as a street tree. The Siberian (*U. pumila*) and Chinese (*U. parvifolia*) elms are much less desirable, although they are resistant to DED. There are many other species, which are all more or less susceptible to Dutch elm disease; the American elm is most susceptible, and the Asiatic elms, in general, are most resistant.

Striking differences in susceptibility are illustrated by the record of mixed plantings in Toronto, Canada. By 1965, all the American (or white) elms were killed, whereas no trees of the English elm were dead. However, the English elm (*U. procera*) performed poorly against the disease in Europe (Gibbs 1978); the earlier report of some resistance may be because a milder strain of the pathogen was present. In general, the European elms are intermediate in resistance, between the American and Asiatic species.

The popularity of elms resulted in much urban planting, often far outside the natural ranges of the genus. Selections were made of the most desirable types, and these were often planted as near monocultures in cities and towns of North America, Europe, and even in central Asia. The genetic homogeneity of these plantings no doubt contributed to the spread and disaster of Dutch elm disease. At one time in North America, there were perhaps a half billion planted elms and a billion or more wild or native elms (Carefoot and Sprott 1967). Total financial losses from the elm disease probably were much greater than those of the great stock market crash of 1929.

Dutch elm disease usually begins in a tree as yellowing or dull appearance of the leaves on a single branch. From this site, the fungus becomes systemic, moving up and down in the water-conducting tissues of the tree. Once the fungus reaches the main bole and roots, the tree is doomed; it often wilts and dies suddenly because the xylem vessels become plugged with tyloses, gums, and gas blockage (a form of embolism) (MacHardy and Beckman 1973). Toxins and enzymes secreted by the fungus may be involved in formation of tyloses and gums. Xylem vessels and surrounding cells are discolored (vascular browning). Systemic spread of the fungus is restricted in the xylem of resistant trees. Recovery of susceptible trees can occur, especially if infected by the less-virulent race; in such cases, the pathogen is restricted, and new healthy tissue develops around the infection.

There have been several studies to determine the physiology of disease development and disease resistance in elms. Natural inoculation occurs through feeding wounds by bark beetles in twigs; from there, the fungus invades sapwood and enters xylem vessels. Yeastlike spores are carried through the xylem to all parts of the tree. In general, resistant trees have shorter xylem-vessel elements; this gives more restriction to spread of the fungus and more time for resistance reactions to occur (Stipes and Campana 1981). There is circumstantial evidence that toxins released by the fungus change the metabolism of elm cells, a systemic response even in cells not in contact with the pathogen (Landis and Hart 1972). A small protein toxin known as cerato-ulmin has been implicated. This toxin is highly active against susceptible elm species but is much less toxic to resistant elms (Takai 1989).

Origin of the causal fungus *Ophiostoma ulmi* (= *Ceratocyctis ulmi*) is not known. Formerly, *O. ulmi* was thought to be endemic in China because the Chinese elms are mostly resistant. This idea was abandoned because the fungus has never been found in China, even after thorough searches by a Japanese mycologist during the 1930s and 1940s (Holmes 1990). One *O. ulmi*–like isolate was found in the Himalaya Mountains; this may have been an ancestral form, but convincing evidence is lacking.

The fungal genus *Ophiostoma* and its close relative *Ceratocystis* are widespread in wood of many tree species, causing stains and vascular disruptions. One such fungus causes the oak wilt disease, which also appeared suddenly and mysteriously. The Dutch elm disease fungus could have evolved from one or another of the related forms, perhaps as a simple mutant. One author (Klein 1979) suggested that the disease may have started centuries ago in Europe, where it was a minor problem, unnoticed; it could have become a major problem following environmental disturbances such as elm monoculture or war disturbances, which led to increased populations of beetle vectors. Also, changes in virulence in an opportunistic fungus could have preceded the early outbreaks; an even more virulent strain of the causal fungus is the cause of a recent second wave of the epidemic.

The elm disease fungus was first isolated and described by Dutch researchers in 1920. It was already widespread in the Netherlands, Belgium, and France by that time, as shown in distribution maps published in the Netherlands. Isolations of the fungus from annual rings of elm trees proved that the disease was in Europe by 1912; circumstantial evidence, based on discoloration of annual rings, indicates that it might have been there as early as 1900. The fungus soon spread over Europe, to Britain (by 1927), Russia (1936), south to Italy and north Africa, and east into Asiatic Uzbekistan by 1939 (Gibbs 1978).

The history of research on the elm disease is of special interest in that all the early work was done by a group of Dutch women scientists. The group was led by Johanna Westerdjik, who in 1917 became the first woman in any field to become a professor in the Netherlands (Holmes and Heybroek 1990). She sponsored and encouraged a group of dedicated young researchers that included Dina Spierenburg, who first became aware of a serious epidemic in elms, described the disease, and isolated what she thought was the causal agent. The work was continued by Marie Schwarz, who generally is credited for first isolating *O. ulmi*. However, the work was not generally accepted until Christine Buisman proved pathogenicity by use of Koch's postulates. Buisman had a short but meteoric career; she traveled extensively and found and identified the elm disease in Ohio in the United States. She was first to use the term *Dutch elm disease*. Buisman also discovered the two mating types in *O. ulmi*, thereby giving a base for later genetic experiments. Other Dutch researchers in the group were Maria Ledeboer, a fungal physiologist, and Johanna Went, who first selected and bred elms for resistance. The famous Christine Buisman elm was one of her accomplishments, named for an old friend, then deceased. The final member of the group of seven was Louise Kerling (Holmes and Heybroek 1990).

All these researchers worked in, or were influenced by, a unique Dutch institution, the Phytopathological Laboratory Willie Commelin Scholten. This institution was associated with three Dutch universities: Utrecht, Amsterdam, and the Free University of Amsterdam (Holmes and Heybroek 1990).

In contrast to the questionable origin of *O. ulmi*, its movement from Europe to North America is well known. Christine Buisman found it in Ohio in 1930, but this site of infestation did not spread. Other early sites of introduction were in Indiana, Ohio, Virginia, Maryland, and New York, always near ports of entry or wood-veneer plants. The introduction into New York became established, and the fungus spread, perhaps because the European bark beetle (a vector) had been there since before 1900, also imported on elm logs. Still another successful introduction occurred later in Quebec (Gibbs 1978).

There were devastating effects when *O. ulmi* spread to the midwestern United States, where many towns had near monocultures of the very susceptible American elm. When the fungus reached Illinois (1955), the rate of infection and mortality skyrocketed; for example, elms in the cities of Champaign and Urbana were almost gone within a few years. No doubt a restricted gene base in American elm, a natural tetraploid, plus high populations of the insect vector and highly virulent strains of the pathogen contributed to this. Bark beetles were numerous because some elms had been killed previously by a phytoplasma disease, elm yellows. Dutch elm disease now occurs throughout the native range of the American elm and beyond into plantations in urban areas to the Pacific coast (Figure 7.4) (Gibbs 1978).

There have been two epidemic waves of the elm disease. In Europe, the first epidemic killed many millions of trees, up to 70 percent in some Dutch communities. This first wave then spread across Europe into Asia, as well as all across North America. The epidemic then subsided, and by 1950 in Europe and by 1960 in North America, there was a decreasing rate of new infections and a low incidence of total cases. Many infected trees appeared to recover; there was optimism that the disease had run its course. There were suggestions that decreases in the rate of infection were caused by a hypovirulence factor such as occurred later with chestnut blight.

The next development for the disease was emergence of a more virulent form of the fungus. The new pathogen first was thought to have originated in mid–North America (Gibbs, Houston, and Smalley 1979). More recent evidence indicates that the new pathogen appeared first in the Romania–Ukraine area before 1950; the new fungus spread east into Asia and west across Europe to North America (Mitchell and Brasier 1994). The destruction of elms started again, in an even more devastating epidemic. The new race was often fatal to the European elms that had been somewhat resistant to the old race. The difference between old and new pathogens was especially apparent on the European elms; American elms were susceptible to both forms, but more so to the new pathogen. The elms in Europe that had been bred for resistance soon melted away. The new, aggressive fungus has now spread as far east in Asia as Iran.

It is rational to suggest that the new pathogen evolved from the older form. If so, the change must have occurred at an early date, because there are many genetic differences in the two forms.

The new elm disease fungus is now considered to be a new species, *O. novi-ulmi* (Brasier 1991). *O. ulmi* killed only older trees; the new species is killing seedlings and young trees. The new and old forms are sexually incompatible (Brasier 1991). Pathogenicity in both is under polygenic control (Brasier 1988), but there are different compatibility groups, which complicate genetic experiments. There also are RNA viruses that infect the fungus, leading to decreased virulence; this is better known in the chestnut blight fungus (see Chapter 7, section 1). *O. novi-ulmi* now occurs over the range of *O. ulmi* and is displacing it (Mitchell and Brasier 1994).

The elm disease fungi have very efficient ways to spread from tree to tree. Roots of elms will graft to the roots of adjacent elms, up to 10 m away. Thus, the fungus moves readily from tree to tree because it is systemic in xylem vessels. A more important movement is via bark beetle vectors; adult insects emerge from brood galleries and move to twigs of healthy trees for feeding. Insects that have emerged from diseased trees are covered with sticky spores of *O. ulmi;* the fungus sporulates abundantly in insect feeding galleries and in bark cracks. The infested insect then flies to a healthy tree, where it chews into the bark, mostly at twig crotches. In doing so, it deposits spores in a fresh wound, the ideal spot for tree inoculation. Twig feeding occurs prior to mating, in early summer, the time that trees are most vulnerable to infection by the fungus.

Mated insects are attracted to boles of trees, especially to those of trees under stress. Eggs are deposited in galleries under the bark; larvae feed in the phloem, making new galleries at right angles to a main gallery. They overwinter as larvae, then pupate, and finally emerge as adults, often covered with spores of *O. ulmi*. Adult insects also move to freshly cut elm wood, which then becomes a reservoir for the fungus. There can be one or two generations of bark beetles per year.

Several species of bark beetles are vectors in Europe and North America. The European bark beetle (*Scolytus multistriatus*) was brought to North America on elm logs before 1900. In North America, the European beetle is the major vector in much of the elm range, but the native American bark beetle (*Hylurgopinus rufipes*) also is involved. In southern Canada, the native beetle is the major vector because the European species does not live through the cold winters. The beetles fly as much as 3.4 km; the disease has spread in Massachusetts at about 16 km per year. About fifty years were required for *O. ulmi* to spread throughout the elm range in North America.

Fairly effective control of the elm disease is achieved by killing bark beetles and by strict sanitation. Complete coverage of trees by one or two spring applications of an insecticide will prevent most new infections. Dead wood on trees must be removed by pruning, and all dying trees must be removed promptly; the wood must be burned or buried, to keep the bark beetle population down. Trenching to eliminate spread by root grafts is effective. Infected trees can be saved, or at least their life prolonged, by injection of fungicides, but this is not practical for other than specimen or special trees. Control in the future may depend on use of trees bred for resistance; another possibility is release of sterile male bark beetles to eliminate the vector species.

Control by use of resistant selections of elm was used in Europe for years, and there are many urban plantings of resistant trees. Unfortunately, most of these trees are susceptible to the new form of the fungus. Nevertheless, there is still hope and progress in breeding resistant elms (Sinclair et al. 1974; Smalley and Guries 1993). Cellular and molecular genetics methods may help (Stichlen and Sherald 1993).

Overall, control of Dutch elm disease is already possible, but it is a political matter because community action is required and costs are high. Cities in the United States that have maintained careful controls still have most of their elms. Cost of control per tree is a small fraction of the cost of removing a dead tree in an urban setting. Even replanting of elms probably is practical as long as effective control is practiced. The American elm populations have decreased in the wild and are being replaced by other species; however, the elm is not likely to become extinct.

Summary

Dutch elm disease (DED) in North America illustrates the destructive effects of an alien pathogen. Origin of the pathogen is unknown, but it may have been present in Europe, unnoticed, for many years. Conclusive evidence from Dutch scientists shows it was in Europe by 1912, and possibly by 1900. Possible explanations of the sudden outbreak of DED in Europe soon after World War I were the development of virulence in an opportunistic pathogen or increased populations of the beetle vector because of war disturbances. DED became very destructive in the Netherlands, Belgium, France, Germany, and Britain. The causal fungus *Ophiostoma ulmi* was taken to North America around 1930 on elm logs; it soon devastated the native elms. The first epidemic waves in Europe and North America were subsiding by 1960, when a new wave was induced by a more virulent form of the fungus.

O. ulmi causes a systemic infection in the xylem of elms. It is carried from tree to tree by infested bark beetles and to adjacent trees by root grafts. Control is possible but expensive because it must be maintained as a community effort. Elm populations have drastically decreased in urban areas and in woodlands of North America.

7.3. White Pine Blister Rust in Europe and North America

White pine blister rust is unusual in several respects. First, it was a new disease on two continents (by 1854 in Russia and in 1906 in America), with different causal circumstances. The inciting fungus (*Cronartium ribicola*) is an obligate parasite, yet it is often lethal because it girdles the trunk and branches of its host (Figure 7.5). The history of the disease is well documented in Europe and North America, but speculations are still required to explain its origin and means of spread into western Europe. Rapid spread of white pine blister rust was alarming for foresters, horticulturists, and the public at large in both Europe and America.

A basic knowledge of the biology of *C. ribicola* is needed to understand this disease. Like many other rust pathogens, this one requires an alternate host for completion of its life cycle (Figure 7.6). Tree hosts are several

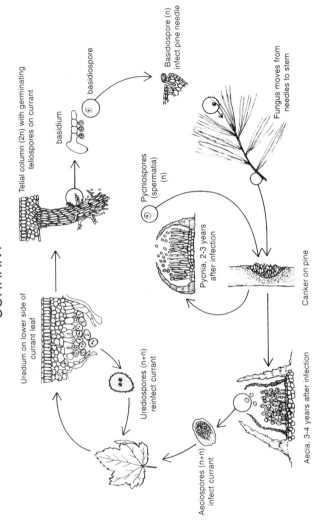

CURRANT

Telial column (2n) with germinating
teliospores on currant

basidium

basidiospore

Basidiospore (n)
infect pine needle

Fungus moves from
needles to stem

Uredium on lower side of
currant leaf

Urediospores (n+n)
reinfect currant

Aeciospores (n+n)
infect currant

Pycniospores
(spermatia)
(n)

Pycnia, 2-3 years
after infection

Canker on pine

Aecia, 3-4 years after infection

PINE

Figure 7.5. Disease cycle of white pine blister rust. (Drawing by Marlene Cameron)

Figure 7.6. Rust lesion in stem of young white pine. Pycniospores and aeciospores are produced in such lesions. (Picture from J. H. Hart)

species of white or five-needle pines (Figure 7.5); alternate hosts are many species of currants (*Ribes*). Both *Ribes* and white pines are adapted to cool regions of the northern hemisphere; both became economically important in Europe before 1725. *C. ribicola* does not spread from pine to pine; hence the alternate host is required. The fungus does not survive on *Ribes* alone because of uncertain overwintering.

There is now general agreement that blister rust originated in northeastern Asia. The endemic host in the area of origin probably was *Pinus cembra* var. *pumila,* which has tolerance to *C. ribicola;* disease-experienced populations usually have tolerance, allowing for survival of both host and pathogen. There is a vast area of Siberia, west of the range of *P. cembra* var. *pumila,* where *P. cembra* var. *siberica* and certain *Ribes* species are endemic; both are susceptible to *C. ribicola* (Spaulding 1922, 1929). Blister rust is thought to have spread gradually westward toward the Ural Mountains between Asia and Europe.

It is strange that the disease on *P. cembra* var. *siberica* was so little noticed or was so slow to reach its westward limit; perhaps the vast distances, scattered host populations in parts of the range, or continental glaciers were involved. At any rate, the fungus apparently did not exist in Europe prior to man's intervention. There was a relict population of relatively resistant *P. cembra* in the Alps, but *C. ribicola* apparently was absent (Spaulding 1922 & 1929; Fischer 1918). Gäumann (1950) believed that *C. ribicola* was endemic in the Swiss Alps, but the basis of his belief is not evident.

Originally, neither host of *C. ribicola* was present in central and western Europe, other than the relict alpine population of *P. cembra* and a small population of *P. peuce* in the Balkans. Currants and gooseberries (*R. reclinatum*) were imported into Europe at an early date, well before 1700. According to Moir (1924) and Spaulding (1929), some of the imported *Ribes* species became naturalized; however, the naturalized population is not believed to have had a role in the European epidemic (Moir 1924; Boyce 1926; Spaulding 1929). Currants became very popular, and *R. nigrum* was planted extensively in northern and central Europe; these plantations were essential for later spread of *C. ribicola.* Beginning in 1705, the white pine of North America (*P. strobus*) was imported into Europe for reforestations (Bean 1921). This pine is well adapted to the climate, and soon was widely planted, especially in Germany and the Baltic States. There were widely scattered plantings to the east across Russia.

There still is uncertainty regarding the spread of blister rust across Russia, although it was known to be common enough in western Russia by 1900 (Mayr 1900). Speculation has it that the disease swept westward across Europe when exotic *P. strobus* and *Ribes* plantations reached the Ural Mountains, where *C. ribicola* was in wait, parasitizing the native white pine and currant species (Gäumann 1950). However, this appears to be unlikely because of scattered planting of both exotic host species in the eastern part of European Russia. A more likely explanation is that *C. ribicola* was brought to western Russia and to other European countries on plants destined for botanical gardens (Spaulding 1929). This occurred during the time (1750–1850) that such gardens were being established throughout Europe, and species of *Pinus* and *Ribes* were popular items. There were several botanical gardens in Russia by 1810 and forty or more in Europe by 1850. Some of the gardens published lists of their acquisitions, which included several species of five-needle pines.

Botanical records show that *C. ribicola* was present in Russia by 1854, in Finland by 1861, and in East Prussia by 1865. In each case, the fungus was parasitizing the exotic white pines and currants, although the alternate host requirement was not known until later (Klebahn 1889). *C. ribicola* was generally distributed over Germany by 1888; apparently, it had swept over northern and western Europe, beginning about 1860. This rapid spread was no doubt facilitated by nurseries that distributed infected seedlings of *P. strobus* for reforestation (Spaulding 1929).

The spread of *C. ribicola* to North America is well documented. It was clearly identified on currants by 1906 at Geneva, New York (Stewart 1906), but there is little doubt that it was present in the northeastern states of the United States as early as 1898. The earliest known infestation was eradicated, but the disease appeared again in 1909. This was the time that reforestation was beginning in America, and seedlings of the desirable white pine were needed. American nurserymen could not supply such seedlings, so foresters turned to Europe, where *P. strobus* was used on a large scale. The German forester C. A. Schenck warned of blister rust danger, but he was ignored (Benedict 1981).

A reliable source for pine seedlings was the nursery of J. H. Söhne of Halstenbeck, Schleswig-Holstein, Germany. Some shipments from this nursery to the United States were clearly identified as being infected by *C. ribicola* (Spaulding 1911). Other European nurseries were involved as well, but political pressures prevented regulation. Thus, *C. ribicola* came to North America and soon spread over the eastern United States and adjacent areas in Canada. It became common wherever highly susceptible *P. strobus* and *Ribes* species were growing. By 1918, the geographic range of *P. strobus* was saturated, and there were huge losses to the forest industry.

There was a separate introduction of *C. ribicola* into the northwestern region of the United States about 1910. Susceptible plants in the Pacific Northwest included the native white pines, especially *P. monticola*, and endemic *Ribes* species. The northwestern introduction of *C. ribicola* was traced to importations of *P. strobus* seedlings from a nursery in Ussy, France (Spaulding 1929). Again, blister rust soon saturated the range of western white pines.

An endemic blister rust (*C. occidentale*) occurs in the western United States; the species affects piñon pine (a five-needle species) and native *Ribes* species (Spaulding 1922). This created confusion regarding the origin of *C. ribicola;* some foresters believed that *C. ribicola* came to Europe from North America. The confusion was soon clarified when mycologists found that the two rusts are distinctly different (Hedgcock et al. 1918). *C. occidentale* and several other endemic rusts on pines in North America ordinarily kill very few trees, although seedling mortality in nurseries can be high. In contrast, exotic *C. ribicola* has drastically reduced white pine populations over large areas (Peterson and Jewell 1968).

Damage to white pine in North America was clear in early reports of the disease (Snell 1928). This damage was most evident on *P. strobus* in the east and on *P. monticola* (western white pine) and *P. lambertiana* (sugar pine) in the west. For example, one report from Vermont indicated that a sixty-year-old stand of *P. strobus,* first exposed to spores from infected currants in 1908, had 24 percent mortality by 1930 and an additional 41 percent that were doomed. In the author's opinion, all trees in this stand would have died but for *Ribes* eradication in the vicinity (Filler 1933). The situation in the Pacific Northwest was even more alarming; for example, after *C. ribicola* entered British Columbia, a stand of *P. monticola,* with trees 7 m or more in height, had greater than 90 percent mortality in eleven years and practically 100 percent by sixteen years. Larger trees had less mortality (Lachmund 1934).

Young seedlings are severely affected, with a high mortality rate, making natural reproduction doubtful in many areas. However, the rust is not likely to exterminate *P. strobus* because the species still survives in many areas, depending on local weather and climate. The disease is most common and damaging in areas with an average July temperature below 22°C, for example in northern Wisconsin. Damage is slight in southern Wisconsin, where the average July temperature is above 22°C (Van Arsdel et al. 1961; Peterson and Jewell 1968).

Control measures for blister rust have had some success. In the United States, eradication of currants for control of pine rust is still feasible but is practical only in areas with a high concentration of valuable pines (Peterson and Jewell 1968). The general rule in the eastern United States is that all currants must be destroyed within at least 280 m if pines are to be protected. In the western states, a greater distance may be required to protect *P. mon-*

ticola, an extremely susceptible species. A recent survey of blister rust in Maine showed that *Ribes* eradication was worthwhile, but not entirely effective. Blister rust appeared in Maine in 1916; *Ribes* eradication was started the following year and was continued. Eradication zones were used statewide, but only in areas near valuable stands of white pine. There was an estimated 50 percent reduction in tree losses in these protected areas (Ostofsky et al. 1988). In Europe, control measures were abandoned, and *P. strobus* is no longer planted.

Summary

Blister rust of white pines is an excellent model that illustrates the danger of both alien pathogens and the introduction of alien plants. The causal fungus *Cronartium ribicola* requires two hosts (white pines and currants) for completion of the life cycle. Currants (*Ribes* species) were introduced into Europe and were widely planted at an early date; *Pinus strobus,* the white pine of eastern America, was introduced about 1705. This pine was planted over western and northern Europe, eastward into Russia. Blister rust was unknown for many years after the two host species were introduced.

The blister rust fungus, thought now to be endemic in eastern Asia, had spread slowly westward through Siberia almost to the Ural Mountains, parasitizing native currants and the endemic pine (*P. cembra* var. *siberica*). Further spread to the west was slowed, possibly by gaps in the range of suitable hosts. The fungus probably came to western Russia and Germany on infected *Ribes* and *Pinus* plants destined for botanical gardens. Spread throughout Europe probably was hastened by trade in infected pine seedlings for use in reforestation. The fungus was then taken from Europe to North America with pine seedlings for reforestation. Importations into the eastern United States occurred by 1898 and to the western United States by 1910. In both areas, the rust fungus met native pines and currants that were highly susceptible.

7.4. Other Alien Pathogens on Native Plants

There are many other examples of the devastating effects of alien pathogens on native plants. Several more examples will be considered briefly: black rot, downy and powdery mildews on grape in Europe, pine wood nematode disease in Japan, and Seiridium blight of cypress in Europe. Fire blight, caused by an alien pathogen in Europe, is discussed in Chapter 8, section 2, as an alien plant that met a native pathogen in North America.

Powdery mildew is endemic in America on *Vitis* species that tolerate the disease; presumably, long association has eliminated susceptible genotypes. The disease on grape was first noticed in Britain in 1845 by Mr. Tucker, a remarkable gardener who also devised a practical remedy. The fungus *Uncinula nector* probably had been carried from America on grape plants for botanical gardens or for study. By 1848, it had spread to France, and it soon spread to all grape-growing areas of Europe, adjacent Asia, and North Africa. European grape, *V. vinifera,* lacked the tolerance of North American grape species. The costly, but necessary, response in Europe was to spray the vines with a lime–sulfur concoction suggested by Tucker. Powdery mildew was the first great plague of Europe's vineyards.

A better control was needed. Therefore, American grapes were imported for breeding, in the hope of producing a grape acceptable for wine and resistant to mildew. This move brought a second and far greater calamity, in the form of *Phylloxera,* a woolly aphid. The wine industry was on the verge of collapse – think what no wine would mean in France and Italy! Again, the *Vitis* species of America were tolerant, responding with small leaf galls of no consequence. In contrast, the insects attacked roots of *V. vinifera* in Europe, which led to decreased vigor and even death. Resourceful farmers soon learned to manage the problem by grafting superior *V. vinifera* cultivars to rootstocks of resistant American species. Many plants and seeds were brought from America for this purpose.

The third scourge of grapes in Europe was downy mildew (*Plasmopora viticola*), an Oomycete (Peronosporaceae). Again, the fungus was of American origin, introduced along with *Vitis* seedlings for rootstocks used to combat *Phylloxera*. The fungus, first described by Schweinitz in 1837, was indigenous in eastern America on native species that are tolerant, although some cultivars of *V. labrusca* are damaged. *P. viticola* was unknown outside North America until 1878, when it was seen in France on seedlings from America. All cultivars of *V. vinifera* were very susceptible, the climate of Europe was favorable, and the disease became a major threat. This time the day was saved by Professor Pierre Millardet, who observed that vines drenched with a copper sulfate–lime mixture (to prevent pilfering) were free of mildew. From this observation, Millardet developed one of our most effective fungicides, the famous Bordeaux mixture (Millardet 1885), still one of the best fungicides for use on plants that will tolerate it. Downy mildew is now found in all humid grape-growing areas, and regular applications of fungicides are required. The disease has not been a problem in California and certain other dry-climate areas.

All literature on the three grape problems, pertinent to the aims of this book, is reviewed in detail elsewhere (Bulit and Lafon 1978; Lafon and Bulit 1981; Viennot-Bourgin 1981). There are popular accounts (Large 1940; Carefoot and Sprott 1967), but relatively few recent publications.

There are other damaging diseases of grapes. Black rot, caused by the fungus *Guignardia bidwellii,* is endemic on native grapes and Virginia creeper (*Parthenocissus quinquefolia*) in eastern North America. Native American grapes are tolerant, but not immune; the European grape (*V. vinifera*) is very susceptible and is not grown in the southeastern United States. Black rot was taken to Europe with rootstocks of *V. labrusca* for *Phylloxera* control. The disease became damaging, especially in wet years. It soon spread to all grape-growing areas of Europe but is not serious in areas with dry climates. Control is by drastic pruning of diseased parts and by use of fungicides (Anderson 1956).

"Dead arm," caused by *Cryptosporella viticola* (a fungus), is another threat; since 1949, it has moved to all areas where grapes are grown (Anderson 1956).

The pine wood nematode disease in Japan is an important case study (Mamiya 1983). Death of pine trees, cause unknown, was reported early in the twentieth century near Nagasaki on Kyushu Island. Trees at the site were promptly burned to prevent spread, but the problem reappeared nearby in 1925; subsequently, it slowly spread northward. Prompt removal of diseased trees delayed the spread until World War II, when control measures were relaxed. The disease then spread rapidly over most of Japan; the rate of spread indicated an introduced pathogen. Native pines of Japan, *Pinus densiflora* and *P. thunbergii,* are very susceptible. Many trees were killed each year; 2.4 million cu. m of timber were lost in one year at the peak of the epidemic in 1979 (Wingfield 1987).

Cause of the pine disease was identified in 1971 as a nematode (*Bursaphelenchus xylophillus*). The nematode spreads over long distances in commercial logs and wood chips and is vectored locally by sawyer beetles (*Monochamus alternalus* and others). The disease was identified in pines in the United States in 1979; the major North American vector is another sawyer beetle (*M. carolinensis*) (Wingfield 1987).

Surveys have shown that pine wood nematode is widely distributed in conifers in North America. Native *Pinus* species carry nematodes but seldom are diseased; only trees that are under stress or are in plantings show symptoms. Exotic species such as Scots (*P. sylvestris*) and Austrian (*P. nigra*) pines are blighted. These observations indicate that pine wood nematode is a native of North America; it probably was taken to Japan in logs.

Import of logs is always risky, for this and other reasons (Goheen 1993); unprocessed North American logs are now prohibited in some European countries where pine wood nematode is not present.

Pine wood nematode feeds on both fungi and wood. It is thought to have originated as a predator of fungi in wood, with later adaptations that allowed it to parasitize trees. Sawyer beetles from infected trees have the nematode in and on their bodies; they introduce the pathogen with feeding wounds in bark and phloem of living trees. Most infections occur in spring and early summer, and symptoms soon show if the weather is warm and dry. Wilting of the whole tree can occur in three to four months.

So far, the nematode is known only in North America (including Mexico), Japan, China, and Taiwan. Partial control in Japan is achieved with aerial application of insecticides for vector control and by removal of infected trees to slow the spread.

Seiridium blight of cypress was first reported in California, and it has now appeared in the Mediterranean region, where it is a serious threat to the native cypress. The causal fungus is thought to be of Asiatic origin, possibly coming to Europe by way of California, with the importation of Monterey cypress (Graniti 1993).

Two more examples will be mentioned. Hop mildew (*Pseudoperonospora humuli*) was first found in Japan in 1901 and in North America by 1909. The fungus was moved from Japan in 1920 to Britain and soon to continental Europe, all the way to Russia. The disease was always very destructive when first introduced, requiring many applications of protective fungicides. The last example is powdery mildew of gooseberry, which was taken from North America to Britain in 1899. It soon spread to the continent, then eastward to Russia. The spread in Europe was rapid; by 1907, at least 70 percent of all gooseberry plantings were thought to be infected (Stakman and Harrar 1957). This by no means completes the list of damaging alien pathogens.

8
Alien Plants,
Native Pathogens

Just as alien pathogens can overwhelm native plants, so can alien plants be overcome by pathogens present in a new area. There are many examples; perhaps most striking is fire blight of pome fruits, which appeared when Eurasian species were brought to North America. Subsequently, the North American pathogen became equally destructive when it spread back to Eurasia, in this case a native plant affected by an alien pathogen. Another example was Pierce's disease, which appeared in European grapes in North America. Sorghum from Africa met an endemic fungus in America, causing the milo disease. Several other examples that differ in circumstances are summarized.

In short, expansion of agriculture into new areas exposed crops to new pathogens. Some of the new pathogens were carried by trade over the world; intensive monoculture allowed them to become destructive. The following discussions emphasize these points.

8.1. Pierce's Disease of the Grape

Pierce's disease is an excellent example of the problems that sometimes overwhelm alien plants. *Vitis vinifera*, the European grape, was introduced into California by the Spanish invaders; it grew well for at least two hundred years. Then suddenly, in 1884, a mysterious epidemic affected the vines around Anaheim, in southern California; some 160,000 ha were killed by 1906, when the epidemic subsided. A second epidemic, which started farther north about 1935 and lasted until 1946, killed more than 20,000 ha of grapes; included were the important vineyards of the Napa Valley. A third epidemic developed after 1957 and continued until 1972. In addition, there were minor outbreaks, locally known as hot spots (Gardner and Hewitt

1974). For many years, the cause was a mystery; there were dozens of theories, but no definitive conclusions were made until much later.

Some background on the grape is needed to understand this devastating disease. There are many *Vitis* species, native to the temperate regions of the northern hemisphere. Several species have been cultivated, but only three are of significance. The most important is *V. vinifera,* which is assumed to be native in the area around the Caspian Sea; during prehistoric times, the vines were taken west to the Mediterranean region and east into Asia. The second grape species of importance is *V. labrusca,* native to the northeastern United States, adjacent Canada, and south through the highlands of Appalachia; one subspecies, var. *subedentata,* grew in the foothills and coastal plains from New York to South Carolina. *V. vinifera* was imported into eastern North America by early colonists, but the vines failed; the colonists turned to *V. labrusca,* which was domesticated. The third species is *V. rotundifolia,* a grape domesticated from a wild species that grows on the Gulf Coast of North America and on the Atlantic coastal plain north to eastern Virginia (Galet 1979). *V. rotundifolia* includes the muscadine and scuppernong types.

Several other wild grapes are native to North America. *V. riparia* is found in southern Canada and the northern United States from the Dakotas to Maine and south to Virginia and Oklahoma. *V. aestavalis* occurs wild from southern Wisconsin east to Massachusetts and south to Arkansas and South Carolina. *V. berlandieri* is found in southern Texas and Mexico. *V. cinerea* is native to Mexico and Texas, north to Illinois, and east to South Carolina and Florida. *V. monticola* is native to Texas; *V. rupestris* is native to the southwestern United States but has been reported to grow east to Mississippi and Tennessee. Some of the native North American species have been hybridized with *V. vinifera* and *V. labrusca,* mostly as sources of disease resistance (Galet 1979).

Pierce's disease was first known as the Anaheim disease, after the area in California where it first appeared. In 1889, the U.S. Department of Agriculture sent Newton B. Pierce to California to study the grape malady. Pierce, who recently had completed graduate studies in Michigan, was in close touch with Erwin F. Smith, the foremost authority on bacterial diseases of plants. Pierce was the first researcher trained in plant pathology to work in California, and he greatly influenced plant science in that state (Gardner and Hewitt 1974). He contributed much to an understanding of the grape disease and its control, but he never identified the causal agent; the basic science was inadequate at that time. Nevertheless, Pierce's contributions were recognized, and his name is still associated with the disease.

During the second epidemic (1934–46), researchers proved that Pierce's disease was contagious, that it was vectored by leafhoppers (Hewitt et al. 1946), and that the same agent caused a disease of alfalfa known as dwarf. A viral causation was assumed but remained unproven. Finally, the true nature of the causal agent was discovered in 1973 (Goheen, Nyland, and Lowe 1973; Hopkins and Mollenhauer 1973); it proved to be a bacterium, as Pierce had suspected.

The first clue to the mysterious and sudden appearance of Pierce's disease in California became evident to W. B. Hewitt, a major contributor to knowledge of this disease. Hewitt reported that the original outbreak in the Napa Valley was correlated with introduction of *Vitis* species native to the southeastern United States; cuttings had been imported to test for resistance to Phylloxera (an insect) and for possible use as rootstocks. Very soon after this introduction, the disease was found nearby, on the Sheehon Ranch near St. Helena, California (Hewitt 1958; Gardner and Hewitt 1974). No such records were found regarding the origin of the outbreak that occurred at Anaheim several years earlier, but the possibility of imported plant materials is evident.

Hewitt, fortunately, recognized the possibilities and was moved to look for Pierce's disease around the Gulf of Mexico, the source of the cuttings from the east. His travels and studies soon led to the conclusion that the disease is endemic in native *Vitis* species and other plants on the Gulf coastal plain. Many of the suscepts, including the native *Vitis* species, carry the causal agent with minimal effects (Hewitt 1958). Furthermore, the only known source of resistance in *Vitis* is in those species native to the southeastern United States, where the plants had lived with the disease for ages. Inoculation experiments proved that the following species, all natives of the southeastern United States and the Gulf coast of Mexico, are resistant to Pierce's disease: *V. aestivales, V. berlandieri, V. candicans, V. monticola, V. rotundifolia,* and *V. rupestris.*

V. vinifera was brought by British colonists to eastern North America; the vine failed in New England, presumably because of the harsh climate. The vines also failed in the southern colonies, for unknown reasons. Hewitt's excellent detective work clarified the mystery. *Vitis* species native to the area are resistant but carry the causal agent, as do many other plants native to the Gulf coast. In contrast, *V. vinifera* had never been exposed and thus lacked any element of resistance; after contact with the causal agent, the results were disastrous. Originally, the pathogen was not present in California, and *V. vinifera* thrived there prior to the unfortunate introductions of *Vitis* species from the southeast. Thus, we have a case of an alien plant and, in California, an alien pathogen as well.

The pathogen is sensitive to low temperature; hence it does not occur north of Virginia or at higher elevations in the Appalachian Mountains. Hewitt's conclusions are supported by the fact that grapes native to the northeastern United States (including *V. labrusca*) also lack resistance to Pierce's disease. The colonists learned at an early date that *V. labrusca,* which grew so well in New England and New York, did not survive along the Carolina coast and farther south. Now we know the reason. The pathogen is known to be endemic only in the southeastern United States and adjacent Mexico. There were earlier reports of its presence in Chile and Argentina, but this was found to be in error (Hewitt 1970).

The true cause of Pierce's disease was still unknown for nearly ninety years from its first appearance in California, in spite of exhaustive work by Pierce, Hewitt, and others. Circumstantial evidence implicated a virus (Hewitt 1970), but, at the time, a virus was the usual answer for an unknown plant disease. Then Hopkins and others reported that symptoms of Pierce's disease in grapevines were suppressed by application of tetracycline, an antibiotic. This reaction is characteristic of phytoplasma and rickettsiae-like bacteria, as had been shown with other plant diseases; viruses are not affected by such treatments. Finally, in 1973, two groups of researchers, one in Florida and the other in California, reported independent work that implicated a xylem-limited, rickettsiae-like bacterium as the cause (Hopkins and Mollenhauer 1973; Goheen, Nyland, and Lowe 1973). These reports were based on electron-microscopic examination of plant tissues, not on isolation of the bacterium in culture. Rickettsiae-like bacteria were notoriously difficult to isolate. It was not until 1978, after ninety-four years of frustration, that the causal bacterium of Pierce's disease was finally isolated and shown to reproduce the disease in inoculation experiments (Davis, Purcell, and Thomson 1978). The bacterium is now known as *Xylella fastidiosa;* there are recent studies on its relationship with the insect vector (Hill and Purcell 1995). A good general summary of the disease can be found in the compendium by Pearson and Goheen (1988).

8.2. Fire Blight of Apple and Pear

The fire blight disease was unknown until colonists brought pear, apple, and quince from Europe to North America. At first, the North American orchards were small, scattered, and were often started from seed; the disease, if present, was not noticed. Fire blight, its cause then unknown, was first reported in the Hudson River valley of New York in 1780. As settlements moved westward, the disease moved as well, either because the planting stocks were contaminated or because the pathogen moved into the new crops from native plants (Van der Zwet and Keil 1979). In all the world, only North America had fire blight until it suddenly appeared in New Zealand, in 1910. Eventually, fire blight reached Europe (1957). The disease has a stormy history.

It is likely that the causal bacterium *Erwinia amylovora* was endemic on roseaceous plants native to North America (hawthorn, mountain ash, shad or *Amelanchier,* and others). The native species of North America are not immune, but generally they have resistance; this is the usual case with endemic host–pathogen combinations of long standing. The related plants from Eurasia (apple, pear, quince, and many ornamentals) are very susceptible, and they soon fell to the North American bacterium. This is not an unusual fate of alien plants.

Fire blight affects more than seventy-five plant species in the family Rosaceae; the disease has never been found in other families. Susceptible genera include *Amelanchier, Cotoneaster, Crategus* (hawthorn), *Cydonia, Fragaria, Prunus, Malus, Pyracantha* (firethorn), *Pyrus, Rosa, Sorbus* (mountain ash), and *Spiraea.* Of these, only *Malus* (apple), *Cydonia* (quince), and *Pyrus* (pear) are of serious concern. Some of the tolerant species and some of the susceptible ornamental plants (*Sorbus, Cotoneaster,* and crab apples) serve as reservoirs of the bacterium, spreading it to fruit crops. This also is the case with hawthorn, a hedgerow shrub in Europe. Fire blight has caused major shifts in fruit production, from highly susceptible cultivars of pears and apples to more resistant ones, and from the eastern United States to the west.

Fire blight became unusually destructive, especially to pear, when the commercial fruit industry was first developed in the eastern United States. By 1850, the disease reached Illinois, where the pear was popular. Millions of trees were killed, and cultivation of the desirable cultivars became commercially impossible; most of the industry eventually moved to dry areas on the west coast. Fire blight reached California about 1880, and by 1910 there were enormous losses, especially in the San Joaquin valley, where pears were a big item. Apples were severely affected in the eastern and midwestern United States (Thomas and Jones 1992), but not as much as pears; apple cultivation was maintained, with some shifts in cultivars, and at great expense.

Fire blight affects all parts of susceptible hosts. Blossom blight usually is noticed first; infected flowers wilt and die suddenly. Twig blight (Figure 8.1) often follows, with dead leaves hanging to the tree. The dead, blackened foliage gives the tree a scorched appearance, hence the name fire blight. Infection can spread downward from blossoms, causing elliptical

Figure 8.1. Fire blight in foliage of apple. Infections often occur in blossoms and move into the wood. (Picture from E. H. Barnes)

Figure 8.2. Fire blight canker in stem of apple tree. Such lesions girdle and kill the infected branch. (Picture from E. H. Barnes)

cankers in the bark of twigs and branches (Figure 8.2). If cankers reach the main stem, the tree will die; this seldom happens in apples but is common in some pear cultivars. Fire blight is, without doubt, the most serious bacterial disease of pears and apples.

The cause of fire blight was a controversial subject for many years. The disease became serious soon after discovery of fungi as causal agents of potato late blight, rusts, and other diseases. Accordingly, a fungus was suspected to cause fire blight, but no one was able to prove it. There were, of course, the traditional guesses: fermented sap, frozen sap, sun scald, divine intervention, and so forth. Finally, Thomas Burrill of Illinois showed that bacteria are involved, although he lacked conclusive proof. Proof was left for later research; Arthur supplied it in 1885, with pure cultures and inoculation experiments (Koch's postulates). Next, Waite proved in 1898 that the bacterium is vectored by insects, including bees. Still, botanists in Europe refused to accept these findings; their skepticism seems to have been based on the preconception that bacteria could not affect plants. Alfred Fischer, a prominent German botanist, was most outspoken in this view.

Erwin F. Smith of the U.S. Department of Agriculture then came into the picture, and the infamous Fischer–Smith controversy erupted (Campbell 1981). Fischer did little more than restate the ideas of the establishment in Europe. Smith, a self-taught linguist, presented unequivocal data in German, and completely demolished Fischer. The encounter was of little scientific value, but it did result in European respect for American plant science, for the first time. Fire blight was the first plant disease known to be caused by a bacterium; proof of bacterial causation for olive knot, cucurbit wilt, crown gall, and others soon followed. Fire blight also was the first plant disease known to be vectored by insects.

At least two remarkable personalities were involved in the fire blight story: Thomas Burrill and Erwin F. Smith. Burrill started as a mathematics teacher in the public schools of Illinois and moved on to superintendent of schools, to university mathematics professor, to professor of botany, and finally to presidency of the University of Illinois. He built a small college into a great university. He was an original thinker and a dedicated man (Carefoot and Sprott 1967). Erwin F. Smith was a different sort, but equally admirable (Campbell 1983). He was born in upstate New York and moved as a child, with his family, to the small village of Hubbardston in rural Michigan. As a young adult, he finally was able to attend high school in Ionia, Michigan. Next, Smith moved to Lansing, where he worked for the state department of health and attended Michigan Agricultural College. His education was completed at the University of Michigan, where he earned B.S., M.S., and Ph.D. degrees; from there, he moved to a career with the U.S. Department of Agriculture in Washington.

Along the way, Smith always had mentors that recognized his talents. These included Ida Holmes, an exceptional grade-school teacher; Charles F. Wheeler, a pioneer Michigan botanist; and Theobold Smith, an eminent bacteriologist. Soon after finishing high school, Smith collaborated with

Wheeler in the publication of the book *The Flora of Michigan.* Erwin F. Smith became an internationally recognized scientist for his work on bacterial diseases of plants, especially the crown gall disease, a form of plant cancer (Chapter 3, section 3). His interests were broad; he was master of many languages, a poet, and a naturalist. At various times, Smith was president of important scientific organizations, including the American Association for the Advancement of Science, the Botanical Society of America, the American Phytopathological Society, and the American Association of Cancer Research. He was a member of the U.S. National Academy of Science and a fellow of the American Academy of Arts and Sciences. He was one of the true greats in plant pathology.

The microorganism that induces fire blight is a rod-shaped, gram-negative, motile bacterium with flagella over the entire surface (peritrichous). It overwinters in the margins of cankers, often also as latent infections in bark, and on alternative hosts. The bacteria in a polysaccharide matrix are exuded as drops from bark pores and are moved by insects and splashing rain to developing flowers. Small strands of bacteria in the matrix can be carried long distances by wind. Pollinating insects, including bees, carry the bacteria from flower to flower; infection occurs through the stigma or the nectaries, leading to blossom blight. Infection also can occur through stomates in leaves and sepals and through minute wounds in flowers, leaves, and bark. Bacteria move through intercellular spaces to the phloem and xylem and down into stems, causing sunken cankers (Figure 8.2).

E. amylovora can live as an epiphyte on flowers, leaves, and fruit of apple, prior to infection or expression of disease. High populations can occur on and around leaf wounds (Van der Zwet and Keil 1979). The bacteria also exist and multiply in xylem vessels and on buds; the pathogen can even exist, for a time, on the bodies of insects. All such populations are potential sources for secondary spread. A phage (virus) that can kill *E. amylovora* exists in aerial parts of apple and pear trees; the significance of this in epidemiology is unknown (Ritchie and Klos 1977). Also, blight lesions contain saprophytic bacteria that compete with *E. amylovora,* but their role and potential as a biocontrol are unknown (Schroth et al. 1974; Van der Zwet and Keil 1979).

Fire blight epidemics are favored by high temperatures during early blossom time, followed by warm, humid weather with rain. Warm weather favors bacterial growth, and humid conditions favor dissemination. Blight also is favored by high nitrogen, which leads to succulent tissue, and by sprinkler irrigation. At one time, the sugar content of nectar was thought to influence infection, but this factor is now given little credence.

E. amylovora has a very efficient means of dissemination via honey bees, discovered by Waite in 1898 (Van der Zwet and Keil 1979); later, it was studied by Hildebrand and Phillips (1936) and, more recently, by Johnson et al. (1993). These studies gave impressive data, showing, among other things, that each bee visits a total of twenty blossoms per hour of foraging; infested bees transmit blight to as many as 70 percent of the flowers visited. The experiment utilized hives arranged so that each bee had to pass through dust that contained labeled strains of the bacterium; the bees then foraged in an area of the orchard enclosed with nets to prevent contamination by outside insects. The bees also were shown to distribute antagonistic bacterial species that could be a biocontrol factor (Johnson et al. 1993). In addition to being dispersed by insects, *E. amylovora* is dispersed by driving rain, wind, and, possibly, by birds (Psallidas 1993).

Fire blight became especially disastrous in California by 1890, notwithstanding early claims that it should not exist there because of the climate. Insect transmission negated climatic limitations. Epidemics in California, from first appearance in the 1880s until 1979, were analyzed by Reil, Moller, and Thomson (1979), and a predictive model for forecasting epidemics was developed. Major predictive factors include an average temperature of 17° to 19°C at blossom time, followed by several days of rain and high humidity. Cool seasons often result in escape or low incidence of fire blight. More recently, a computer-driven predictive model (MARYBLYT) was devised, but its use has been limited because of great variation in epidemiological factors from place to place (Schroth et al. 1974; Jones 1992). The model is useful in West Virginia and Maryland (Van der Zwet et al. 1994).

Fire blight was first seen in California near Chico about 1888, and millions of pear trees were soon killed. The blight probably was brought over the mountain and desert barriers on infected plants. The pear industry was eliminated in some areas, notably the San Joaquin valley, where pear cultivation never recovered (Reil, Moller, and Thomson 1979). Some of the best pear cultivars were taken off the market. Pierce (of grape-disease fame) confirmed the identity of bacterial isolates from all over California in 1902. The disease soon spread northward to Oregon, Washington, and British Columbia.

Fire blight moved to England in 1957, probably on imported plants or contaminated fruit boxes. Next, it moved to the European continent, in a northeast direction, presumably with bird migrations or on imported plants (Psallidas 1993). Once on the continent, fire blight slowly expanded its range, to the Netherlands by 1966, to Belgium and Denmark by 1968, to

Germany by 1971, to France by 1972 (Van der Zwet 1968; Van der Zwet and Keil 1979), and to Egypt by 1982; it spread to Cyprus in 1984 and, finally, to Israel and Turkey by 1985, presumably from Egypt (Momol and Zeller 1992).

Europe was not prepared for this disaster; the affected countries reacted slowly with quarantines and attempts at eradication, but it was too late. Fire blight was firmly established, and it will stay. In part, Europe's difficulties can be traced to the hawthorn hedgerows that serve as bacterial reservoirs and to an understandable reluctance to eradicate the hedgerows. Nevertheless, many hectares of orchards were destroyed in the attempt at eradication, along with 200,000 m of hawthorn hedge and perhaps a half million cotoneaster plants. Spread of *E. amylovora* was scarcely slowed (Roosje 1979).

In earlier times, fire blight was controlled, in part, by careful pruning to remove diseased tissue and by use of copper-containing sprays at blossom time. Copper has disadvantages, including phytotoxicity and russeting of fruit; it was replaced by an antibiotic, streptomycin, which became a control mainstay for years. *E. amylovora* now has resistance to streptomycin in many parts of the United States especially in California, where resistance to the antibiotic was first reported in 1971. This resistance posed serious problems. Streptomycin resistance is spreading rapidly, probably because the genes that control resistance are passed readily among bacterial populations. Chiou and Jones (1991) isolated streptomycin-resistant *E. amylovora* from an apple orchard in Michigan and from crab apple trees nearby. These findings have been confirmed and expanded (McManus and Jones 1994).

There are at least two types of resistance to streptomycin (McManus and Jones 1994): One type is carried on a plasmid, an extrachromosomal inheritance that is transferred readily from cell to cell; the other type of resistance is chromosomally inherited. The latter has been found in California, Oregon, Michigan, and Washington. To date, plasmid-carried resistance in *E. amylovora* has been found only in Michigan (Chiou and Jones 1991), although the same plasmid is present in other bacterial species from streptomycin-sprayed orchards in New York. Such resistance likely will be transferred to *E. amylovora* in time (Burr 1993). Genetic transfers between *Erwinia* and related genera are well known; these transfers include genes for resistance to antibiotics (Chatterjee 1972).

There were unsuccessful attempts in the mid-1970s to detect streptomycin resistance in *E. amylovora* in New York and Michigan (Sutton and Jones 1975; Beer and Norelli 1976). However, streptomycin resistance was soon found in another pathogenic bacterial species (*Pseudomonas syringae* pv. *papulans*) in New York and Michigan orchards that were sprayed with streptomycin (Burr et al. 1988; Norelli et al. 1991; Jones, Norelli, and Ehret 1991). Streptomycin resistance was then found in *E. amylovora* in Michigan, in isolates from streptomycin-sprayed orchards. Earlier failures to detect streptomycin resistance suggest a recent introduction into *E. amylovora,* probably from other bacterial species. These findings indicate the importance of using as few antibiotic sprays as possible, interspersed with other toxicants, to delay development of resistance. This places greater emphasis on the use of computer models to predict outbreaks of fire blight.

We now have a better understanding of antibiotic resistance in *E. amylovora* than in any other plant pathogen, thanks to the use of modern techniques of molecular biology. The gene for streptomycin resistance (Michigan and New York type) is carried on a conjugative plasmid (Burr et al. 1988), first found in several bacterial species other than *E. amylovora;* much of this research was on *Pseudomonas syringae* pv. *papulans,* which is widespread in orchards (Norelli et al. 1991; Sobiczewski, Chiou, and Jones 1991; Jones, Norelli, and Ehret 1991). The gene was then found in *E. amylovora* and was conclusively identified by transposon analysis and DNA hybridization experiments, using a probe from a streptomycin-resistant strain of *P. syringae* pv. *papulans.* The probe hybridized with 2.7-kb restriction fragments of DNA digests (Ava 1 digestion) from all streptomycin-resistant isolates, but not from streptomycin-susceptible isolates. These hybridizations did not occur with DNA from *E. amylovora* (streptomycin-resistant) isolates from Washington, showing that the resistances in Washington and Michigan isolates are different.

A 33-kb plasmid was in all Michigan and New York isolates with streptomycin resistance, but was not in the streptomycin-sensitive isolates or in the resistant isolates from Washington. The plasmid was transferred in mating experiments to streptomycin-susceptible strains of *E. amylovora,* which then became resistant (Chiou and Jones 1991, 1993; Norelli et al. 1991; Sobiczewski, Chiou, and Jones 1991).

Other molecular genetics work with *E. amylovora* has been notable, making fire blight one of our best understood diseases. The bacterium carries *hrp* genes that confer virulence or avirulence. A significant product of *hrp* is harpin, a protein associated with the bacterial cell surface. Harpin creates susceptibility in some host cultivars and induces cell collapse in resistant cultivars (Wei et al. 1992, 1993). This is discussed further in Chapter 3, section 5.

Summary

Fire blight is a disease of apples and pears that appeared with the introduction of Eurasian plants to North America. The disease is caused by a bacterium (*Erwinia amylovora*) that is endemic on rosaceous plants that are native to eastern North America. The disease was restricted to North America until the early twentieth century, when the causal bacterium was found in New Zealand. Later, the disease appeared in Europe (1957) and elsewhere, as an alien. It is expected to move still farther into Asia, Africa, and Australia.

Fire blight became destructive in the United States with the development of commercial apple and pear orchards. It moved west with the movement of agriculture. Fire blight was the first plant disease known to be caused by a bacterium and the first known to be vectored by insects. The disease was controlled by the use of antibiotic (streptomycin) sprays, until the bacterium developed resistance to streptomycin. This was among the first reported examples of antibiotic resistance. Streptomycin resistance was studied by modern methods of molecular biology, leading to a much better understanding of antibiotic resistance. The molecular basis of pathogenicity is now apparent, making fire blight one of the best understood of all plant diseases.

8.3. The Milo Disease of Grain Sorghum in the Southwestern United States

The milo disease, caused by *Periconia circinata,* became a major problem because of several factors that are now known to encourage epidemics. First and foremost, sorghum was a new introduction into North America; the introduced grain sorghum (*Sorghum bicolor*) had a narrow genetic base and was planted as a monoculture over a large area. Milo sorghum soon encountered a native fungus for which the crop had no resistance.

Grain sorghum was brought to the United States from Africa about 1885, to fill the need for a crop adapted to an area too dry for maize. The original introduction probably was a land race of the Giant Milo type (Karper and Quinby 1946). During the early years after introduction, chance selections of mutants and new genetic combinations multiplied the number of cultivars available. The new cultivars included the dwarf and double-dwarf types that proved to be well adapted to machine harvesting and to the agriculture of the dry southern plains. Selection of dwarf types led to further shrinkage of the gene pool and to large areas planted to genetically uniform types. These events set the stage for a disease disaster.

The milo disease was first observed in 1924 at Chillicothe, Texas, and in 1926 at Garden City, Kansas, in fields cropped to sorghum for several years (Leukel 1948). Ten years later, the disease was widespread and creating damage all over the southern plains and west to California and north to Nebraska. For many years, the causal organism was thought to be a *Pythium* species, even though *Pythium* lacked the cultivar specificity evident for the disease. Finally, Leukel (1948) isolated *Periconia circinata,* a soil fungus that had the proper cultivar selectivity and that reproduced the disease in inoculation experiments. The earlier confusion was caused by slow growth of *P. circinata* in culture and by rapid growth of ubiquitous, opportunistic forms of *Pythium* and *Fusarium* that overran the cultures. The genus *Periconia* was not well known to plant pathologists at that time; its species are now known as saprophytes and as opportunistic, low-grade pathogens on roots of various plants (Domsch, Gams, and Anderson 1980). *P. circinata* was first described as a soil saprophyte growing on wheat roots in France (Mangin 1899).

The pathogenic form of *P. circinata* invades the roots and the lower culm of sorghum plants. Infected roots have the usual symptoms of root rot, including cortex decay and red discoloration of the central cylinder; culm bases also turn red and die. Leaves and upper stem are not invaded, but leaves on older plants may roll, wilt, turn yellow, and become blighted. Infected young seedlings have a scalded appearance (Quinby and Karper 1949). These manifestations suggest systemic toxemia.

There are several hosts of *P. circinata,* all closely related: the milo grain sorghums; milo derivatives, including some Darso sorghums; one cultivar of sweet sorghum or sorgo (*S. vulgare* var. *saccharatum* cv. Extra Early Sumac, of unknown heritage) (Quinby and Karper 1949); and shattercane (*S. bicolor*), a weed sorghum in the western United States (Dunkle 1979). Susceptible genotypes of grain sorghum were found in accessions from India, Sudan, Nigeria, and China (Dunkle 1981). The original sorghum brought from Africa probably was a suscept (Karper and Quinby 1946; Dunkle 1981). All other sorghums, including most of the international sorghum lines, are resistant or immune, as far as is known. Even before the causal fungus was identified, susceptibility was shown to be under the control of a single dominant gene. Resistant plants are thought to arise by mutations; the disease in grain and sweet sorghums was soon controlled by use of resistant selections that had survived in infested fields. The milo disease is unknown outside the United States and affects no genera other than *Sorghum.*

There has been little speculation on the origin of the virulent, host-selective form of *P. circinata.* However, it probably originated from opportunistic forms of *Periconia,* which are widespread (Sprague 1950). The virulent form may have started as a mutant or some other genetic recombination. Virulent *P. circinata* has been found with certainty only in the United States, even though susceptible sorghum is planted elsewhere (Dunkle 1981). We might have expected to find the disease in Africa or India, where sorghum is a significant crop; however, there have been no conclusive reports. It is of special interest that shattercane, a weed sorghum, also carries the gene for susceptibility; 35 to 40 percent of shattercane seeds from Nebraska were susceptible (Dunkle 1979). This finding does not prove that shattercane was an endemic host of the fungus; weed and cultivated sorghums are interfertile, and the gene for susceptibility in shattercane may have been from cultivated sorghum. The weed host may be significant as a reservoir for the specialized fungus or as a maintenance source for the gene for susceptibility.

A host-specific toxin as a determinant in the milo disease was first suggested by Leukel (1948) in his work on the causal role of *P. circinata*. Leukel seeded sorghum in pans of sand that were heavily infested with the fungus. After susceptible seedlings were killed, he eliminated the fungus by steam pasteurization and then reseeded; again, susceptible seedlings were killed but resistant control seedlings were not, implicating a residual, heat-tolerant toxin.

The involvement of such a toxin was demonstrated unequivocally in later work (Scheffer and Pringle 1961). *P. circinata* was grown in liquid cultures, and sterile filtrates were tested against germinated seeds and cuttings from sorghum plants. Filtrates from some isolates completely inhibited root growth in susceptible seedlings at both low and high dilutions, whereas growth of resistant seedlings was not affected. Other isolates produced filtrates without selective toxicity; these isolates were not pathogenic in inoculation tests. Shoot cuttings from susceptible plants were killed by dilute solutions of toxin-containing filtrates, whereas cuttings from resistant sorghum and other species were not affected. Toxin was isolated and purified by solvent extraction and chromatography (Pringle and Scheffer 1963). The toxin, originally called PC toxin, was finally characterized in 1992 and given the trivial name peritoxin. It is a low-mol-weight compound (m/z 575) containing aspartic acid, cyclic hydroxylysine, chlorine, and cyclopropane derivatives (Figure 8.3) (Macko et al. 1992).

All cultivars of sorghum that are susceptible to *P. circinata* are sensitive to peritoxin, and all cultivars that are resistant to the fungus are insensitive to the toxin (Figure 8.4). Other researchers confirmed that susceptibility and resistance to peritoxin and to the fungus are controlled by the same gene locus, that heterozygotes are intermediate in susceptibility to the fungus and in sensitivity to toxin, and that resistant mutants appear frequently in susceptible cultivars (Schertz and Tai 1969). Toxin was shown to be reliable for screening for disease resistance. Resistant sorghum, with the *pc* gene, will tolerate at least 20,000-times-higher concentrations of peritoxin than will sorghum with the *PC* gene for susceptibility. The postulated site of action for toxin is the plasma membrane (Scheffer and Livingston 1984).

The ability of *P. circinata* to produce peritoxin had ecological conse-

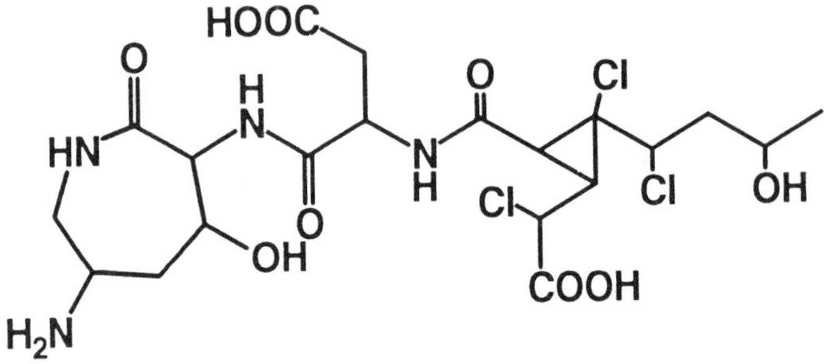

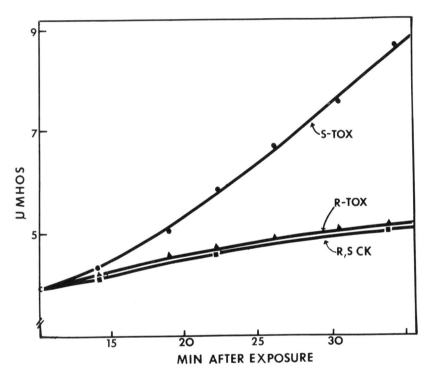

Figure 8.3. Chemical structure of the host-specific toxin from *Periconia circinata*.

Figure 8.4. Host-specific effects of peritoxin on electrolyte leakage from susceptible sorghum tissues. Leaf sections were exposed to toxin, and conductivities of leaching solutions were determined at intervals. ●–● = susceptible toxin-treated tissues; ▲–▲ = resistant toxin-treated tissues; and ■–■ = controls (resistant and susceptible) without toxin. Losses from susceptible tissues indicate cell-membrane damage, but the effect may be secondary to a primary lesion. (Gardner, Mansour, and Scheffer 1972)

quences. First, the ability to produce toxin led to great increases in populations of the fungus and to a devastating epidemic in the crop. Also, seasonal incidence of the disease is a function of toxin reactions because heat affects sensitivity of sorghum tissues. Sorghum is a high-temperature plant, growing well at 35°C; when plants are held at this temperature for several hours, they become insensitive to peritoxin, perhaps because a protein receptor for toxin is altered. When the temperature is lowered to 20°C, plants regain sensitivity in two to three days, perhaps because new receptors are synthesized (Bronson and Scheffer 1977; Scheffer 1991).

These heat responses fit with observations of disease incidence in the field and in plants held under controlled conditions. In the field, Periconia blight appeared in spring, decreased or disappeared during the heat of summer, and reappeared in fall (Quinby and Karper 1949). Inoculated plants under controlled conditions grew well with no symptoms at 35°C but were quickly blighted when temperature was lowered to 20°C (Bronson and Scheffer 1977). Similar heat effects are evident with *Bipolaris*, which affects sugarcane (a toxin producer; see Chapter 12, section 4), and with the pear disease, which is caused by an *Alternaria* species (Chapter 13, section 1) (Scheffer 1989a). The parallel between toxin sensitivity and disease incidence is further proof of the key roles of host-specific toxins as determinants in certain disease. Still other plant diseases, not now known to be mediated by toxins, have comparable responses to heat.

Summary and Conclusions

The major factor in the epidemiology of Periconia blight of sorghum was the introduction of an alien crop with a narrow genetic base. A new pathogen (*P. circinata*) was encountered in the United States about forty years after the introduction. The situation was complicated further when the genetic base was narrowed again by selection of dwarf genotypes and by large-scale monoculture. This is typical of many other diseases that appeared suddenly following changes in agriculture.

P. circinata probably evolved from a nonspecialized, opportunistic, or low-grade pathogen to a virulent, host-specialized form capable of attacking metabolically active plant tissue. The change occurred when the fungus gained the ability to produce a highly active, host-specific toxin (peritoxin). The most obvious ecological consequence was a devastating epidemic in an important crop. Toxin sensitivity of sorghum tissues is affected by heat, and this controls the seasonal incidence of the disease. The epidemic was brought under control by the use of resistance genes, common in sorghum from various parts of the world. A dominant gene gives susceptibility to *P.*

circinata and sensitivity to its toxin. Plant breeders can screen for resistance by exposing germinated seeds to toxin.

8.4. Alien Pines, Native Pathogens

Dothistroma needle blight of Monterey pine (*Pinus radiata*) became destructive following movement of the tree to new areas and planting in monoculture. *Dothistroma* also has damaged ponderosa (*P. resinosa*) and Austrian (*P. nigra*) pines that were planted outside their native habitats. Summer rains are needed for serious infections by *Dothistroma;* without such rains, the fungus is present but of no consequence. Epidemics have been explosive on *P. radiata* in the southern hemisphere, but the disease is not the ecological disaster that was caused by chestnut blight. In a sense, Dothistroma blight is a disease mediated by attempts at maximum production.

The Monterey pine has a very limited natural range. It was restricted to a narrow, 200-km zone of the California coast, from San Francisco south to San Luis Obispo. The area has a mild Mediterraneanlike climate, with rain primarily in winter and with much fog in summer. *P. radiata* is not an imposing or fast-growing tree in its native area. The climate of coastal California appears to be suboptimal for growth but optimal for long-term survival, perhaps, in part, because pests and pathogens are missing. The species is limited elsewhere by frost, midsummer heat, disease, and insect pests (Hepting 1971).

Monterey pine was imported by other countries at the end of the nineteenth century. The tree did well and became very important in the southern hemisphere, where there are no native pines. Millions of hectares were planted in Chile, New Zealand, Australia, and Africa. The tree produced good timber and grew more rapidly in these areas than it did in its native habitat. In New Zealand, Monterey pine accounts for more than 90 percent of harvested wood (Chou 1989).

The needle blight fungus (*Dothistroma septospora,* =*D. pini;* teleomorph, *Mycosphaerella pini,* an Ascomycete) was first described in Idaho (United States) in 1917 and in Georgia (USSR) in 1919. The disease became a worldwide problem during the next sixty years, sometimes causing heavy mortality of young trees in plantations. The pathogen probably was distributed, in part, with pine planting stocks (Sinclair, Lyon, and Johnson 1987). However, there are different races in Africa, South America, and North America (Gibson 1972); it is possible that virulent forms originated separately from a parental, opportunistic type.

Dothistroma blight occurs in North America in Newfoundland, south to

Virginia, and west to the Pacific coast. It is common, but causes little damage except in disturbed areas such as parks, shelterbelts, and Christmas-tree plantations, and in planted forests. Plantations of Austrian and ponderosa pines outside their native ranges are seriously affected (Peterson 1973). The disease has been destructive on ponderosa plantations in the mid–United States (Petersen 1967, 1973; Eldridge et al. 1980). Disease is first evident as killing of needles, often with reddish spots. Seedlings and saplings frequently are killed, but the major damage is from reduced growth (Peterson 1967). The rates of spread in plantations have been rapid.

Plantations of Monterey pine in the southern hemisphere remained free of serious problems prior to about 1957, when Dothistroma blight suddenly appeared in New Zealand, Africa, and Chile. In Africa, it was first noticed in Tanzania, but herbarium specimens indicated that it was in central Africa, unrecognized, as early as 1940. The disease was evident in all major African plantings by 1964. Dothistroma blight also became destructive in North America in plantations of ponderosa and Austrian pines, at about the same time that trouble started in the southern hemisphere (Gibson 1972). This suggests, again, new and more virulent races of the pathogen. But how could this have occurred worldwide, in such a brief time?

There have been no serious outbreaks in natural forests in the western hemisphere. Also, Dothistroma blight was not a problem in Spain, South Africa, or northern Chile before 1972 (Gibson 1972), and not in Australia until about 1980 (Eldridge et al. 1980). In California, Dothistroma blight had little effect on ponderosa pine, but it has been severe on Monterey pine in plantations north of the native range of the tree (Cobb, Uhrenholdt, and Krohn 1969).

The Dothistroma blight fungus affects more than thirty *Pinus* species; Monterey, Austrian, and ponderosa are most susceptible, whereas Mexican yellow (*P. patula*) is resistant. Infections sometimes occur in fir (*Pseudotsuga*) and larch (*Larix*) trees growing near infected pines (Gibson 1972). The sexual stage of the fungus is found in scattered areas; the role of ascospores in dissemination is unknown.

Resistance is known in some individuals of each species; these include ponderosa pines from certain areas of the western United States (Peterson 1984) and Austrian pines from Yugoslavia (Peterson and Read 1971; Sinclair, Lyon, and Johnson 1987). *P. radiata* tends to become tolerant with age. A form of induced resistance is said to be systemic and graft transmissible (Gibson 1972).

D. pini enters needles through stomates and colonizes mesophyll tissue (Gibson 1972). A toxin (dothistromin) may be involved, which may be responsible for red spots on needles (Shain and Franich 1981). Symptoms are evident about four months after initial penetration. Light seems to be needed for symptoms to develop; shaded trees seldom show good symptoms (Gibson 1972).

Fungicidal sprays with certain copper formulations, including Bordeaux mixture, give adequate control with two properly timed applications per year (Peterson 1967). This method of control may not be economical, except for seedling nurseries and young plantations in some areas. It has been used in New Zealand, with application by aircraft, to protect young plantings until they reach a tolerant stage (Manion 1981).

Conifer diseases similar to Dothistroma blight include Diplodia tip blight, Elytroderma needle blight, Lophodermella blight, and brown spot needle blight caused by *Scirrhia acicola* (Manion 1981; Sinclair, Lyon, and Johnson 1987).

Two alien pines have been widely planted in North America: the Scots (*P. sylvestris*) and the Austrian (*P. nigra*). Scots pine was brought to the United States from Europe at an early date; it was a favored conifer and was widely planted by 1870, even on Kansas prairies. It had variable success in North America, depending, in part, on the European source of seeds. Sphaeropsis canker, spittlebug, pine-wood nematode, and woodgate rust (*Endocronartium harkensii*) are common causes of failure. Austrian pine in North America is damaged, often seriously, by pine-wood nematodes, by blights caused by *Diplodia* and *Dothistroma* species, and by other diseases (Hepting 1971).

8.5. Other Alien Plants, Native Pathogens

African mosaic virus of cassava is an important example of a native pathogen meeting with an alien crop. Cassava, a major source of carbohydrates, is indigenous to the tropics of South America. It was moved by the Portuguese to West Africa in the sixteenth century; since then, it was planted throughout the humid tropics. Although cassava is the most important food plant in Africa, it is almost entirely a subsistence crop; there are industrial or export uses only in Brazil and Thailand. Potentially, cassava can yield eighty tons per hectare, but the average is much lower. Cassava is propagated by stem cuttings; as usual with such a practice, viruses soon became a problem. Cassava is the source of tapioca.

Cassava mosaic virus, unknown in South America, emerged about 1929 and became a major problem in Africa; it is now found throughout Africa and Madagascar east to India and Java. The virus is found in more than 80 percent of cassava plants in Africa, where overall yields are estimated to be reduced by 50 percent. However, farmers generally are unaware of the losses, because the disease is not recognized or is taken for granted. The virus probably was endemic in native African weeds and then spread to cassava when it was widely planted.

Cassava mosaic virus is transmitted by whiteflies and by infected cuttings used in propagation. There are several strains, all serologically related to other whitefly-transmitted viruses. Infected cassava, rather than weed hosts, is now the major reservoir of both the virus and vector. The disease can be controlled by eradication of infected plants and by use of seedlings to establish disease-free stocks in isolated plantations. The virus is not seed-borne. Some cultivars are tolerant (Thurston 1984; Farquet and Farquet 1990).

Several diseases of beets (*Beta vulgaris*) are instructive models. Beets are of Eurasian origin. The yellow wilt disease appeared in Argentina in 1929, soon after beet cultivation began there; the disease eliminated beet growing in the Rio Negro valley of central Argentina and restricted the area of profitable production elsewhere in Argentina and in parts of Chile. The causal bacterium, a rickettsiae-like organism, is vectored by leafhoppers and probably came from a native host plant on which it is of no consequence; if beets had not been introduced, yellow wilt pathogen would never have been known. The disease is now a potential threat worldwide (Bennett and Munck 1946; Bennett et al. 1967). We do not know why it has not, to date, spread from South America. Perhaps a narrow host range, propagation by seed, or limited distribution of the vector has prevented spread.

The viral-induced pseudo-yellows disease of beet has a different natural history but also illustrates problems for alien plants. The virus is transmitted by whiteflies, as is the cassava mosaic virus. Beet pseudo-yellows was first found by Duffus (1965) in a localized area of California. The virus has a wide host range and thus is a potential threat to many crops other than beet; indeed, the virus now occurs on cucumbers and melons in Ohio, Spain, Japan, and other areas. The disease is increasing in importance and is spreading rapidly (Liu and Duffus 1990).

Beets have several other virus diseases of interest, including yellows, curly-top, and yellow vein or rhizomania disease (Richards and Tamada 1992), which is becoming a serious threat. All originated in lands where beet is an alien.

Peach (*Prunus persica*), native to China, was brought to the United States during colonial days. Several new virus and phytoplasma diseases appeared in North America, including peach yellows (described in 1791), little peach (caused by a strain of yellows phytoplasma, which appeared in 1898), phony peach (about 1885), X-disease (1913), and peach mosaic (1931). These diseases are caused either by new forms of endemic pathogens or by native pathogens passed directly from native hosts to peach.

X-disease of peach and cherry was first found in 1931 in California and in 1933 in Connecticut, where it apparently had been present for several years. Within a few years, it had spread to most North American peach- and cherry-growing areas. Presumably, the source of X-disease was the native chokecherry (*P. virginiana*) and related species. The disease appears as yellowing and stunting on chokecherry, which commonly carries the X-pathogen, even in areas with no peach or cherry cultivation. The disease in peach shows as irregular red or yellow blotches on upward-rolling leaves, which soon become ragged and pale; fruit on infected trees is small, often dropped, and has a bitter flavor. Transmission is mostly by grafts and leafhoppers. The causal agent is a phytoplasma. Control is by sanitation, eradication of nearby chokecherries, and use of insecticides to eliminate vectors (Gilmer and Blodgett 1976).

X-disease appears and disappears in cycles. Observations indicate that the microbe spreads in chokecherry populations until they collapse. New plants develop from abundant dormant seeds; chokecherry populations are then restored, only to be eliminated again by the X-phytoplasma. Such fluctuations are thought to affect the level of X-disease in peach and cherry orchards. There is little spread from plant to plant in the orchard because leafhopper vectors prefer wild chokecherry (Gilmer and Blodgett 1976 and unpublished).

The origin of brown rot of peach is not known, but there is enough information for some interesting speculations. Two causal and easily distinguished fungi are involved: *Sclerotinia fructigena* is the usual cause of peach rot in Europe, whereas *S. fructicola* is the North American version. In recent years, the two fungal species have invaded new areas, to a limited extent. *S. fructicola* is thought to have been endemic in wild plums of North America (Anderson 1956); if true, then peaches must have been brought to the New World without the European fungus, allowing free rein to the native pathogen. The peach is endemic in the mountains of western China and Tibet, where wild peaches still grow today; the crop was brought to the Mediterranean area about 300 B.C. There are no known records of diseases during early times; it is possible that peach met *S. fructigena* when the crop was moved to the West. Our modern peach is a delectable but fragile product; it is difficult to protect from disease.

Phony peach appeared in Georgia in the United States, about 1885, causing dwarfing of trees and fruit. By 1932, it had affected 35 million cultivated peach trees over the southeastern United States. All peach cultivars are susceptible, as are apricots and certain plums. Since 1932, phony peach has been eradicated in some states and controlled in others by sanitation, insecticides for vector control, and removal of wild plums. Phony peach is caused by a bacterium (*Xylella fastidiosa*), transmitted by grafting and by several species of insects (*Cicadellidae*). There were cycles in the prevalence of phony peach; it reached a peak in the period 1940–45, when half the trees in new plantings sometimes were infected by year five, and 99 percent by year ten. The disease is of significance today only in local areas without adequate care (Cochran and Hutchins 1976). The source host of phony peach is unknown, but apparently it is in North America, probably a wild plum.

The decline disease of pear (*Pyrus communis*) was first discovered in British Columbia in 1948; the pathogen appears to be a native that moved to pear, an alien plant. The cause is a phytoplasma vectored by an insect (psylla) (Hibino, Kaloostian, and Schneider 1971) that was introduced from Europe into the eastern United States in 1882 (Jensen et al. 1964). Pear decline did not appear in the eastern United States for almost one hundred years after psylla was introduced. However, the insect spread to the west; pear decline appeared suddenly and became destructive in the Pacific Northwest soon after the psylla became common there. Source of the causal phytoplasma is uncertain, but it evidently existed in the native flora of the Pacific region and moved into pear orchards via the vector. Pear decline later appeared in Europe (Behnke, Schaper, and Seëmuller 1980) and in the eastern United States (McIntyre et al. 1978). It is now found in Israel, Argentina, and possibly elsewhere. Pear decline most likely spread with trade in plant materials. Pear decline quickly became destructive, killing millions of pear trees in the Pacific area of the United States and Canada after 1948. The disease moved into California by 1959; at least ten thousand trees soon became infected and were in decline (Jensen et al. 1964). The disease also spread rapidly once it reached the eastern United States and Europe.

Pears usually are grown on grafted trees, using the best available rootstocks. Decline often is associated with graft failure (Schneider 1970). The disease is especially destructive on pear cv. Bartlett grafted to rootstocks of *P. serotina*, *P. ressuriensis*, or *Cydonia oblongata* (quince). The causal phytoplasma is vectored to tolerant scions and moves in phloem to the hypersensitive roots, resulting in necrosis below the graft union. This leads to quick decline of trees under various stresses and to slow decline under more favorable conditions. Nongrafted plants of *P. serotina* and *C. oblongata* are very susceptible.

Cacao (*Theobroma cacao*), the source of chocolate, is a small tree endemic in the rain forest of the upper Amazon basin. Several diseases occur on wild *Theobroma* species in South America, but they were never problems prior to domestication and monoculture. After domestication, cacao was moved to west Africa, where extensive plantings were made and where more diseases appeared. One of these was cacao swollen shoot, caused by a complex of virus strains. The disease was first found in Ghana in 1936; it now occurs in other countries of west Africa, causing serious losses. Millions of trees were destroyed in attempts to eradicate the viruses. The pathogens are not found in the western hemisphere, and every effort should be made to keep them out. There is little question that the viruses came from plant species endemic in the African rain forest; such plants are still reservoirs of the viruses, which are transmitted by mealybugs. It is hoped that resistant cultivars will be forthcoming (Thurston 1984).

Chrysanthemum morifolium, the florist "mum," originated in eastern Asia and is now planted worldwide in temperate areas. The plant has picked up several diseases from various sources over the world. Ray blight, caused by *Ascochyta chrysanthemi,* was found first in North Carolina in the United States. The disease quickly spread over the southeastern United States and west to California; it now occurs worldwide. A major factor in its spread was the distribution of infected cuttings by a single nursery in California, which sold a hundred million rooted cuttings in 1958 (Baker, Dimock, and Davis 1961).

Snapdragon (*Antirrhinum majus*), native to the Mediterranean region, is now cultivated worldwide. It, too, has picked up various diseases, including rust (*Puccinia antirrhini*), first seen in California in 1897. The subsequent spread of rust is well documented: to Oregon by 1909, Illinois by 1912, New England by 1915, Canada by 1916, Bermuda by 1922, France by 1931, throughout western Europe and Egypt by 1936, Russia by 1937, and Rhodesia by 1941. The fungus was distributed with infected plants, cuttings, seeds, and, locally, by wind-borne spores. Only uredial and telial stages are known (McClellan 1953).

Sugarcane (*Saccharum officinarum*), a plant of Asiatic origin, is now grown throughout the tropics. One of the new diseases in this alien plant was ratoon stunting, first found in Australia in 1914. The disease is now worldwide where sugarcane is grown and is said to be the most damaging disease in the crop. Its cause is a bacterium (*Clavibacter xyli*) that is spread in infected pieces of cane used for planting and by knives used to cut seed pieces and to harvest. Several weed grasses are hosts. Control is the use of heat in preparation of seed pieces (Thurston 1984).

Many other diseases have appeared when plants were moved to new areas. The potato, a native of the Andes, was moved north by native people; presumably, it met the blight fungus from other *Solanum* species in Mexico. Europeans moved the plant over the world, and many other diseases appeared; the potato is now host to more than three hundred pathogens (Hooker 1981). Douglas fir (*Pseudotsuga Douglasii*) was taken from America to Europe about 1850; this useful tree grew well until 1911, when a leaf disease caused by *Rhabdocline pseudotsuga* appeared in Scotland. The disease became epidemic by 1922 and spread throughout Europe (Gäumann 1950). *Ribes* spp. have mildew (*Sphaerotheca ovae*), which is endemic in North America; when European gooseberries (*Ribes*) were planted in North America, they were destroyed. The mildew reached Europe by 1900 and caused a disastrous epidemic in gooseberry (Gäumann 1950).

9

Pathogens Overtake Movement
of Crop Plants

There are well-known examples of crops that were moved to new areas without their native pathogens. Good model cases include late blight of potato, coffee rust, maize rust in Africa, rust and smut of sugarcane, banana leafspot in the Americas, and apple scab in Japan. In general, resistance evolves where a pathogen is endemic, and often it is lost without the selective pressure of the pathogen. Loss of resistance may be pushed further by agriculturists who tend to select only for yield (Harlan 1977). In all these model cases, the pathogen finally overtook the highly susceptible crop in its new area, with serious consequences.

9.1. Late Blight of Potato

Late blight of potato, one of the world's most destructive plant diseases (Cobb and Niederhauser 1959; Cox and Large 1960), is best known as the cause of a great famine in Ireland (Woodham-Smith 1962). Primary emphasis here will be on late blight as an example of a pathogenic fungus that, in time, overtook its host, which had been moved by people to new geographic areas. Movements of host and pathogen were not the only factors involved in the destructive consequences; clearly, adaptability by the fungus (*Phytophthora infestans*) and genetic uniformity in the potato were additional factors.

Our cultivated potato (*Solanum tuberosum*) originated in the Lake Titicaca region high in the Andes Mountains of Peru and Bolivia. It has been cultivated and was a major food crop in the Andes for at least two thousand years. *S. tuberosum* is one of six *Solanum* species that form underground tubers; other tuberous species are endemic in the southwestern United States, Mexico, Central America, and South America. Around 1570, the Spanish conquistadors carried the potato to Europe, where it was maintained

124

for years as a curiosity, not accepted as food. There were various reasons for this; among them, a belief that the potato was poisonous and evil, that it caused leprosy, and that it was not mentioned in the Bible and therefore was not meant for mankind. Even its reputation as an aphrodisiac did not make it acceptable. However, it was fed without troubles to farm animals, and this practice finally led to its acceptance as food for people (Klein 1979).

The potato became a regular part of the human diet in Europe by 1800. Easy cultivation and high yields in the cool climate led to major dependence on the potato in northern Europe, especially in Germany and Ireland. Use of the potato in Ireland was encouraged by the nobility as a means of saving grain for export and for use by themselves, the elite. The economic, social, and political situations finally made the population of Ireland dependent on the potato – a recipe for disaster (Salaman 1949; Klein 1979).

The potato in Europe always had problems; nevertheless, the crop thrived for almost three hundred years. Then in 1845–7, devastation hit throughout northern Europe, in the form of a blight not previously seen. The cause was obscure; thus, many theories were generated, many that, in retrospect, were absurd. The problem had appeared, with less impact, in the United States several years earlier; in Boston, an amateur scientist, J. B. Teschemacher, suggested that the culprit was a microorganism (Bourke 1991).

Devastation from the blight in northern Europe was completely unexpected; a great famine occurred in Ireland, where the population was most dependent on the potato. However, there was general hardship throughout northern Europe, especially in Germany (Large 1940). The story of the potato famine has been repeated many times, and I will not belabor it here; suffice it to say that the population of Ireland was reduced drastically. The best available estimate is a reduction from 8.2 million in 1841 to 6.5 million in 1851 (Hampson 1992). Considering births over this ten-year period, the population loss was approximately 2.5 million, of which 1.5 million perished from famine and accompanying disease, and at least 1 million emigrated, mostly to North America. There are other estimates; one that is frequently quoted is that 2 million persons died and 2 million emigrated. In more recent times, potato blight caused many deaths in Germany in 1916 (Carefoot and Sprott 1967) because fungicides were not available for control during World War I.

The blight fungus was first thought to have been brought to Europe from South America with the potato. This hypothesis did not explain why blight had remained dormant in Europe for many years. Therefore, the idea that Central America was the source was suggested; this hypothesis became popular after 1928, with claims that *P. infestans* was unknown in the Andean

countries until recent times. The Mexican or Central American origin became generally accepted. However, careful perusal of historical records has
provided convincing evidence that the fungus was in the Andes when Europeans arrived. Further, there are no early records of infection of wild potatoes in Central America and no indications that potatoes were cultivated
there by the Amerindians (Abad and Abad 1995).

   P. infestans probably was taken to Europe in the 1830s, with tubers
brought to Belgium from Peru for experimental purposes (Abad and Abad
1995). The disease was first noticed in Belgium in July of 1845, and it soon
spread over Europe; by mid-August, it was in England and France; by mid-
September, in Germany, Denmark, and Ireland; and by mid-October, in
Scandinavia, Scotland, Austria, and Italy (Bourke 1991).

   There was much debate in Europe over the cause of the new blight. The
established political and scientific authorities held that the blight resulted
from excess water that the plant could not expel; certainly, the blight years
were unusually cool and wet. Others (including Berkeley in England, von
Martius in Bavaria, Martens and Morren in Belgium, and Montagne in
France) on various occasions had proposed a fungal cause. Berkeley, who
had made careful microscopic studies of a fungus in the diseased plants, was
the most steadfast proponent (Large 1940). The controversy over fungal causation became heated and was vitriolic for years. Those persons who supported a fungal cause were ridiculed in the popular press and in scientific
circles; there were even popular ballads that derided the fungal theory and its
weird supporters.

   Finally, in 1861, the great German biologist Anton deBary (Figure 9.1)
presented conclusive proof for the role of a fungus that he named *Phytophthora,* meaning plant destroyer. DeBary gave evidence that no rational
person could dispute. That ended the fun; the scoffers no longer had that
stupid fungal theory to kick around. All this is remarkable because it occurred before the famous work of Louis Pasteur, which led to general acceptance of the germ theory of disease. We generally acknowledge deBary's
work as the true beginning for the scientific study of plant disease.

   The late blight disease affects all parts of the potato plant. On leaves, the
first symptoms are brownish to purple or black spreading lesions; under
moist conditions, the spots enlarge rapidly, and there is a characteristic pun-
gent odor. A white, mildewlike growth may show on the lesions; under the
microscope, the white growth is recognized as sporangiophores (spore-bearing branches) and sporangiospores that usually protrude through stomatal
openings. Tubers are infected in the soil or at harvest time; at first, the
infected area shows as a brown to purplish discoloration on the surface,

Figure 9.1. Anton deBary (1831–88). (Courtesy of American Phytopathological Society)

followed by brown dry rot penetrating to about 12 mm. Wet rot often follows, a result of secondary invasion by bacteria.

Mycelial structures of the late blight fungus indicate that it is an Oomycete (see Chapter 2). However, there was a mystery for many years because the sexual stage with oospores was never found. Finally, in 1956, oospores were found in the highlands of Mexico (Gallegly and Galindo

1958). The Mexican isolates had two mating types, classified as A1 and A2, that were needed for sexual union and oospore production. Surprisingly, the A2 mating types occurred only in Mexico; isolates from elsewhere in the world were all of the A1 type. Evidently, the fungus that was first moved to Europe was entirely the A1 type. This explained why sexually produced oospores were never found until the fungus in Mexico was studied seriously (Niederhauser 1991; Fry et al. 1992). Mexican populations of *P. infestans* contained many more genotypes than were found elsewhere (Goodwin et al. 1992).

A second migration of *P. infestans* probably started in the late 1970s (Spielman et al. 1991); this migration amounted to a reconquest of the potato crop. The A2 mating type was found in Europe in 1981. It probably was taken to Europe by 1976, as dormant infections in tubers that were imported from Mexico for breeding purposes; otherwise, trade in potatoes between Mexico and the rest of the world had been minimal. The new A2 races in Europe were surprisingly diverse, and new and more virulent races of A1 types also were found. The new races appear to be displacing the old A1 races. DNA fingerprinting, along with other genetic and biochemical evidence, clearly shows that the new races are identical to those in the highlands of Mexico. Isolates of the A2 type have since been found over most of the world where potatoes are grown, with the exception to date of Australia. Spread of the new races over the world may have been via export of seed potatoes from Europe, which is a major source of virus-free stocks. In the United States, A2 races could have come from Mexico (Deahl et al. 1991; Goodwin et al. 1994) via infected tubers or with movement of tomato fruit and plants. The tomato is another host of *P. infestans*.

Theoretically, there is a possible source of A2 isolates other than Mexico. Rare oospore isolates of the A1 type can be induced by hormones to produce both A1 and A2 progeny; the same is true for A2 isolates (Ko 1994). However, the evidence from molecular studies shows that the new A1 and A2 races in Europe and the United States are identical to races found in Mexico.

Molecular biology techniques have confirmed and increased our knowledge of lineages and movements of *P. infestans* over the world. Historical isolates taken before 1971 from Europe were of a single type, as are isolates from areas not yet affected by the 1970s migrations. This was predominantly the case for the United States as well, although data from DNA fingerprinting and allozyme alleles suggest somewhat greater diversity than was found in Europe. There is now evidence that sexual recombinations are occurring between mating types of *P. infestans* in North America (north of Mexico). The first report of such recombinations came from potato fields in British

Columbia. DNA data show the presence of new genotypes that could have arisen from matings of the most common A1 and A2 genotypes (Goodwin et al. 1995).

An earlier hypothesis was that *P. infestans* was moved from Mexico to the United States about 1842 and from the United States to Europe about 1845 (Niederhauser 1991). However, transport directly from Mexico to Europe was not ruled out. There may have been only a single incident of *P. infestans* movement to Europe during the nineteenth century, based on uniformity of populations. More recently, historical documents indicate that the early movement to Europe was from Peru (Abad and Abad 1995). The more recent introductions of races A1 and A2 have greatly increased worldwide genetic diversity in *P. infestans* populations. The new A1 and A2 races may be displacing the old A1 races rather than hybridizing with them (Spielman et al. 1991; Fry et al. 1993).

The data and claims regarding the source of *P. infestans* are still confusing. The endemic potato species in the highlands of Mexico are resistant to *P. infestans;* potato populations in the lowlands and in dry areas of Mexico are not resistant. This fits Vavilov's theory (1926) that pathogen–plant associations of long standing have resistant genotypes as survivors. The potato taken to United States and Europe was from the Andes of South America, where *P. infestans* was thought to be absent until recently; however, new evidence suggests that the fungus may have been in Peru for a long time, possibly brought from Mexico with trade by the Amerindians. The potato in Europe was selected over centuries for yield and quality, in the absence of *P. infestans;* the result was loss of resistance.

The fungus in the highlands of Mexico has greater genetic diversity than that found anywhere else in the world (Niederhauser, Cervantes, and Servin 1954); this also fits Vavilov's theory regarding endemic areas of origin. Both A1 and A2 mating types, giving a complete fungus, occurred only in Mexico until recently. Finally, recent genetic, biochemical, and molecular evidence supports the origin of *P. infestans* in Mexico and its spread from there throughout the world. However, the fungus spread to Peru much earlier than once thought (Abad and Abad 1995).

There are predictable consequences for the worldwide presence of the new A2 races, as well as other possible results that cannot be foreseen. Among the immediate consequences: Many of the A2 races carry resistance to metalaxyl, a fungicide that has become a mainstay for control. Metalaxyl resistance was rare in the old A1 races. Metalaxyl must be replaced with other fungicides that are either less effective or not yet developed. Also, some A2 races are pathogenic to many of the currently used potato cultivars that carry

R genes for resistance to the old A1 races. Other A2 races may also have higher virulence on cultivars with general or field resistance (Ingram and Williams 1991). The ability of *P. infestans* to reproduce sexually increases chances for development of new races (Goodwin et al. 1995), although even the old asexual forms had no apparent difficulty with this. Certainly there is a greater diversity of races in Mexico, where the sexual phase had been operating for thousands of years. Oospores offer another means for the fungus to survive in the soil from season to season; this may lead to changes in epidemiology because it could allow infections to occur earlier in the growing season.

All these factors probably mean increased reliance on the use of fungicides, at just the time that we had hoped for decreases in such use. More dependence on costly fungicides will be especially difficult for the less-developed nations that rely so heavily on potato as a main carbohydrate source. Development of new cultivars with general resistance to all races of *P. infestans* must become a world priority. An international program to develop resistant cultivars is now in place (Niederhauser 1993); it includes the International Potato Center in Peru, which has been functional for years.

There are two types of resistance to *P. infestans* in potato (Niederhauser, Cervantes, and Servin 1954). One type is dependent on the so-called *R* genes, mostly from a Mexican species (*Solanum demissum*) that crosses readily with *S. tubersum*. Other tuberous *Solanum* species have, to date, been little used as sources of resistance. The *R* genes are specific to individual races of the fungus; there are many races and many race-specific *R* genes. The race situation is unusually complex in Mexico, where both A1 and A2 mating types were always present. The *R* genes are the potato plant's part of a gene-for-gene complex, a theory originally formulated by Flor from his work with flax rust (see Chapter 6, section 1). An *R* gene is easy to incorporate into new potato cultivars, but it also is easy for the fungus to counter with new races that infect plants with the *R* gene. This has been the general situation with late blight, even before the two mating types (A1 and A2) became widespread. Previously, new races in places other than Mexico presumably arose via mutations or parasexual mechanisms.

The second type of resistance is a general resistance that is effective against all races of *P. infestans*. Many genes are involved, making breeding much more difficult. Plants carrying such resistance are not immune, but the resistance is reliable and has held for years in Mexico. It is the hope for the future. General (or multigene, or horizontal) resistance is being incorporated into acceptable cultivars in Mexico, where many races of *P. infestans* occur

Figure 9.2. John Niederhauser. (Courtesy of American Phytopathological Society)

and where the conditions favor disease. Researchers in Mexico have selected the potato lines that are most resistant to all races, have crossed them with other lines to select for quality, and have released surviving selections for worldwide testing. The results are promising. However, some fungicides will be needed for high yields under severe blight conditions because general resistance is not immunity.

John Niederhauser (Figure 9.2) received the World Food Prize in 1990 for his work on potato blight in Mexico. Niederhauser was cited for increasing potato production in less-developed countries. Such efforts are essential if we hope to feed ever-increasing populations in the world.

Summary

Late blight of potato, cause of famines, had a major impact on human affairs. The potato was moved from its endemic home in South America to Europe by the Spanish conquistadors in the 1500s. Late blight may not have been endemic in South America; instead, it could have been part of the Mexican flora. By 1841, the causal fungus *Phytophthora infestans* was finally moved with trade to the United States and Europe. The result was catastrophic because the potato outside Mexico was very susceptible and because Europe had become so dependent on this one crop.

The devastation by late blight stimulated early studies on plant disease. Among the important early scholars were the Reverend Miles Berkeley in England and Professor Anton deBary in Germany. DeBary's work was the beginning of the science of phytopathology.

P. infestans that had spread prior to 1980 consisted of only one mating type (A1), and the fungus did not reproduce sexually except in Mexico. Modern techniques, including DNA monitoring, confirmed the ideas about movements of *P. infestans*. A1 forms had moved to Europe and the United States by 1841, and the great famines started in 1845. All isolates of *P. infestans* outside Mexico were A1 types until recently. New A1 plus A2 races were taken by 1979 to Europe with imported tubers; the new races soon spread over the world. Copresence of A1 and A2 races could enhance development of new races and may pose a threat to food supplies in underdeveloped countries.

9.2. Coffee Rust in Africa, Asia, and South America

Coffee rust is a well-known example of a plant disease that, in time, overtook movement of its host throughout the tropics. Another important factor in the serious effects of coffee rust was monoculture of the host; observations on this disease gave the first notice of problems with monoculture.

Coffee rust generally is conceded to be endemic in the mountains of Ethiopia, also the origin of several coffee species. Rust also occurs on wild coffee in the lowlands around Lake Victoria, in Uganda and Kenya. In fact, the fungus *Hemileia vastatrix* was first discovered in 1861 by a British explorer, on wild coffee plants near Lake Victoria (Wellman 1970; Wellman and Echandi 1981; Thurston 1984).

Coffee has been used by humans since prehistoric times, and its use is the basis of interesting speculations. The belief is that hunters first found coffee trees in the forest, and the seeds, or "beans," were consumed as dry food.

Tribal chiefs "owned" jungle trees, which were given holy names. Weeds and brush were cleared around the holy plants; this practice could have been a forerunner of cultivation. Primitive cultivators discovered that roasting and steeping the beans resulted in a good drink, which led them to true cultivation of coffee, including monoculture. Wild trees scattered through the forest occasionally were infected with rust fungi, but with no serious consequences. Monoculture changed the situation; rust was able to spread easily from plant to plant, and the disease became a problem (Wellman 1961; Wellman and Echandi 1981).

At least sixty species of coffee are known; of these, *Coffea arabica* (Arabian coffee) is, by far, the most important. *C. arabica* still grows as a wild plant in the southwestern mountains of Ethiopia. *C. canephora* (robusta coffee), endemic in lowland forests of east Africa, is next in importance, but it accounts for no more than 10 percent of the crop. There is a little production of *C. liberica* and *C. dewevrei* for local consumption in central and west Africa.

The coffee crop usually is started from seed, but sometimes it is propagated by cuttings. Plants started from seed begin to yield a crop in three or four years, and if conditions are favorable, they will continue to produce for fifty years. *C. arabica,* the major species of commerce, is a tetraploid with forty-four chromosomes and is self-fertile; the other species are diploid (twenty-two chromosomes) and are self-sterile. Another group of coffee species, known as the para coffee types, are native to India, Sri Lanka (Ceylon), and southeast Asia; all others are native to Africa. The coffee bush, an evergreen, produces a drupe, or berry, containing large seeds. The seeds are extracted from the pulp and then dried to become the coffee beans of commerce.

C. arabica was taken from Africa to Yemen on the Arabian peninsula at an early date, possibly by A.D. 575. Yemen then became the major center for further distribution: hence, "Arabian" coffee. A plant of *C. arabica* was taken to the botanical garden of Amsterdam; from this plant, a single seedling was taken to Java in 1600. Another seedling offspring was taken from Amsterdam to the botanical garden in Paris in 1713. Later, coffee seeds were moved from Paris to Martinique and from there to Central America and Colombia. Thus, all coffee plantings in Ceylon (Sri Lanka), India, and the Americas are descendants of a single seedling line, an extremely narrow genetic base for a vast crop. Passage through the botanical gardens did give a temporary advantage: Coffee was moved to Ceylon, the Americas, and other areas without its rust parasite, which was left behind in Africa (Rodrigues, Betancourt, and Rifo 1975). However, the limited genetic base plus self-fertility and selections over many years without the rust disease created a dangerous situation.

Coffee was cultivated in Ceylon since 1700. When the British took over the island, they developed a large coffee industry and became a major world supplier. Coffee was planted over most of the island, with more than 162,000 ha devoted exclusively to the crop. This planting set the stage for rust invasion, first observed in 1868. Within five years, the new disease covered the whole island, and production quickly dropped to half its former level (Monaco 1977). Forty-five million kg were produced in 1870; by 1889, there were only 2.25 million kg. The cultivation of coffee was soon abandoned, replaced by that of tea and rubber (Wellman 1961; Schumann 1991). Rust specimens from Ceylon were sent in 1869 to Berkeley (of potato blight fame; see Chapter 9, section 1), who named the species. Berkeley warned of the danger but was unheeded; five years later, coffee production in Ceylon was overwhelmed by rust.

Coffee was taken to Brazil in 1727; it became a major crop by 1850. When coffee cultivation was abandoned in Ceylon, Brazil became the major source of coffee. Coffee is now grown in an almost continuous zone, with at least 280 million trees, from the Brazilian state of Paraná to Bahia, west to the Amazon basin, and in scattered plantings elsewhere (Monaco 1977). The Andean countries, Central America, and Mexico also became important producers. Coffee production is the economic base in more than fifty countries in the tropics (Rodrigues, Betancourt, and Rifo 1975).

There are two species of coffee rust: *Hemileia vastatrix* and *H. coffeicola*. The genus affects only plants in the family *Rubiaceae*, which includes *Gardenia* spp. Only *H. vastatrix* and *H. coffeicola* infect *Coffea* spp. (Rodrigues, Betancourt, and Rifo 1975). Both rust species appear to have originated and evolved in Africa. *H. vastatrix* soon became common in east and central Africa, and it is believed to have spread, by 1868, from the horn of Africa via prevailing monsoon winds to Ceylon (Rayner 1960). Rust has since spread throughout tropical Asia, east to the Philippines, with dire results. In Indonesia, coffee was replaced by another exotic crop, rubber.

When rust threatened the important coffee industry in Ceylon, the British government sent an energetic young mycologist to study the problem and recommend a solution. Harry Marshall Ward, age twenty-five, had studied at Cambridge and was well aware of the work of Anton deBary. Ward soon discredited the prevailing theories on the coffee disease; he proved, by inoculation experiments, that Berkeley's *H. vastatrix* was the culprit. He found that the fungus, a Basidiomycete, is disseminated entirely by urediospores; he described the teliospore and the basidiospore but could attribute no function to them (Ward 1882; Wellman 1970). To this day, the complete rust life cycle with aeciospores and pycnospores is unknown; if there is an alternate host, it should be found in east Africa.

Ward was first to elucidate two major principles in plant pathology. One concerns the importance of monoculture in epidemiology. Ward observed that rust occurs in wild coffee in the forest but is not readily disseminated to host plants interspersed with other trees. Thus, the disease is of no significance. All this changed when coffee was cultivated in dense monoculture. Ward warned of the worldwide danger of monoculture in other crops, but no one paid attention.

Ward's second significant idea also was simple. He clearly stated, for the first time, that the spore during germination is the vulnerable point in a fungal life cycle. At this stage, the sporeling is not protected within host tissue. Thus, it became clear that protective chemicals must be present in the vicinity of germinating spores (Large 1940; Schumann 1991). Such was the only way for control of plant disease by chemicals until curative systemic fungicides were developed many years later (see Chapter 6, section 2). Ward failed to recommend a magic potion for practical control of rust in Ceylon; safe, effective fungicides were not yet available. However, protective sprays were later used to save the coffee industry in other places.

Coffee rust first appeared in the western hemisphere in 1903, when it was brought to Puerto Rico on seedlings for experimental purposes. Rust at this site was soon eradicated (Thurston 1984; Wellman and Echandi 1981), providing what should have been a useful lesson. Subsequently, coffee rust was excluded from the western hemisphere by strict quarantines until 1970, when it appeared in the state of Bahia in Brazil. There is good circumstantial evidence that coffee rust was carried from west Africa to Brazil with prevailing winds. The Brazilian introduction followed recent movement and buildup of rust race II into Angola, West Africa, in 1966; this most-prevalent race in Africa was first to appear in Brazil. There are estimates from data on prevailing surface winds that rust urediospores could have been carried from West Africa to Bahia in five to seven days, well within the viability limit for such spores (Schieber 1972). Coffee rust urediospores were trapped 5,400 m above the southern Atlantic Ocean (Bawden, Gregory, and Johnson 1971). However, there is still a possibility that rust could have been brought to Brazil on infected seedlings or via airplane traffic.

Wellman (1952) warned that coffee rust sooner or later would come to the Americas; his warning was echoed by D'Oliveira in 1954 and by Rayner in 1960. All warnings went unheeded. Eradication on a massive scale was attempted after rust was found in Brazil, but the area of infestation was too great and the attempts were too late. There was precedence for eradication because rust had been eradicated in 1903 in Puerto Rico, and again in 1965 in Papua, New Guinea. It is ironic that Professor A. J. Riker of Wisconsin had applied for a grant in 1963 to develop a detection-and-eradication plan for Brazil, to use when the expected rust invasion occurred; the proposal was denied. If enacted, this plan, at a cost of only $190,000, could have prevented coffee rust in Latin America or at least prolonged the rust-free time. Such plans have, to date, saved the rubber industry in southeastern Asia from the South American blight of the *Hevea* rubber tree (see Chapter 10, section 1). Presently, many millions of dollars are spent each year on research to control coffee rust (Thurston 1984).

The climate and the vast monoculture of coffee in Brazil were ideal for the rust disease. In that country, urediospores of *H. vasatrix* clearly are dispersed by wind (Becker, Mulinge, and Krantz 1975; Martinez et al. 1975). Viable spores have been caught 3,000 m above areas at least 150 km from known areas of infection (Schieber 1972; Martinez et al. 1975). Rust has spread with prevailing winds in a southwesterly direction from Bahia to São Paulo to Paraná in Brazil, on to Paraguay and Argentina; this movement is thoroughly documented (Schieber 1975). Locally, rust spores also are disseminated by splashing rain and by insects. Rust was moved to Bolivia (1978) and Peru (1979) with infected seedlings (Schieber and Zentmyer 1984), a method previously known for rust movements from island to island in the Pacific and Indian oceans (Schieber 1972). By 1983, rust had spread to El Salvador, Honduras, Colombia, and Mexico, in some cases probably by spores carried on workers in coffee plantations (Fulton 1984; Schieber and Zentmyer 1984).

Coffee rust is primarily a leaf disease, showing as yellow to orange spore masses on the underside and as an oily spot on the upper surface. Spots often are surrounded by pale haloes, and eventually infected areas become necrotic. Major damage is from defoliation; a single spot can cause a leaf to drop. After successive defoliations, the plant is weakened and often dies. Infections on the berries are rare (Thurston 1984).

Coffee is the base of the economy in at least eleven countries in Latin America, and rust has been costly. First, there is the direct cost of fungicides for control, which was estimated to be 20 percent of the costs of production (Monaco 1977). Fungicides are now required everywhere. Next, there is the cost of replacing highly susceptible cultivars with more resistant ones. Also, rust has required changes in agricultural practices. More spacious planting is necessary for the use of spray equipment and for the creation of conditions less favorable for disease; these changes result in decreased yields per unit of land. There are costly effects on commerce and related industries and other services. Hardest hit have been the small growers, who often give up, joining the ranks of the unemployed. Overall, rust has been devastating, but coffee production will survive, thanks to the use of new control procedures. On the positive side, every coffee-producing country has increased its support of research that can affect coffee and other agricultural industries.

Fungicides are now one of the main controls for coffee rust. Fungicides are costly but are essential for the industry; if properly applied, fungicides can increase yields two- to threefold. Three to five applications per year are made in Brazil, from October to April. In the past, several different fungicides have been used, mostly copper and organic compounds, as protective sprays. More recently, systemic compounds, such as pyracarbolid and tridimefon, have been used to give curative effects (Fulton 1984; Schieber and Zentmyer 1984; Kushalappa and Eskes 1989).

More than thirty races of *H. vastatrix* are now known, creating a major problem for resistance breeding. Races I, II, III, and XV are worldwide and are most frequently seen (Schieber and Zentmyer 1984). All were first found in Africa; they are the only races identified in Ethiopia, the endemic home of *C. arabica*. However, genes for resistance to these and other races were found in Ethiopia, suggesting that other races have occurred there. Race I overwhelmed coffee cv. Kent, the first popular rust-resistant cultivar. Race II, the first race to reach Brazil, is the most prevalent worldwide. Race III was first found in east Africa; it moved to west Africa by 1966 and to Brazil in 1973. There are several virulence genes in *H. vastatrix,* and these, in different combinations, determine the race of the fungus.

Since 1970, a whole series of new races has appeared in Brazil, the place of the greatest coffee and *H. vastatrix* populations. New races continue to overcome new resistant cultivars, comparable to but more slowly than is the case with stem rust of wheat (Chapter 11, section 1), which has a functional sexual stage. New races of *H. vastatrix* apparently arise via mutations and heterokaryosis (Schieber 1972, 1975; Betancourt and Rifo 1975; Schieber and Zentmyer 1984; Kushalappa and Eskes 1989).

Races have been identified and worldwide research has been coordinated at a coffee rust research center (Centro de Investigacao das Ferrugens do Cafeeiro, Oeiras, Portugal). This institution was organized in 1955 by Wellman and D'Oliveira and directed by D'Oliveira. These two gentlemen, along with H. M. Ward, were the individuals most associated with research on coffee rust. Wellman brought a collection of resistant coffees to Turrialba, Costa Rica, in 1952; these coffees were propagated by the U.S. Department of Agriculture and were made available to breeders.

Control of rust by breeding and use of resistant coffee cultivars has been successful, and the outlook is promising, even with the problem of rust races. There are breeding programs in Angola, Brazil, Colombia, Costa Rica, India, Kenya, Portugal, and Tanzania. Resistance genes have been found in *C. arabica, C. coffeicola, C. canephora,* and other species. Five dominant genes for resistance were soon identified, and there are others; major genes (high resistance to specified races) and minor (tolerance) genes are available. There are now nine known resistance factors ($V_1–V_9$) from several *Coffea* species (Kushalappa and Eskes 1989). A gene-for-gene system of virulence and resistance is apparent.

When rust came to Brazil, D'Oliveira showed that all coffee cultivars in the Americas were susceptible (Rodrigues, Betancourt, and Rifo 1975). So far, there are no known cultivars that are resistant to all rust races; however, some Ethiopian selections have multigenic resistance to many races. Such resistance is needed in good coffee cultivars.

Cv. Kent was the first useful resistant cultivar; it was developed on the Kent estate in Mysore, India. Kent was widely planted but eventually was destroyed by rust race I. The cv. Hibrido de Timor was another early source of resistance; Hibrido de Timor is a natural hybrid of *C. arabica* × *C. canephora,* which was discovered by D'Oliveira. From this hybrid, D'Oliveira produced cv. Catrimor, a cultivar much used in breeding programs (D'Oliveira 1954; Schieber and Zentmyer 1984). Another early resistant cultivar was Tipica, also frequently used in breeding programs. Several resistant but little used cultivars of *C. arabica* were found in Ethiopia.

C. liberica, C. canephora, and other minor coffee species are, in general, resistant to rust, but all have susceptible segregating individuals. Hybrids with *C. arabica* occur naturally, and others are produced by plant breeders. Most such hybrids are triploid and unusable because of sterile seed, although they can be propagated vegetatively. Some matings give fertile seed, notably the *arabica* × *canephora* hybrids (Rodrigues, Betancourt, and Rifo 1975).

The second rust species, *Hemileia coffeicola,* poses another danger for the Americas. *H. coffeicola* is present in eastern and central Africa and is spreading west; eventually, it may reach South America. Coffee cultivars now used in the Americas are all susceptible to *H. coffeicola.*

Summary

Arabian coffee (*Coffea arabica*), the main species of commerce, grows wild in the highlands of Ethiopia. Rust disease occurs in wild coffee but is of no consequence because of tolerance in the host and poor spore dissemination among coffee plants that are scattered in the forest. Robusta coffee (*C. canephora*) is endemic, along with rust, in lowland forests around Lake Victoria

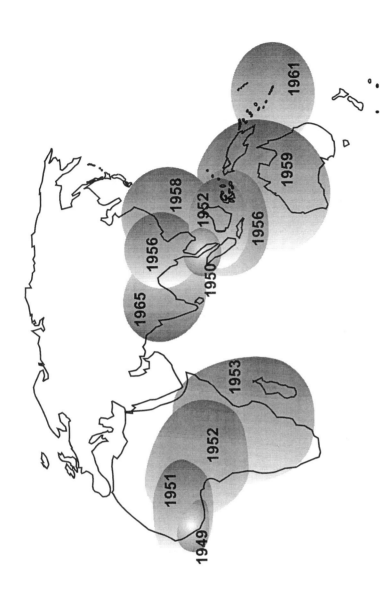

Figure 9.3. Spread of maize rust in the eastern hemisphere, 1949–65. (Adapted from Gregory 1973)

in Africa; the disease occurs in natural populations of robusta coffee but is of no significance. Great rust epidemics occurred when the fungus overtook earlier movements of the coffee crop that was planted in monoculture over the tropics.

A large coffee industry was developed by the British during the 1800s in Sri Lanka (then known as Ceylon). In 1868, rust reached Ceylon. The industry was destroyed and was replaced by tea, thus changing caffeine consumption in the whole British empire from coffee to tea.

Rust finally reached the great coffee plantations in the Americas in 1970. At present, with the use of fungicides, resistant cultivars, and changes in crop management, the disease is under control, but at great cost. The fungus now has more than thirty races, and coffee has several known genes for resistance to the most prevalent races.

9.3. Other Pathogens with Delayed Movement

Rust on maize is an easily understood example of a pathogen that overtakes movement of its host. Both the host and the causal fungi (*Puccinia polyspora* and *P. sorghi*) are native to the Americas. The disease caused little damage in the Americas, although sometimes recently introduced cultivars were affected; long association has eliminated highly susceptible genotypes. Maize was taken to Africa at an early time, probably before the seventeenth century, without the rust; maize became a widely planted and important food crop. Over time, maize became highly susceptible to rust in Africa, where there was no exposure to the disease. In 1969, maize rust appeared in Sierra Leone (West Africa), presumably via wind-borne spores from South America (Gregory 1973). *P. polyspora* soon spread over all of Africa to southern Asia and, finally, to Australia and New Zealand (Figure 9.3). The disease became very destructive. Since the initial epidemic, rust has declined for several reasons. Resistant cultivars were brought from Central and South America, and the most susceptible land races of African maize were eliminated (Commack 1961; Leonard 1969; Harlan 1977).

Sugarcane (*Saccharum officinarum* and related species) was moved throughout the tropics and subtropics from an endemic home in New Guinea or southeast Asia. Rust, primarily *Puccinia melanocephala,* was a minor disease in southeastern Asia; the disease was absent for years in many new areas of cultivation. Rust apparently spread to west Africa at some time prior to 1978. In July 1978, rust was found in the Dominican Republic, the first occurrence in the western hemisphere. Rust spores are thought to have been carried to the West Indies by prevailing winds from west Africa, as happened with coffee rust (Chapter 9, section 2). The evidence for this is circumstantial but convincing, and other possibilities were eliminated. From the Dominican Republic rust spread within a year, throughout the Caribbean area south to Venezuela and west to Florida, Louisiana, Mexico, and Central America. Australia also was invaded for the first time in 1978, presumably via urediospores carried on monsoon winds. Rust was very damaging in its new range, in part because prevalent sugarcane cultivars were very susceptible. For example, Mexico had estimated 50 percent losses (12.6 tons per hectare) in some cultivars. Only the uredial and telial stages of the fungus are known, and there are no known alternate hosts. Several grasses related to *Saccharum* also are infected (Purdy, Krupa, and Dean 1985; Nagarajan and Singh 1990).

Smut (*Ustilago scitaminea*) of sugarcane is another disease that spread to new areas long after sugarcane was introduced. A plant with smut develops a long, curved "whip," with masses of powdery spores. The disease, one of the most damaging to sugarcane, was first found in South Africa in 1877. Smut spread to Argentina and Brazil by 1940, possibly with planting material for sugarcane improvement. The disease then spread north via windborne spores through South America. Since then, its spread is well documented: to Hawaii in 1971, Florida in 1978, Louisiana in 1981, and Texas in 1982. Cultivars in use in the newly affected areas were very susceptible; the so-called Noble canes, originally from the East Indies, are resistant to smut (Sreeramulu and Vittal 1972; Lee-Lovick 1978; Hoy and Grisham 1988).

Banana leaf spot, or Sigatoka, is now a major disease (Stover 1962a, 1972, 1980; Meredith 1970; Thurston 1984). The probable center of its origin was tropical southeastern Asia (Stover 1977), where banana also was endemic. Bananas were moved long ago, without this disease, to most of the tropics. The disease spread to the Pacific Islands at some early date, causing damage in the Sigatoka valley of Fiji by 1912; a black-leaf-streak variant of the disease was found in the Sigatoka valley in 1964. The Sigatoka disease reached Africa at some unknown time. The next significant development was intercontinental spread over a line from Australia to African Cameroon to the Caribbean area. The hypothesis is that this occurred via spores, with a "fortuitous combination of disease, climate, and air turbulences superimposed on the belt of easterlies" (winds) (Stover 1962a).

In 1972, Sigatoka finally reached continental America (Stover 1980). Black Sigatoka, caused by a variant of the black-streak fungus, was found in Honduras; black Sigatoka became the most damaging of the leaf spot diseases of banana. The whole banana industry in Central America was threatened until controls were developed.

Figure 9.4. Scab on pear. (Picture by R. P. Scheffer)

Three related fungi (Ascomycetes) are involved in Sigatoka: *Mycosphaerella musicola, M. fujiensis,* and *M. fujiensis* var. *difformis.* The anamorph is *Cercospora musae.* These three, respectively, cause Sigatoka, black leaf streak, and black Sigatoka; collectively, they are known as Sigatoka. The spores are disseminated by wind, infected fruit and plants, and clothing of workers.

Sigatoka is controlled by fungicidal sprays, often applied by aircraft (Stover 1980). At one time, the fungicide program of United Fruit Company was said to be the largest such operation in the world. Unfortunately, the Sigatoka fungi tend to develop tolerance to the most effective fungicides, requiring use of oils plus fungicides for control, and this is expensive. Many small and subsistence farmers could not afford the necessary equipment and, consequently, abandoned banana farming. The standard export cultivar, Cavendish, is very susceptible; resistance has been found in some wild diploid bananas, but breeding is slow, expensive, and retarded. There has been little effort to provide resistant plantains (cooking bananas), most needed by subsistence farmers.

Many common diseases present today in various parts of the world probably were brought with imported crop plants or soon overtook them. In most cases, there is not enough information to consider them here. Worthy of mention are apple and pear scabs, caused by *Venturia inaequalis* and *V. piriana* (Figure 9.4), said to have been brought to North America from Europe about 1830, and cherry leafspot, caused by *Coccomyces hiemalis,* perhaps brought from Europe about 1890 (Waterworth 1993); however, the documentation is uncertain. Apple scab is known with more certainty in Japan, where it first appeared in 1952; it is thought to have been brought from the United States with apples for the Allied Forces (Kiyosawa 1977).

10

Monoculture:
Removal of Ecological Restraints

We know from the history of agriculture that plant diseases became more severe with the development of monoculture; this probably has occurred with most of our crops, but at such early dates that we have no records. It is the down side of an essential practice that has many advantages.

Recorded experiences during recent times show that wild plants usually develop unexpected diseases when grown in monoculture. The reasons are removal of the ecological restraints present in natural systems, easier dissemination of pathogens, and loss of genetic diversity. Several examples will be discussed; the most dramatic is a leaf disease of *Hevea* rubber trees in Amazonia. There are recent examples in North America with two diseases of *Populus* species that appeared when the trees were grown in plantations. When wild rice was domesticated in Minnesota, *Bipolaris oryzae* and *Cochliobolus sativus* became limiting factors. When cacao was cultivated in the tropics, pod rot became a problem. Dothistroma blight became damaging in plantation pines (Chapter 8, section 4), especially in those pines outside the native range. There are many other examples.

10.1. South American Leaf Blight of the Rubber Tree

Leaf blight of *Hevea* rubber illustrates very well the central problem of monoculture. The causal fungus *Microcyclus ulei* is endemic but of no significance on scattered *Hevea* trees in the forests of Amazonia. The tree *H. brasiliensis* was taken to southeastern Asia without the blight fungus; large plantations were established that, to date, are still free of the disease because of strict quarantines. In contrast, blight has foiled all attempts at plantation cultivation in tropical America. Prior to 1900, most rubber came from Amazonia, harvested from jungle trees by the native Amerindians.

Rubber can be extracted from the latex of several plant species. These

144

include several species of *Hevea, Taraxacum* spp. (dandelions), *Ficus elastica* (a fig), guayule (*Parthenium argentatum,* a Mexican weed), *Landolphia* (an African vine), various *Euphorbia* spp., *Manihot,* and *Castilla elastica* (a tropical American tree). Of these, *H. brasiliensis* is by far the best and most important producer of rubber. It is a large tree, to 40 m in height, endemic in the upper Amazon basin of Brazil and Peru, where it still grows wild, as scattered trees throughout the jungle. *H. benthamiana,* another good tree for rubber, is native in the northern part of South America, up to the Caribbean Sea. *H. benthamiana* is used for hybridization with *H. brasiliensis.*

The blight fungus *M. ulei* is restricted to several *Hevea* species. In native forests, the disease does not reach epidemic levels because there are only a few scattered host trees per hectare; thus, dissemination by spores is not efficient (Tollenaar 1959; Thurston 1984). When plantations were established in South America, the fungus spread readily, and the crop always failed.

Some knowledge of the history of rubber as a commercial product is needed to understand the leaf blight disease. Rubber was first used by native people of tropical America to make toys and other small items. The Amerindians long ago learned to collect and process the latex from tapped trees. Columbus and, later, Cortez found the natives occupied with a vigorous game that involved the use of a large rubber ball; the object was to knock the ball, with bounces from the player's head and body, through a small hoop at the end of a large court. Early explorers sent samples of rubber to Europe, where it became an item of curiosity. Priestly, a famous English chemist, gave us the name rubber because it could be used to "rub" away pencil marks. Botanical explorers were soon sent from France to Brazil to find sources of rubber for possible commercial use. Rubber boots had been sold in the United States as early as 1820, but they were impractical because rubber melted in summer and cracked in winter.

The commercial use of rubber began with Charles Goodyear and the process of vulcanization. Goodyear, born in 1800 in Connecticut, became a rubber zealot and spent his life trying to improve the product. In 1839, he found that cured latex plus sulfur plus heat (vulcanization) gave a stable rubber that was pliable in the cold and did not melt in the heat of summer. Thomas Hancock of England developed a similar process; Hancock also learned to coat fabrics with rubber, making possible the pneumatic tire. The pneumatic tire was first used for bicycles, later for the newly popular automobile, soon leading to tremendous demands for rubber. The demand stimulated exploitation of native peoples of South America and, later, of Africa;

great wealth was foreseen, and it soon accumulated. Akron, Ohio, was the home of Harvey Firestone's nonskid tire; Akron became known as rubber city by 1910, when it used half the world's supply of rubber (Klein 1979).

The Brazilian rubber boom resulted from demand in North America and Europe. Rubber was harvested from the forest by native Amerindians, who became virtual slaves; thousands died from bullets and disease, and all existed in misery. The Brazilian city of Manáos, in the heart of rubber country, became a center of wealth, home of the rubber barons. The boom in Brazil lasted from 1860 to 1910 (Collier 1968).

Rubber frenzy soon affected Africa. The Congo basin, the private fiefdom of King Leopold of Belgium, was also the endemic home of a latex-producing vine, *Landolphia* spp. Leopold hired a famous explorer, H. M. Stanley (the man who said "Dr. Livingston, I presume" when he rescued the old missionary in Africa), to explore the Congo for rubber. Soon there were destructive harvests of rubber vines by natives (Klein 1979), who, in effect, became slaves. All rebellious Africans were killed by a black mercenary army; many others died from white man's diseases. The population of Congo is said to have dropped from 25 million to 10 million in ten years, all for sixty thousand tons of rubber. Leopold became rich, but the world was repulsed by the greed and carnage ("one life = 8 lbs. of rubber"). The same exploitation occurred in other parts of the African rain forest, until all readily available vines were gone by 1914. The rubber booms in South America and Africa had ended even earlier, with development of competition from southeastern Asia.

In 1875, Henry Wickham, a planter in Brazil, smuggled seventy thousand seeds of *H. brasiliensis* to the Kew Gardens in England, without permission from Brazilian authorities. We now recognize this as one of the most significant smuggling operations in history. *Hevea* seeds are fragile, but some germinated at Kew, and the seedlings were promptly sent to Ceylon (Sri Lanka) and Singapore. The great rubber industry of southeastern Asia developed from the two thousand plants from Kew, plus a few from a second smuggling operation. Obviously this was a very restricted genetic base for a major crop; nevertheless, the plantations grew rapidly. By 1908, there were more than 300,000 ha of rubber trees in the Asian plantations, and the quantity of rubber from there soon surpassed that from Amazonia and Africa. Destructive and inefficient harvests had ruined wild rubber; Brazil, the major producer until 1900, soon imported most of its rubber (Carefoot and Sprott 1967; Klein 1979; Thurston 1984; Schumann 1991).

The great smuggling through Kew had one strong plus: It allowed the movement of *Hevea* to southeastern Asia without the leaf blight fungus. *M.*

ulei apparently is not dispersed with seed, and its spores are not durable enough to be carried long distances by winds.

M. ulei (= *Dothidella ulei*) was first discovered in Brazil in 1904 (Thurston 1984). The fungus is an Ascomycete, with sexual spores produced in small vaselike structures (perithecia). *M. ulei* also produces asexual or vegetative spores (conidia), which are the main means of dissemination. There is a third spore type, the pycnidiospore; its function is unknown but probably is involved in sexual mating. Infections from germinating conidia can occur on leaves, green stems, flowers, and young fruit. Infected young leaves develop gray-black lesions, soon covered with new spores; the leaves often are distorted, and defoliation usually follows. When older leaves are infected, distortion is unlikely; the leaf does not fall, but a "shot hole" develops. Still older leaves become immune. Several successive defoliations usually result in death of the whole plant. *M. ulei* occurs only in the western hemisphere, now present wherever *Hevea* grows in tropical America (Figure 10.1). To date, *M. ulei* has not reached the great rubber plantations in Asia. There are strict quarantines to keep it out, with personnel trained to eliminate the fungus if it appears (Rao 1973).

The rubber industry in Asia, first controlled by the British, soon was shared by the Dutch in Indonesia. By 1935, the world depended largely on southeastern Asia for its rubber, although there was some cultivation in west Africa. Success in plantations led to efforts, early in the twentieth century, to cultivate *H. brasiliensis* in tropical America. The efforts were encouraged by the United States and Brazil, to break the Asian monopoly; financing was provided by the great rubber companies of the United States and by Henry Ford, the automobile magnate. Ford, for example, started a large plantation at Fordlandia, on the Tapajós River, near the site of Wickham's seed collections in Brazil. The venture failed because of the leaf blight disease. A second attempt was made downriver at Belterra, with a still larger plantation of *H. brasiliensis*. Many of the trees at Belterra had resistant tops (scions) grafted to good latex-producing roots. The Belterra effort fared better than that at Fordlandia, but eventually it fell to a new race of *M. ulei* (Russell 1942).

Hevea was then planted in Central America, Mexico, and the Caribbean islands. These areas were beyond the native range of *M. ulei*, and, for a time, the plantations thrived. Eventually, *M. ulei* reached most of them, with the usual result. The fungus was inadvertently introduced into new areas with planting materials and by farm workers. Spread to all Central American and Mexican plantings was complete by 1952 (Figure 10.1), when *M. ulei* was found in Honduras (Waite and Dunlop 1953). The fungus has

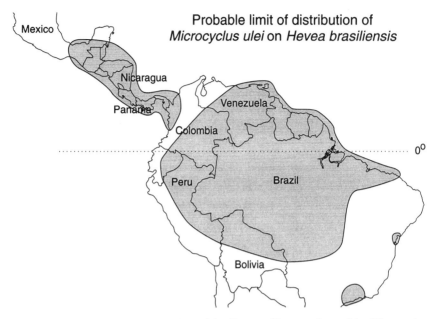

Figure 10.1. Distribution of leaf blight of the *Hevea* rubber tree (caused by *Microcyclus ulei*) in South and Central America. (Map by Marlene Cameron; source: Holliday 1970)

not reached certain Caribbean islands, but plantations there have not been a success because of low yields of latex (Carefoot and Sprott 1967; Klein 1979; Thurston 1984; Schumann 1991). Unfortunately, planting mistakes are still being made in western Brazil, although there are recent reports of successful cultivation in southern Brazil.

When World War II began, the Japanese gained control of most of the world's rubber supply. This left the allies in a desperate situation because modern war vehicles roll on rubber. The allies turned to several possible solutions: Rubber was reclaimed; further planting of *Hevea* was subsidized; and there were attempts to extract rubber from other plants. These efforts helped, but the final solution was the development of a synthetic-rubber industry, using a petroleum base. Earlier, synthetic rubber had been used by the Germans. Synthetic rubber is still the major source for industry, but demand for natural rubber has held. At present, the Asian plantations are reviving, along with some in west Africa.

There is a well-developed program of breeding for resistance to *M. ulei*, but it is complicated by new races of the fungus. The first resistant clones soon fell to a second race of *M. ulei,* as indicated previously for Ford's plantations at Belterra. Races are designated by number, according to the dates of discovery; the original race is now known as race 1. Race 2 is found in South and Central America. *Hevea* clones with F 4542 parentage (from *H. benthamiana*) were relatively resistant to all races known in 1965, although such clones were affected more by race 2 than by race 1. Clones lacking this heritage are equally susceptible to both races (Langdon 1965).

Next, a new fungal race appeared, with virulence to all progeny of F 4542. A third race infected many Ford clones at Belterra that previously were resistant. Differential clones were then used to identify races, and two more races were soon found (Miller 1966). Generally, plant breeders have found that susceptibility is linked to high yields, adding to the difficulties in exploiting resistance genes. Grafting resistant scions to susceptible rootstocks has been used in Asia, in anticipation of the appearance of *M. ulei.* The *Hevea* clones in use in Asia are highly susceptible to certain known races of *M. ulei;* Malayan scientists are now in South America, breeding for resistance.

So far, control of rubber leaf blight in Asia is dependent on strict quarantines (Berg 1970). Grafting resistant scions to high-latex stocks has had limited success. Chemical control is not practical, except in nurseries. Genes for resistance to specific races are well known but, so far, have had limited use.

Summary
Included herein is a brief history of rubber, including accounts of rubber as a product, harvests of native rubber from the forests of South America and Africa, cultivation of the *Hevea* rubber tree, and the devastating effects of blight.

The fungus *Microcyclus ulei,* cause of South American leaf blight, is endemic in the Amazon basin on native *Hevea* trees. The disease is of no consequence in the native forest, where there are few host trees per hectare. Demand for rubber led to attempts to establish plantations in tropical America, but the plantations always failed, primarily because of monoculture and blight.

H. brasiliensis, the rubber tree, was moved in 1875 by seed from Brazil to Kew Gardens in England and from there to southeastern Asia. The move left behind the leaf blight fungus, making possible the great rubber-producing industry of Asia. To date, *M. ulei* has been excluded from Asia by strict quarantines. The potential for disaster is great.

10.2. Mycosphaerella (Septoria) Canker and Rust
in Poplar Plantations

North American *Populus* species include aspens (*P. tremuloides* and *P. grandidentata*), cottonwood (*P. deltoides*), and balsam poplar (*P. balsamifera*). These species and certain interspecific *Populus* hybrids grow rapidly on good sites, are easily propagated, and are planted in the northeastern and north central United States for wood fiber production. Diseases have limited the plantations; the fastest-growing hybrids and several introduced European poplars are especially vulnerable to an endemic canker-producing fungus, *Mycosphaerella populorum* (*anamorph = Septoria musiva*). The fungus is present in natural stands of *Populus,* causing an inconsequential leafspot, but seldom causing cankers (Thompson 1941). *M. populorum* soon became destructive in seedling nurseries and young plantations (Filer et al. 1971). *Populus* species vary greatly in relative susceptibility, as do clones (ramets) within species. Breeding for resistance is feasible, although the fungus no doubt is adaptable.

M. populorum is present throughout North America where poplars are growing. The fungus overwinters on fallen leaves, producing infectious ascospores in spring (Luley and McNabb 1989). New infections occur in spring and summer. Conidia are produced on the perennial cankers. Infection often leads to leaf drop, but the major damage is from stem cankers, which often girdle and kill young trees (Manion 1981; Sinclair, Lyon, and Johnson 1987).

Poplar rust, caused by *Melampsora medusae* (= *M. albertensis*), is one of the most serious diseases in intensely managed plantations of several *Populus* species and hybrids (Schipper and Dawson 1974; Sinclair, Lyon, and Johnson 1987). The fungus occurs worldwide, wherever *Populus* and certain conifers grow together. Alternate hosts are conifers, but the rust is of little consequence to them. The rust also is of little significance on poplars in natural stands; native trees have a wide genetic base, and the fungus is not efficiently disseminated. This situation changed when poplars were selected for fast growth and managed in solid stands; the selections usually were highly susceptible, and the fungus spread more readily. Rust became serious, often causing much defoliation and reduced growth (Hamelin et al. 1993; Newcombe and Chastagner 1993). Poplars planted outside their natural range are especially vulnerable; some shelterbelt plantings of poplars in the western United States were destroyed.

There are many species of *Melampsora* specialized to many different hosts; important examples are the rusts on Douglas fir, poplars, and onion. *M. medusae* is a long-cycle rust (see Chapter 2), with spermagonial and aecial stages on several conifers (fir, larches, several pines, and others). Aeciospores released in spring infect poplars; there, the fungus produces golden-yellow urediospores, which repeat on poplar, and brown, crustlike telia with black teliospores. Telia overwinter in fallen leaves; in spring, teliospores form basidiospores that infect conifers. In areas with mild winters, the fungus lives over as uredinal mycelium in poplar cankers; thus, the fungus can persist in the southern United States without the conifer host. In the north, the conifer host is required.

A similar rust fungus, *M. occidentalis,* occurs along with *M. medusae* in the western United States. Another similar species, *M. populnea,* was introduced into the United States from Europe.

Poplars are genetically diverse (Thielges and Adams 1975; Thielges et al. 1988), and resistant selections are easy to find. However, the fungus varies in virulence and is adaptable (Ziller 1965; Prakash and Thielges 1987), as are rusts in general.

The canker disease caused by *Hypoxylon mammatum* (Figure 10.2) is common on poplars, primarily P. *tremuloides* (trembling aspen). Hypoxylon canker kills an estimated 10 percent of trembling aspens in plantations and in natural stands in the Great Lakes area of North America. This endemic disease often decreases the density of stands.

10.3. Other Examples of Removal of Restraints

Cacao (*Threobroma cacao*), a small tree native to the upper Amazon basin, is the source of chocolate. Demand for chocolate, which is extracted from pods, led to large-scale cultivation in South and Central America and in Africa. Two serious diseases developed with cultivation: black pod rot, caused by *Phytophthora* species, and Monilia pod rot, caused by *Monilia roreri*. There also is a threatening wilt disease caused by *Ceratocystis fimbriata* and, in Africa, the devastating swollen shoot disease (Chapter 8, section 5) (Thurston 1984).

The Monilia pod rot fungus is native in the Amazon rain forest, but has little effect there. Plantations in Colombia, eastern Ecuador, and Costa Rica often suffer 30 to 40 percent losses, sometimes leading to abandoned plantations. Dissemination is by wind-borne spores. The disease has spread to new areas of cultivation (Thurston 1984).

Wild rice (*Zizania palustris*), a native grass in the wetlands of the north-

Figure 10.2. Cankers in aspen *(Populus tremuloides,* caused by *Hypoxylon mammatum)* can girdle the tree and often leads to breakage. (Picture by R. P. Scheffer)

ern United States and southern Canada, was used by the Amerindians as food; in recent years, it has become a gourmet item. High prices encouraged cultivation, which began in Minnesota and now occurs in adjacent states and Canada. This was the first and only domestication of a native small grain in North America. Disease problems now limit cultivation; the major disease is fungal brown spot, caused by *Bipolaris oryzae* (anamorph, *Cochliobolus miyabheanus*) and by *B. sorokiniana* (anamorph, *C. sativus*). The first cultivated field of *Z. palustris* in Minnesota was destroyed in its second year; the disease is now destructive elsewhere in the Great Lakes area. In 1973 and 1974, there were complete losses in many fields in Minnesota. Resistant cultivars are unknown; the new industry is in peril in the Great Lakes region and may move to regions with a dry climate and irrigation (Johnson and Percich 1992).

The European larch (*Larix decidera*), a very productive and useful tree, was brought to North America. Plantations are severely limited by the fungus *Mycosphaerella laricina,* first reported in 1982 (Ostry, Pijut, and Skilling 1991).

Bunt or stinking smut of wheat, an important disease that was strongly influenced by monoculture, was well known in ancient times and was severe in Europe during the Middle Ages. Bunt, important in the historical development of plant pathology, was studied by many early mycologists, including Tillet, Prévost, the Tulasnes, deBary, Kuhn, and Brefeld. Tillet was first to demonstrate the infectious nature of a disease. Prévost, in 1807, showed clearly that a microbe causes bunt (Walker 1969); this was a first for any disease, plant or animal, but Prévost has never been given proper credit by biologists in general.

A number of important plant diseases have not been categorized and discussed in this book, although monoculture clearly is a factor in their incidence and severity. Included are smut and powdery mildew diseases of cereals, bacterial blight of rice, anthracnose of bean, cedar-apple rust, citrus canker, apple scab, downy mildews of grasses, sugarcane mosaic, ergot of grains, and the root knot nematode disease of many species. These and other diseases are difficult to categorize, for several reasons. In some, numerous epidemiological and anthropogenic factors are involved, making classification difficult. Many of these diseases probably were problems in agriculture from its beginnings, but there are no records of their appearance and spread. Agricultural management practices are involved in the prevalence of some, such as loose smut of small grains. Some crops were greatly altered by selection and breeding; resistance to disease often was lost.

11
Monoculture:
Pathogen Adaptability

Intensely managed and genetically manipulated crops soon lose genetic diversity. Any pathogen that can attack such uniform genotypes has few limitations; this requires further breeding to introduce resistance. Unfortunately, many pathogens are adaptable; they change by mutation, new gene combinations, or selection and spread of existing but minor pathotypes. New strains that can attack recently developed resistant plants have few restrictions. This has happened frequently, time after time, in our most important crops. Constant monitoring and genetic manipulation are required to keep ahead of the microorganisms and insects. Several outstanding models, differing in natural history, will be discussed. The situation is not completely hopeless.

11.1. Stem Rust of Wheat

Black stem rust of wheat is a classic example of the consequences of genetic adaptability in pathogens. There are, of course, other factors that affect prevalence of rust, especially large-scale monoculture of genetically uniform plants and seasonal reintroduction of the causal fungus, but pathogen adaptability is the overriding factor in wheat rust epidemics.

Small grains, including wheat (*Triticum aestivum*) and rice, are among the most important crops in the modern world. Wheat and barley as crops, along with their rust diseases, originated in western Asia, probably in the so-called fertile crescent of Mesopotamia, before the dawn of recorded history. These grains and their rusts have been with us ever since. Wheat was soon taken to the Mediterranean basin, where it became the staple crop. Later, European colonists took wheat to the Americas and to Australia.

Wheat is grown on a vast area in mid–North America, in an almost continuous belt from northern Mexico and Texas through the midwestern plains

154

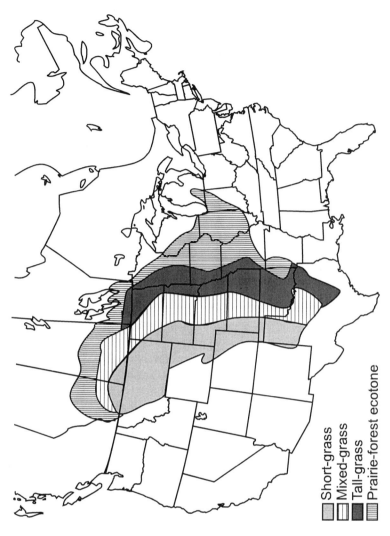

Short-grass

Mixed-grass

Tall-grass

Prairie-forest ecotone

Figure 11.1. Natural grasslands of North America, prior to disturbance by cultivation. Great herds of ungulates, including bison, elk, and antelope, formerly grazed on these plains. The present-day wheat belt of the midwestern United States is on the area of short-grass prairie and the western part of the long-grass prairie. The maize belt is centered on long-grass areas. (Map by Marlene Cameron, from various sources)

of the United States to southern Canada. Before Europeans arrived, this area
was grassland, remnants of which still survive (Figure 11.1). Blue grama
grass (*Bouteloua gracilis*) is an important native species; it has rust diseases,
and it offers an interesting comparison with cultivated wheat.

Wheat stem rust is among the most important of plant diseases, causing
hunger, hardship, and great financial losses during recurrent epidemic years.
Rust is found wherever wheat is grown; it is most severe in moist climates
and during years with above average rainfall. Wheat rust in North America
is most damaging in the spring wheat area of the upper Mississippi valley
and Great Lakes regions.

Rust is an ancient problem for the wheat crop. Rust spores were found on
seeds that are three thousand years old. Rust diseases are described in the
earliest writings, including the Bible. Greek philosophers wrote about rust
problems. During classical times, rust was of no consequence on the Italian
peninsula until barberry (*Berberis vulgaris*) was introduced. Rust then became very serious in the whole Mediterranean area, so serious, indeed, that
Romans had a pair of rust gods (Robigo and Robigus), along with a reli-
gious ceremony, the Robigalia. The ceremony was observed for more than
fifteen hundred years (Schumann 1991).

Farmers soon noticed a relationship between barberry and wheat stem
rust. Laws against growing barberry were enacted in France by 1660, followed by similar actions in other countries of Europe and America. However, the biological connection between barberry and grain rust was not
understood until Anton deBary (Figure 9.1), a German biologist, showed
that barberry is an essential alternate host in the rust life cycle. The common
barberry is not to be confused with the Japanese barberry, which is not
susceptible enough to be of concern.

European colonists brought wheat to the western hemisphere; they also
brought barberry because its wood was used for tool handles and its berries
were good for jellies and dyes. Wheat rust probably was brought to North
and South America with barberry plants. The barberry was soon naturalized
and became a weed throughout the northeast and west to the plains. Anti-
barberry laws were passed in New England by 1726, but the laws were soon
disregarded.

Two people were especially important in early studies of rust diseases:
Anton deBary of Germany (Figure 9.1) and E. C. Stakman (Figure 11.2) of
Minnesota. Important later work on wheat was that of Norman Borlaug (Figure 11.3), who developed high-yielding wheat, which led to the "green revolution"; Borlaug received the Nobel Peace Prize for this work in 1970.

Stakman studied the several subspecies (*forma specialis*) of *P. graminis*
when he was a graduate student at the University of Minnesota. All the

Figure 11.2. Elvin Charles Stakman (1885–1979). (Courtesy of American Phytopathological Society)

subspecies have the same morphology, but each is highly virulent or specialized to one major host species. Stakman's subspecies were *tritici* (specialized on wheat), *hordei* on barley, *secalis* on rye, *avenae* on oats, and
others. Later, the f. sp. *hordei* was dropped, but still others were added; the
new forms are specialized on various grasses. Stakman showed that each
subspecies can infect other host grasses to a limited degree, leading to ref-
utation of the theory of bridging hosts (Christensen 1984). The theory was

Figure 11.3. Norman Borlaug. (Courtesy of American Phytopathological Society)

an earlier attempt to explain how rusts adapt to new host species, by first infecting hosts with moderate resistance.

Stakman pioneered in three areas of rust research, each of major significance. He was first to show the importance of rust races or genotypes, further subdivisions of *P. graminis tritici*. At present, some three hundred races of *P. graminis tritici* are recognized; the races are specialized to different cultivars of wheat. Many other plant pathogens are now known to have races, all following the pattern established by Stakman in his pioneering work. Stakman's attention was next directed to the alternate host of wheat rust. He supervised the eradication of the common barberry in the main grain belt of the United States, beginning in 1918; the task was essentially completed by 1928. Eradication was important in the epidemiology of wheat rust (Stakman 1934); eliminating the alternate host decreased the generation of new races, but it did not eliminate rust, as had been hoped.

Figure 11.4. The *Puccinia* pathway in central North America. Rust spores are moved north each spring and summer with prevailing winds. In autumn, spores are moved back to the south, infecting newly planted wheat. Other organisms are aided by these air movements. (Map by Marlene Cameron; source, Schumann 1991)

Stakman's next major contribution was in aerobiology; he was first to prove long-distance dispersal of rust spores. In North America, spore movements follow the "*Puccinia* pathway" (Figure 11.4), from Texas north to Canada in spring and back to Texas in fall (Christensen 1984). The pathway is utilized by other pathogens and insects, including aphids (see Chapter 15, section 1) and monarch butterflies. Stakman's major achievements will be discussed further, but first, a brief description of the rust fungus is needed.

The wheat stem rust fungus is pleomorphic, which means that it has many different spore forms. There is a dark-colored overwintering spore known as the teliospore, which is produced on infected wheat plants as they mature (see Figure 2.1). The teliospore germinates, forming basidiospores that infect barberry, not wheat. Two spore types are formed on barberry: pycniospores, which appear in pale yellow, slightly erumpant areas on the upper leaf surface; and yellow aeciospores, which appear in cuplike structures on the undersides of leaves. Aeciospores are carried by wind and infect wheat, not barberry. Rust-colored spores known as urediospores are produced on infected wheat plants; urediospores are known as repeating spores because they spread the rust from wheat to wheat. Stages of the rust fungi and their host selectivities were discovered and described by Anton deBary by 1865; deBary first studied bean rust and later turned to wheat.

The function of pycniospores was obscure until 1927, when J. H. Craigie, of the rust laboratory in Winnipeg, found that pycniospores function as spermatia in sexual fertilization (Littlefield 1981). This discovery led to an understanding of how new races are formed. The pycniospores (haploid) are moved by insects or rain to receptive hyphae of the opposite mating type of the fungus. A vegetative fusion occurs, and the fungus becomes binucleate (N + N nuclei). Nuclear fusion is delayed until teliospores are formed; the fungus then becomes a true diploid (2N). Haploid (N) and binucleate (N + N) spores infect grains, and diploid (2N) spores infect barberry (Figure 2.1).

We now know that *P. graminis tritici* contains many different races or genotypes, thanks to the pioneering work of E. C. Stakman in 1916. How do we know that this is the case? By painstaking experiments with a specified group of wheat cultivars and species that were inoculated with each of many collections of the rust fungus, taken from many locations over many seasons. Each race is defined by the wheat cultivars that it attacks; for example, a given race may attack wheat cultivar A, but not B; another race may not attack cultivars A and B, but attacks C, and so on. Races have been assigned numbers, based on the order in which each was recognized. Wheat breeding for rust resistance requires knowledge of rust races.

Stakman selected twelve different cultivars of *Triticum* (wheat) that differ in susceptibility to the various races of the rust fungus; each race has a specific spectrum in the differential wheat cultivars that it will infect. Stakman's cultivars were from five different *Triticum* species; the original selections were cultivars Little Club, Marquis, Reliance, Kota, Arnautka, Mindum, Spelmer, Kubanka, Acme, Einkorn, Vernal, and Khapli (Stakman et al. 1944). Race determination is further complicated because there are several different reaction types in wheat to various rust races. The reaction types are assigned numbers for convenience. The types are $0 =$ no reaction (immune); $1 =$ and $2 =$ resistance, with flecking (development of small necrotic spots, the hypersensitive response) at the point of fungal penetration; $3 =$ and $4 =$ susceptible, with spreading lesions; and type $X =$ indeterminate, with both resistant and susceptible reactions on the same leaf (Stakman et al. 1944; Roelfs and Groth 1988).

Stakman's system is still used, with modifications based on modern knowledge of the rust genes involved (Littlefield 1981). Stakman's code numbers were dropped in the new system, and code letters are used to designate specific virulence genes in the races of *P. graminis tritici*. The revised system helps plant breeders because the specific genes to be dealt with are identified.

Once the rust race situation was clarified, Stakman turned his attention to barberry, the alternate host and "race breeding station" for *P. graminis*. Stakman was aware that barberry eradication had relegated stem rust to a minor disease in northwestern Europe. After the rust epidemic of 1916 and its effects on supplies for World War I in Europe, Stakman was able to convince the U.S. Department of Agriculture and the U.S. Congress of the urgent need to eradicate barberry. Several state and federal laws against barberry were passed by 1917. The eradication program, based on solid information showing that local epidemics started early in the season near barberry bushes (Stakman 1934), was under way by 1918, with Stakman as director (Christensen 1984). This was the first big battle in the war against stem rust. Eradication was essentially completed in the spring wheat area by 1928.

In the northern spring wheat region, the rust uredial stage cannot live over winter on wheat stubble. Rust does survive as infections on barberry and as teliospores; the basidiospore offspring of the latter infect barberry, not wheat. Thus, eradication of barberry should break the cycle, or so it was thought. In fact, eradication was successful, in part, because it eliminated many early season sources of infection and removed the breeding station for development of new races by *P. graminis* (Stakman 1934). Unfortunately, the eradication program was not a complete success, for reasons that soon became apparent.

Stakman turned his attention to another rust problem when it became evident that rust would still be with us after barberry eradication (Christensen

1984). Rust was showing up in areas far from barberry bushes, sometimes to a damaging degree. How could this be? Stakman was able to trap urediospores on sticky slides in these areas, prior to the appearance of rusted wheat. Stakman then started extensive aerobiological studies, another field that he pioneered. The weapons in the research were spore traps, sticky slides, airplanes equipped to trap spores at high altitudes, and continental surveys of rust races. These detailed studies of airborne rust spores were continued for thirty years; rust development in wheat was followed across the continent, from south to north and from north to south. The final results: Airborne spores carry rust each year from northern Mexico and Texas across the Great Plains to Canada; winds allow the fungus to survive and cover the continent, even though rust cannot overwinter in the north or oversummer in the south; in fall, prevailing winds carry rust spores south to Texas, leading to infection of the new southern crop that is seeded in the fall (Christensen 1951; Hamilton and Stakman 1967).

Stakman had not expected these results because rust races prevalent on the southern plains did not occur in wheat cultivars that were planted in the north. However, the situation changed with new wheat cultivars that were planted in both south and north and with barberry eradication. Soon it was evident that southern races of rust were moving into northern wheat. When the season favors rust in Texas, vast clouds of urediospores are carried upward by convection currents and northward by prevailing winds. The rust advances northward, from a few kilometers to as many as 1,000 km or more at a time. This movement occurs as wheat matures from south to north with the season. In the spring wheat area, spores that arrive early establish initial infections, and local infections supply inoculum for damaging epidemics (Rowell and Romig 1966; Christensen 1984). Damaging epidemics do not happen every year, but usually there is at least a little rusting.

All these conclusions were backed by massive data. Barberry eradication resulted in fewer severe epidemics and less chance for the fungus to develop new and virulent races (Roelfs and Groth 1980; Roelfs 1982, 1985). The final battle in the rust war was development and deployment of new resistant wheat cultivars for the spring wheat area; cultivars must resist the rust races found in the south. This step has reduced the chances of rust epidemics even more; there have been no major epidemics of stem rust in wheat in America since 1955 (Roelfs 1982; Roelfs, Long, and Roberts 1993). However, constant vigilance is required to monitor new rust races, which can occur by mutation. New and resistant wheat cultivars must always be ready to replace those that may fall to rust.

During this century, severe epidemics of wheat stem rust occurred in the spring wheat area of the United States in 1904, 1916, 1935, 1950–1, and 1953–4; heavy losses also occurred in 1923 and 1927. These epidemics happened when new races of *P. graminis* became prevalent and when weather conditions were conducive to rust (Van der Plank 1963; Roelfs 1982).

Historical records show that as each new resistant wheat cultivar was introduced and widely planted, the fungus responded with a new and virulent race that attacked it. The wheat cv. Marquis was by far the most popular variety until 1916, when it was eliminated by several rust races. The 1916 epidemic led to the great barberry eradication program. Cv. Marquis was replaced by the durum wheats, which endured for several years because they were resistant to rust races prevalent at the time. The durums were eliminated by rust race 11 and several others with minor roles (races 12–32). Next, cv. Marquillo was widely planted, to be taken out, in turn, by new rust races. Following Marquillo, cv. Kanred fell and was replaced by cv. Reliance, which fell in turn (Roelfs and Groth 1980; Van der Plank 1984; Roelfs 1985).

The first popular cultivar that was bred specifically for rust resistance was Ceres. Resistance in Ceres was based on a single gene, a type of resistance that became popular by 1930. Ceres was devastated by rust race 56 in 1935. Subsequently, cv. Hope H44 had a short life, and cv. Newhatch took its place; Newhatch lasted until the great rust epidemic of 1950–1, when it was destroyed, from Texas to Canada, by rust race 15B. This race had been discovered as a minor race in 1940 near barberry bushes that escaped eradication. Race 15B was virulent to all wheat cultivars known in 1950 and caused another epidemic in 1953–4 (Van der Plank 1984). Several new cultivars resistant to race 15B are now grown, and there have been no major epidemics in recent years (Roelfs 1982; Roelfs, Long, and Roberts 1993).

Rust losses still occur every year but are relatively minor. The fungus still overwinters in Mexico and Texas and sometimes on the Gulf coast of Louisiana and Alabama. In 1992, wheat rust race Pgt-TPMK (new nomenclature system) was most common, accounting for 53 percent of the total rust collections (Roelfs, Long, and Roberts 1993). This race lacks virulence on wheat with any one of fourteen different genes for rust resistance that are present in major cultivars now grown in the spring wheat area.

Aerial dissemination of rust occurs in parts of the world other than North America. There has been less study in other areas, and the dispersions appear to be less dramatic. Nevertheless, a rust track was found in western Europe and another in eastern Europe. The western track starts in Morocco and Iberia and moves north to Scandinavia. The eastern track is Turkey to

Romania to Scandinavia (Santiago 1968; Littlefield 1981). Both tracks also carry spores of tobacco blue mold (Chapter 15, section 2) and other pathogens. The data for Australia are limited, but there seem to be spore movements from north to south, occasional movements from southern Queensland to southwestern Australia, and even some from Australia to New Zealand. In India, there are movements of *P. graminis* and *P. recondita* from south to north in spring and from north to south in fall (Littlefield 1981). However, the situation in India is complex; urediospores are always present because of double-cropping of wheat in the foothills of the Himalayas.

Spores of sugarcane and coffee rusts, as well as other pathogens, have moved, on occasion, for long distances over the Atlantic, Pacific, and Indian Oceans (Nagarajan and Singh 1990).

Now compare wheat with its predecessor grasses in mid–North America. Prior to the coming of Europeans, blue grama grass (*Bouteloua gracilis*) existed on the high plains from Mexico to Canada (Figure 11.1). East of the high plains, where conditions are more humid, little bluestem grass (*Andropogon scoparius*) and western wheatgrass (*Agropyron smithii*) grew from south to north, but in interrupted stands mixed with other grasses and forbs. All three of these native grasses have their rust diseases; blue grama has *Puccinia bartholamaei,* little bluestem has *P. andropogonis,* and wheatgrass has *P. graminis* and *P. rubigo-vera.* Usually, there are scattered pustules, but rust never severely damages the plants. This restriction no doubt results from genetic variability, which is enormously complex in these grasses; for example, blue grama has chromosome races with diploids of 20, 40, 60, 42, and 84 chromosomes. Such genetic diversity is at least an order of magnitude greater than that of cultivated wheat. Rust fungi never have a chance to develop specialized and virulent races; instead, genetic diversity and coevolution of host and pathogen led to a balance. All these observations further confirm that epidemics of plant diseases are largely caused by human activities, either directly or indirectly (Harlan 1976).

Another interesting spin-off from rust studies was the gene-for-gene theory, which has greatly influenced thinking in plant pathology (Chapter 6, section 1).

Summary
Pathogen adaptability is the major factor in development of wheat rust epidemics. E. C. Stakman showed that the rust fungus has many races, each differing in virulence to specific cultivars of wheat; this was the first known example of such in a major plant pathogen. New races frequently arise by genetic recombination in the sexual stage, which occurs on barberry, the

alternate host. Virulent new races appeared promptly after the release of each new resistant cultivar of wheat. Stakman advocated and supervised eradication of barberry in the major wheat belt of the United States. Eradication decreased the frequency of appearance of new rust races and the frequency of major epidemics. Thus, the disease situation was stabilized, giving plant breeders time to introduce new resistance. Eradication did not eliminate wheat rust.

The wheat rust fungus only survives winter in the northern United States and Canada in barberry. Furthermore, rust seldom survives summer in Texas and Mexico, where barberry does not grow; prevalence of rust in these places was a mystery. Stakman's aerobiological studies proved that rust urediospores are carried each year by prevailing winds north to Canada. In fall, rust is carried by winds back to Texas, allowing the fungus to survive without its alternate host. There have been no major rust epidemics since 1955 because of barberry eradication and the cultivation of resistant wheat from south to north. There is still a threat because *P. graminis* has the potential to develop new races by mutation.

11.2. Phytophthora Root and Stem Rot of Soybean

The *Phytophthora* root disease of soybean is an excellent model to illustrate that genetic variability and adaptability of microbes can lead to new and destructive epidemics in agriculture. There are, of course, other interrelated and contributory factors, including the basic and obvious prerequisite of monoculture of a genetically uniform plant. Also of possible significance is the fact that the soybean was an alien plant of relatively recent introduction from eastern Asia into the United States. Large-scale cultivation over a large area of the United States was even more recent; at present, more than half of world production is in the United States.

The soybean has an interesting history. It was an important crop in China and Japan for ages. Seeds were brought to Germany in 1692, but soybean never became more than a curiosity in botanical gardens and large estates. Seeds were brought to the United States as early as 1804, and again from Japan in 1854 by Commodore Perry. However, the crop did not become a success until after World War I; then, production increased from 80 million kg in 1920 to approximately 2.7 billion kg by 1970. The world soybean production lead now held by the United States will be challenged by Brazil, where production was increased with a Japanese subsidy that began when United States exports to Japan were halted in 1973–4 (Klein 1979).

The *Phytophthora* disease was first found in 1948 in Indiana (United

States). Since then, *Phytophthora* has become a major problem in soybean production in all areas. The outbreak in the Maumee River basin of northeastern Indiana and adjacent Ohio was soon followed by appearance of the disease throughout soybean-growing areas of the United States and Canada. Its spread was alarmingly rapid, possibly too fast to be traced to a single point of origin, when we consider that this is primarily a soil-borne pathogen. More recently, the disease was found in Europe, Japan, and Australia (Schmitthenner 1985). The causal organism was not conclusively identified until 1955, when the disease appeared in North Carolina.

Origin of the causal fungus is not known with certainty. The fungus (*Phytophthora megasperma* f. sp. *glycinea*, also known as *P. megasperma* var. *sojae* and *P. sojae*), a specialized and virulent pathogen, is assumed to have arisen as a variant of *P. megasperma*, an unspecialized and opportunistic pathogen that affects many plant species (Schmitthenner 1988). The presumed parental form is common throughout the area of soybean production and elsewhere. The soybean pathogen also affects lupines that are native in the soybean-growing regions.

Knowledge of dispersal of *P. megasperma* is required to understand epidemiology of the disease and biology of the pathogen. A few *Phytophthora* species are disseminated by wind and blowing rain, notably *P. infestans*, which affects potato (Chapter 9, section 1). In contrast, *P. megasperma* and most of the other *Phytophthora* species are soil-borne and are not known to spread rapidly over long distances by natural means. They are moved to new areas by human activities, including commerce in plant materials, especially nursery crops.

Limitation in means of dispersal sometimes leads to physical and genetic isolation; chance changes in ploidy, known for *P. megasperma*, also contribute to genetic isolation, which can result in rapid speciation (microevolution). Genetic variants that give virulence or host specialization are expected to increase rapidly in extensive plantings of genetically uniform crops, at least locally. Host specialization adds further to genetic isolation and to speciation. *P. megasperma* and other species of the genus are well known for variability and ecological adaptability (Erwin 1983) and are notorious for spawning new form species and races.

P. megasperma includes a varied collection of genotypes, and new forms continue to appear. All the races and subspecies (f. sp.) have the same morphology as the ecologically generalized form, which is presumed to be ancestral. Genetic relationships of parental and specialized forms of the species are evident in data on DNA homology, protein profiles, and chromosome numbers, powerful tools in determining genetic relationships (Hansen 1987; Brasier 1992). Other fungal genera appear to have similar patterns of microevolution.

A source of resistance in soybean to the new disease was reported during the same year that pathogenicity of *P. megasperma* was clarified. Soybean cultivars Blackhawk and Monroe, already in use in the midwestern United States, were found to carry a single gene for resistance. This gene, now known as *Rps-1*, was quickly incorporated into other cultivars, including Harosoy, the most popular cultivar in the northern zone of soybean cultivation. *Rps-1* was used throughout the area of soybean production, and it controlled the disease for years. Race 2 was discovered in 1965, but it was found only on soybean breeding lines in the south, and it never became a threat. Race 3 appeared in the northern zone of soybean culture and caused heavy losses in 1972.

Since 1972, the situation with fungal races and host genes for resistance has become incredibly complex, rivaling the situation with stem rust of wheat. From 1972 to 1982, a number of new races appeared, each with virulence to soybean carrying the *Rps-1* gene; these races became common in all areas where such soybeans were planted. However, genetic research again saved the day. *Rps-1* was shown to be a locus with an allelic series (*Rps-1, Rps-1b, Rps-1c,* etc.), each giving resistance to one or another of several races of the fungus. Next, additional loci for resistance were found (*Rps-2* through *Rps-7*), each controlling resistance to different races of *P. megasperma* f. sp. *glycinea*. At least twenty-five races of the fungus were known by 1986 (Sinclair and Backman 1989), although relatively few of them (races 1–9) were important. Races are defined by selective pathogenicity to eight different cultivars of soybeans. At least fourteen *Rps* loci and alleles giving resistance are now known. Gene *Rps-1,* which gave resistance to a number of fungal races (1, 2, 3, 6, 11, 13, 15, 17, 21, and 23), lost effectiveness when race 4 of the fungus became prevalent. This is only an example of the continuous war between new races of *P. megasperma* f. sp. *glycinea* and new genes that breeders have used for resistance.

Genetic control of resistance to various races of *P. megasperma* f. sp. *glycinea* obviously is complex. In a series of experiments, resistant and susceptible soybeans of several cultivars were mated, and F_2 and F_3 progenies were tested for resistance. Several independent, dominant genes were found to control resistance to various races (1–9), but the pattern was confusing. For examples: Two genes gave resistance to races 1, 2, and 3; another gene conferred resistance to races 4, 5, 7, 8, and 9; a fourth gene gave resistance to all races except 6; and a fifth gene gave resistance to all races except 2 (Athow, Laviolette, and Wilcox 1979; Athow et al. 1980; Athow and Laviolette 1982).

Stable control by single gene resistance appears to be unlikely (Ryley and Obst 1992). Potentially, there are still untapped genes for resistance, but the fungus appears to have the capacity to respond by changes in pathogenicity. This capacity to adapt is evident when we consider that there were three races in 1972, nine in 1977, and twenty in 1982 (Sinclair and Backman 1989). There were twenty-six known races in 1994, with indications of still more that were not numbered (Schmitthenner, Hobe, and Bhat 1994). Race nomenclature has become so complicated that a better system is needed, along with better designations of genes for resistance. The problem becomes increasingly difficult for the plant breeder because new cultivars need many genes for resistance; there are examples of resistant genotypes that became obsolete (susceptible) before they were released. Presently, there are races that can defeat all known *Rps* genes and gene combinations (Schmitthenner, Hobe, and Bhat 1994). These data are a testament to the tremendous capacity for genetic variation and recombination in *P. megasperma*.

Several characteristics contribute to the extraordinary ecological and pathogenic adaptability of *P. megasperma*. The vegetative stage is diploid (Long and Keen 1977b), in contrast to the situation in most fungi, and probably is heterozygous. Thus, recessive genes may be better able to survive. Diploidy, combined with the ability to hybridize by heterokaryosis (vegetative rather than sexual fusion) *in planta* (Layton and Kuhn 1990) and *in vitro* (Long and Keen 1977a), is a powerful combination for adaptation and no doubt accounts for the formation of many new races and form species. Another possible contributory factor in adaptation may be the hormonal control of heterothallism versus homothallism (Ko 1988), although the nature of such involvement is not yet clear.

Heterokaryotic hybridization was studied by the use of fungal isolates that differed in resistance to certain toxicants. Plants were inoculated simultaneously with two isolates of *P. megasperma* f. sp. *glycinea*, each with resistance to a different toxicant (Layton and Kuhn 1988a). After an incubation period, the fungus was reisolated from the infected plants. Fungal isolates with dual resistance to the toxicants were recovered from 4 to 17 percent of the plants (recovery varied from one experiment to another), proving that heterokaryosis had occurred *in planta*. Comparable results were obtained *in vitro*, following protoplast fusions of the two toxicant-resistant isolates. The recombinants (*in planta* and *in vitro*) also differed from parental types in virulence and pathogenicity. Avirulence was said to be dominant over virulence, as expected if the host–pathogen relationship is gene for gene (see Chapter 6, section 1); that is, if genes for virulence are matched by genes for resistance in the host (Layton and Kuhn 1988b). However, other data for Phytophthora root rot do not fit the gene-for-gene hypothesis.

Variability data must be considered in relation to the gene-for-gene hypothesis. A wide range of races that differ in pathogenicity was isolated from single soybean fields (Anderson and Buzzell 1992) and from mass transfer cultures. This clearly affirms that single gene control of virulence in *P. megasperma* f. sp. *glycinea* in soybean will always be transitory. Completely avirulent isolates were obtained after prolonged growth and transfers in culture; for example, an avirulent isolate was obtained from cultures of race 6. The completely avirulent isolates were incompatible with all genotypes of the host; these isolates penetrated plant surfaces and caused hypersensitive reactions that accompany resistance, but they did not develop further. According to the gene-for-gene hypothesis, such isolates should carry many genes for avirulence, which is dominant over virulence. This situation appears to be improbable for the soybean disease because it requires an excessive genetic load for the fungus; thus, there are some doubts that soybean *P. megasperma* fits the classical gene-for-gene situation (Schmitthenner 1988).

Genetic variability in *P. megasperma* was further confirmed and was strikingly evident in the work of Long et al. (1975). These workers began with four field isolates, representing three races of the fungus, and selected a number of single-zoospore lines. The single spore lines varied greatly in pathogenicity and in several other characters. The isolate lines were repeatedly selfed, and single zoospore isolates were grown after each selfing; each transfer generation was selected for virulence, pathogenicity, spore production, and other characters. The fifth generation transfer lines were stable for all characters, were said to be wild-type in pathogenicity, and were deemed to be suitable for genetic studies.

The approach just described was developed further by Rutherford, Ward, and Buzzell (1985), who made repeated single-zoospore isolates of races 3 and 6. The resulting isolate lines were used in inoculation experiments with several cultivars of soybean, each with a different *Rps* gene. Results showed both gains and losses in virulence and pathogenicity of the fungus over successive generations, as compared with the parental types. Mass transfer isolate lines gave similar results but were less variable than were the single spore lines. In one case, Rutherford et al. started with race 6 and ended with race 3 after a series of single-zoospore transfer generations. It is unlikely that such a change in pathogenicity could be the result of mutations, because at least seven loci should have been involved. The parental isolates must have carried masked potentials for pathogenicity and virulence. These kinds of changes could occur in nature and could account for the well-known adaptability of *P. megasperma* in its many subspecies.

There is some hope for future control of *Phytophthora* by use of "field resistance" or tolerance in the soybean plant. Such resistance is multigene-controlled and appears to be based on entirely different biochemical mechanisms than is single-gene resistance. Field resistance is not always apparent in laboratory experiments and may be evident only in large-scale, long-term field trials. Obviously, such resistance usually is less striking and less complete than is single-gene resistance, but field resistance is not fungal race-specific and should be more stable (Schmitthenner 1988; Tooley, Grau, and Stough 1986). Field resistance does not place the same population pressures for change in the pathogen because it and the host plant can coexist. The price is lower crop yields under some conditions. Another handicap is that multigenic control makes field resistance a more difficult problem for the plant breeder; nevertheless, efforts are under way. The biochemistry is poorly understood in both types of resistance.

When control of a plant disease by use of resistance genes is not successful, the usual response is to try fungicides. The use of fungicides can be effective for soybeans, especially when combined with use of tolerant or field-resistant cultivars, but it is costly. Damping-off of tolerant seedlings has been controlled by use of fungicidal seed treatments. The systemic fungicide metalaxyl is effective for root rot control when used either as a seed treatment for damping-off or as a soil treatment with granules in the furrow (Schmitthenner 1985). Metalaxyl is most effective when used with disease-tolerant cultivars. Unfortunately, there are limitations to its use, including cost, unreliability, and reduced yields under some conditions. A major potential problem is the buildup of resistance to metalaxyl, a problem that is well known for this and other site-specific fungicides. There are also potential ecological dangers in the use of a fungicide on a crop with the enormous acreage of soybean.

Several cultural and management practices also can decrease the threat of *P. megasperma* to the soybean crop, but these are beyond the scope of this treatise.

11.3. Rice Blast

Worldwide, blast is the most damaging disease of rice, one of the world's most important crops. The causal fungus, an Ascomycete (*Magnaporthe grisea*, teleomorph of *Pyricularia grisea*), is perhaps the most variable and adaptable of all major plant pathogens. Scientists working with rice have been amazed and dismayed by the rapidity and regularity with which resistant cultivars suddenly became susceptible.

Rice feeds a third of the world's people; the total crop yields more than 400 million metric tons, with 90 percent in Asia. Other important producing areas are in Latin America, Africa, Australia, and the United States. Rice is a dependable and productive crop; Australian farmers have yields of more than 1,000 kg per hectare (100 bu per acre), although the average for Asia has been about 275 kg per hectare. The crop is cultivated in flooded fields (paddy fields) and in upland fields that are rain fed. Rice has two major types: Short-grain rice (*Oryza sativa* f. *japonica*) is grown in temperate regions of China and Japan, as far north as the island of Hokkaido and into central Europe; long-grain rice (*O. sativa* f. *indica*) is grown mostly in the tropics. Thousands of cultivars have been developed by farmers over the centuries; more have been added in recent times by plant breeders.

The agriculture of southeastern Asia centers on rice production (Carefoot and Sprott 1967). Flooded paddy fields (unpolished rice in Asia is known as paddy) are used to grow fish such as the golden carp, which eats algae, fungi, weeds, and insects in the water. The fish help to control mosquitoes that carry malaria, and they provide a necessary protein supplement to a rice diet. Estimates indicate that 500 million tons of paddy fish are produced in Asia. Still, deficiency diseases such as beriberi are common in southeastern Asia because of the strong preference for polished rice as food. In Japan, the significance of rice is symbolized by the fact that the word *rice* means a meal; *no rice* means nothing to eat. Overall, rice is so important in the world that anything that affects rice agriculture and rice distribution affects the whole world in one way or another.

Cultivated rice presumably was first developed from a native species during the Stone Age, perhaps eight thousand years ago. The first rice is thought to have been indigenous to the wetlands of southeastern Asia. Rice cultivation is mentioned in the earliest Chinese and Hindu writings, more than three thousand years ago. Rice was moved to Japan about one thousand years later, and to Europe about 300 B.C., when Romans brought it from Egypt. However, rice cultivation was of no significance in Europe until the Moors came to Spain, beginning in A.D. 711. Rice was grown in Italy during the fifteenth century, and it was soon taken to Brazil by the Portuguese. In 1718, rice was brought to North America, where cultivation flourished in the Carolinas. Arkansas and California, now the major producing areas in United States, started rice cultivation in 1905 and 1912, respectively. Most rice is grown in the monsoon zone, and most of the rice grown in the monsoon areas of Asia is eaten by the growers (Carefoot and Sprott 1967; Thurston 1984).

Rice is the only grain that has been grown in the same fields for century after century without exhausting the soil. The soil remained fertile because rice is grown in flooded fields, where nitrogen is fixed by *Anabaena,* a blue-green alga that has a symbiotic relationship with a water fern, *Azolla.* The Chinese have exploited this relationship for centuries. *Azolla* grows as a floating mat on open water; the fern and its alga can be skimmed off and used as fertilizer when the field is puddled in preparation for transplanting rice seedlings. *Azolla* and alga also grow and fix nitrogen in flooded fields during growth of the rice crop.

Estimates indicate that about 20 percent of the rice crop is lost to pests and diseases each year. Among the major destructive agents are the blast disease and the brown spot disease, caused by *Helminthosporium oryzae;* these and other problems (insects, drought, flooding, political instability, and lack of distribution systems) have caused major famines in Asia over the centuries. Rice blast caused local famines in Japan as late as 1934 and 1941. India has had at least 70 major famines during its recorded history, the latest in 1966. China may have had as many as 1,800 famines since 100 B.C., but most of them were local.

The major modern development in rice agriculture was the "green revolution." This revolution was created by plant breeders who developed short-stemmed cultivars that can tolerate heavy fertilization, especially increased nitrogen. Use of the new cultivars, plus more fertilizers, can give a fivefold or more increase in yields of Asian paddy fields. The new cultivars mature rapidly, allowing two crops each year in some areas that formerly grew only one. Also, the development and use of new storage methods have reduced spoilage and loss. Rice production in Asia increased 1.4 percent per year between 1955 and 1970. Unfortunately, population in the same regions of Asia increased 2.5 percent per year.

Blast disease is our major focus. The term *blast* apparently was first used at the beginning of the twentieth century to describe the problem in the Carolinas, where the disease was common as early as 1876. Infected seedlings have a scalded appearance, and this was seen as "blast"; there are other descriptive names. The disease was described in Chinese literature as early as 1637 and in Japan by 1704. The causal fungus was first described in Italy in 1891, and the first pathogenicity tests were reported in 1908 (Thurston 1984).

Blast shows on rice leaves as spindle-shaped spots with gray centers and brown margins. If conditions are favorable, the spots enlarge and coalesce, thus killing (blighting) the leaves. Severe losses occur when infections are on the stem of the panicle (the flower and grain-bearing shoot); these infections lead to breakage of the stem and grain loss. If a panicle infection occurs early, there is no grain; if the infection occurs late, the grain quality is poor. The fungus can infect and damage wheat, barley, maize, sugarcane, and other crops; forage grasses often are infected but seldom are damaged.

Nomenclature of the causal *Pyricularia* species has been controversial, even the spelling (*Piricularia*). Certain forms are specialized and highly virulent on rice, and other forms primarily affect certain other gramineous plants. The rice-infecting forms soon became known as *P. oryzae* and the grass-specialized forms as *P. grisea*. Discovery of the sexual stage (*Magnaporthe*) in the laboratory made genetic experiments possible. The sexual or perfect stage has not been found in the field. Rice and grass-specialized forms are morphologically identical and sexually compatible; thus, they appear to be specialized forms of the same species. At present, researchers are increasingly inclined to label the whole group as *M. grisea*. In retrospect, the situation with *Magnaporthe* appears to be similar to that with *Puccinia graminis* and its several form species, such as *tritici* and *avenae* (Chapter 11, section 1).

M. grisea lives from season to season as spores and mycelium in debris and in seed. Conidia are carried by wind and splashing rain. Spores have been found at altitudes of 2,200 m, but they are too fragile to be carried far. Infection occurs on wet leaves; cool, wet weather favors infection, as does high nitrogen in the soil and plant. At higher temperatures, plants become more resistant. The disease is more damaging to upland rice than to rice in flooded fields, in both the tropic and temperate zones. In fact, holding the rice in flooded fields to near maturity is a significant control method. Other control measures are the use of resistance genes, fungicides, and antibiotics. Chemical controls are costly, and their use is more or less limited to Japan.

M. grisea occurs worldwide on cereals and forage grasses belonging to at least thirty genera; a few nongrass hosts are sometimes affected. The fungus has many races or genotypes that differ in virulence and pathogenicity to different cultivars of rice; this finding was first reported in Japan in 1922 (Ou 1972). The race situation greatly complicates breeding of rice for resistance and is responsible for contradictory results from different areas. Breeding for resistance has been especially frustrating because of the ability of *M. grisea* to spawn new and virulent races that soon eliminate the newly developed resistant cultivars of rice.

Complexity of the race situation is illustrated well by some data from Ou and associates (Ou and Ayad 1968; Ou 1980). These workers picked single spores (conidia) from a single lesion on one rice leaf; the spores were germinated, grown in culture, and used to inoculate a series of rice plants of different cultivars. Results showed that the conidia fell into fourteen different races, as shown by selective pathogenicity to different rice cultivars. To illustrate hypothetically, race A infected only rice cv. 1, whereas race B was avirulent on cv. 1, but infected other cultivars, and so on. Single spores from the inoculated plants were grown in culture and analyzed again on new plants of the same cultivar series. Some of the second generation isolates were identical to the parents, but others had to be classified as different races.

The findings of Ou and Ayad (1968) have been confirmed many times by other researchers who felt that there is no limit to the number of fungal races that appear from one generation to the next. Several groups of researchers came to the conclusion that available cytological and genetic knowledge does not explain such variability, and that it can be understood only if the fungus has many nuclei per cell. Later electron microscope work indicated that conidia and vegetative cells sometimes are multinucleate but, in general, are not (Giatgong and Frederiksen 1969).

Mating experiments have shown that avirulence and virulence in *M. grisea* to rice generally are inherited, as expected (Kolmer and Ellingboe 1988; Ellingboe, Wu, and Robertson 1990; Ellingboe 1992). Resistance in rice usually is dominant to susceptibility, with one or a few genes controlling resistance to any given race of the fungus. Crosses between various fungal races and form species gave progeny that were more fertile and mated more readily than usual. Mating type, female fertility, and pathogenicity to differential rice cultivars segregated randomly, but progenies virulent on rice were female sterile and all of mating type A; there was no corresponding mating type a. A single mating type virulent on rice confirms data from other sources (Itoi et al. 1983) and explains why rice isolates usually mate only with isolates from goosegrass or other grasses.

What is the explanation of the extreme variability in *M. grisea?* Several mechanisms appear to be involved, in addition to expected sexual union with segregation. Observations have shown an active parasexual cycle (vegetative fusion between races), at least in culture; whether or not this occurs in nature is not known (Leung and Taga 1988). The two mating types (A and a) are needed for both vegetative fusion and sexual union. However, many isolates throughout the world are hermaphroditic (carry both A and a potential). Also, many isolates from rice in Japan are female sterile and act as males only when mated with isolates from grasses (Itoi et al. 1983); this finding has been reported from other areas. It appears to be based on a single-gene mutation. Chromosomes that lag in the formation of ascospores were observed during both meiosis and mitosis. The frequency of lagging in spore development varies from 13 percent to 76 percent, depending on the fungal strain (Ou 1980). Lags can result in differences in chromosome numbers and may be an important source of variation. Some cells are multinucleate, which also may affect variability.

Molecular genetics methods are now being used to clarify race and variability problems in *M. grisea*. DNA fingerprint profiles were obtained for isolates from various areas in North and South America; profiles can differentiate clonal lineages and linkages. Each lineage (genotype, or race) gives a distinctive DNA profile, and linkage relationships between lineages become evident. There appear to be a limited number of clonal lineages that have been stable for years. The same distinct lineages and clonal structures were found for both continents. The data suggest repeated migrations of *M. grisea* over all rice-growing regions that were examined (McDermot and McDonald 1993).

Cloning experiments with host specificity genes show that *M. grisea* follows expected avirulence–virulence patterns found for several other pathogens (see Chapter 4, section 4). Two avirulence genes in rice-infecting races were identified and cloned; one precludes infection of certain rice cultivars that are resistant, and the other precludes infection of certain other grass species. Some genetic instability is thought to result from location of host specificity genes near the end of a chromosome (Valent and Chumley 1991; Valent et al. 1994).

Some fungal races infect most rice cultivars, and some rice cultivars are resistant to most races. This occurs because there are many virulence genes in the fungus and many resistance genes in rice. The important point to remember is that *M. grisea* is extremely adaptable, changing constantly. Worldwide tests have uncovered a few rice cultivars with resistance to many races of *M. orisea;* these cultivars carry a generalized resistance controlled by many genes. Such cultivars have a low level of resistance; many fungal races can attack them, but few races are virulent, and the plants are not severely damaged (Ou 1980). Wild species of *Oryzae* vary in the resistance-to-susceptibility scale comparable to that seen in cultivated rice.

Breeding for resistance to control blast began more than sixty years ago. Many new and highly resistant cultivars were developed, but they were never effective for more than a few years. The complex race situation is evident in that a cultivar resistant in one geographical region is susceptible in another. In worldwide tests, no cultivars were highly resistant in all locations. Many tropical (*indica*) cultivars are temporarily resistant in temperate regions, and many temperate (*japonica*) cultivars are resistant for a while in the tropics; however, new and virulent races of *M. grisea* soon overcome them. Many resistance genes (more than twenty) are known in different rice cultivars, and there are dozens of known pathogenic races of the fungus. A gene-for-gene relationship is indicated (see Chapters 4 and 6, section 1).

Fungicides are used to control blast in Japan. However, *M. grisea* is as efficient in development of resistance to systemic fungicides and antibiotics as it is in overcoming new resistant cultivars of rice. Japanese researchers are constantly searching for new antibiotics.

Summary

Blast of rice, caused by the fungus *Magnaporthe grisea*, has been a major problem in Asia and worldwide for many years. The disease has been a significant factor in famines in Asia. *M. grisea* has several subspecies, with selective or specialized pathogenicity to different gramineous plants, and many races that vary in virulence to different cultivars of rice. The fungus spawns new races that attack each new rice cultivar that is bred for resistance. Variability is a major problem because use of genetic resistance is the best control for diseases of this type.

Many factors seem to be involved in the variability and adaptability of the blast fungus, but full understanding is still elusive. Observed mechanisms that could contribute include anastomosis (vegetative fusion), which gives hybrid races, heterokaryosis (many nuclei per cell), sexual mating and segregation, and mutation. Lagging chromosomes in *M. grisea* could account for differences in chromosome numbers and thus contribute to variability. The variability problem is being clarified by use of molecular genetics techniques.

11.4. Other Examples of Pathogen Adaptability

Monoculture is a factor in most, if not all, important plant diseases in modern agriculture, and adaptability by pathogens is involved in many of them. These factors apply to Fusarium wilt of banana, described in Chapter 14, section 2, as a disease facilitated by propagation methods, although monoculture and adaptability are significant. Genetic uniformity of crops and pathogen adaptability affect epidemics by the *Alternaria* and *Cochlibolus* diseases, but these are treated elsewhere (Chapters 12 and 13) because of unique circumstances.

Several other Fusarium wilt diseases are significant and have had much

Figure 11.5. Fusarium wilt of tomato. (Scheffer and Walker 1953)

study. The pathogens are host-specific form species of *F. oxysporum;* among them are f. sp. *lycopersici,* affecting tomato (Figure 11.5); *conglutinans* (cabbage); *vasinfectum* (cotton); and at least fifty others that affect various hosts. The pathogens are morphologically identical (Bosland 1988) and have in common a number of vegetative mating types as described for the *Fusarium* that affects banana (Chapter 14, section 2). The tomato form has received most study and will be considered here as an example of the Fusarium wilts.

Three races of *F. oxysporum* f. sp. *lycopersici* are now known. The original (race 1) was virulent on tomato cultivars that were widely used by market and home gardeners. A gene for resistance was found in *Lycopersicon pimpinellifolium* from Peru; plant breeders succeeded in incorporating this gene into a number of good tomato cultivars that were used successfully in areas where race 1 was present and cultivation was intense. Then, a second race of the pathogen appeared; it was virulent on cultivars that were resistant to race 1. Races 1 and 2 are now present in most regions of market gardening. Next, a gene for resistance to race 2 was found and put to use; this was followed by appearance of race 3, now present in a few places in Australia, California, and Florida (Elias and Schneider 1992).

New races apparently develop by random mutations or by genetic recombinations via parasexual mechanisms; the fungus has no known sexual stage. The form species *lycopersici* appears to have arisen from at least two progenitor populations, as indicated by molecular biology studies (Elias and Schneider 1992). The progenitors may have been opportunistic and nonspecialized or saprophytic forms of *F. oxysporum*, which are widely distributed.

Other Fusarium wilt diseases have much in common with those caused by *F. oxysporum* f. sp. *lycopersici* and f. sp. *cubense* (Chapter 14, section 2), the models for physiological, biochemical, molecular, and epidemiological research (Beckman 1987). A number of other form species of the fungus are known to have races specialized to specific host cultivars.

Puccinia striiformis, cause of stripe rust of wheat, is an instructive example of the power of adaptability in pathogens. A new resistant cultivar was introduced in the Netherlands in 1950; it was planted on 14 percent of the wheat acreage the first year, on 43 percent the second, and on 81 percent by the third year. A small locus of stripe rust infection was found in the breeder's plots during the second year; by year three, there was infection in all fields planted to the new cultivar, and by year six, 70 percent of the winter-wheat crop in the Netherlands was destroyed (ten Houten 1974).

12

Monoculture:
Cochliobolus Diseases with Toxins

Certain plant diseases have become leading models for study; among these are diseases caused by toxin-producing fungi. Natural histories and recent molecular studies on several toxin-mediated diseases will be discussed.

Several pathogenic species or races in the fungal genera *Cochliobolus* and *Alternaria* are well known for their ability to produce host-specific toxins, compounds that affect only hosts of the producer (Figure 12.1) (Scheffer and Livingston 1984). Host-specific toxins were thought, for many years, to be unusual accidents of nature because only two were known. We gradually are discovering more such toxins; at present, seventeen or more are known from at least six fungal genera. The toxin producers and the diseases they cause are, in most respects, typical of many other pathogens and plant diseases.

Some toxin-mediated diseases have, in recent years, become possibly the best understood of all plant diseases, in regard to the biochemistry of disease development and disease resistance. Potentially, infectious plant diseases mediated by toxins are useful models for study of plant diseases in general. This does not imply that toxins are involved in all plant diseases; there are advanced studies with diseases not mediated by toxins.

Each disease caused by the several toxin-producing Cochlioboli and Alternariae differs from the others in many important ways. The diseases considered here have one common feature: Susceptibility and resistance in hosts are determined by the selectively toxic compounds that are released by the pathogen. There may be other features in common. The known host-specific toxin producers appeared following genetic manipulation and widespread planting of new genotypes of crop plants. The new genotypes of crops, in some cases, began with genetic mutations (Wise, Pring, and Gengenbach 1987). Endemic microorganisms soon adapted to the new host genotypes, creating new diseases and damaging epidemics.

179

Figure 12.1. Selective toxicity of victorin to susceptible oat seedlings. Equal amounts of toxin were added to the nutrient solutions that contained roots of susceptible (center) and resistant (right) plants. Control seedlings without toxin are shown on the left. (Scheffer and Yoder 1972)

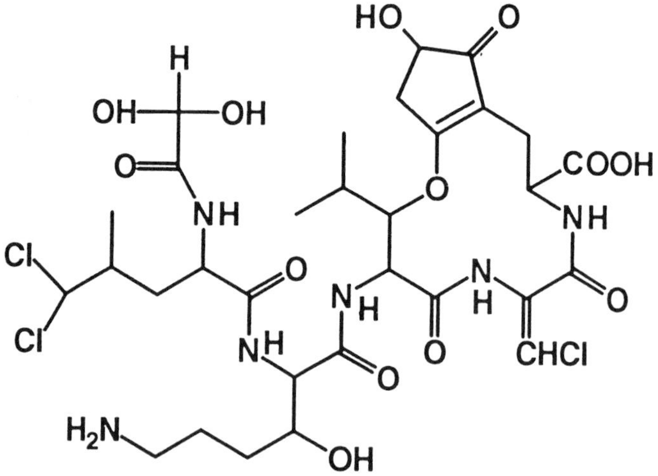

Figure 12.2. Chemical structure of victorin C, the major form of the toxin.

12.1. Victoria Blight of Oats

Victoria blight was the first *Cochliobolus* disease known to be mediated by a toxin. History of the disease began with introduction of oat cv. Victoria from Argentina to the United States, to be used in plant breeding. The cultivar had a gene (*Vb*) for resistance to many races of crown rust and to smut. New cultivars carrying the *Vb* gene were released in Iowa in 1942; they were so good that, within five years, 98 percent of oat acreage in Iowa and 50 percent in all of North America was planted to them. Oat is self-pollinated and therefore maintains genetic uniformity. Victoria blight appeared soon after the new cultivars were established; all cultivars carrying the *Vb* gene had to be discarded (reviewed by Pringle and Scheffer 1964). The causal fungus was described as a new species, *Helminthosporium victoriae* (=*Bipolaris victoriae*); later, the sexual stage was found, and the fungus became known as *Cochliobolus victoriae* (an Ascomycete).

C. victoriae was soon found to produce in culture a toxin that affects only plants with the *Vb* gene (reviewed by Pringle and Scheffer 1964) (Figure 12.1). However, the toxin was not characterized chemically for many years because of its complexity and lability. Finally, Macko et al. (1985) characterized it as a cyclic peptide, mol weight 814, containing glyoxylic acid, several unusual amino acids, and chlorinated entities (Figure 12.2). Several forms of the toxin vary slightly in structure. The compound, often referred to as HV-toxin, is now known as victorin, following the usual custom of using a name chosen by the characterizer.

Victorin is one of the most active natural products known; it is toxic to oats with the *Vb* gene at 37pM; resistant oats and other species are not affected by a million-times-higher concentration (Scheffer and Livingston 1984). Toxin production by *C. victoriae* is controlled by a single Mendelian gene that also controls selective pathogenicity to oats with the *Vb* gene (Scheffer, Nelson, and Ullstrup 1967). The toxin gene allowed *C. victoriae* to cause a continental epidemic of oats in North America in 1946–8.

C. victoriae is closely related to *C. sativus,* as was discussed earlier (Chapter 5, section 2). The two species are similar in morphology, and some isolates have been reported to be sexually compatible. Still another indication of relatedness is the ability to produce a compound known as victoxinine, a sativane with a known structure. All *C. victoriae* isolates produce victoxinine, as do some isolates of *C. sativus;* no other fungi examined produce it, including a number of *Helminthosporium* species (Pringle 1976). This finding supports the hypothesis that *C. victoriae* was derived from a strain of *C. sativum,* perhaps as a mutation or other genetic rearrangement.

This origin was postulated many years ago (reviewed by Scheffer 1989a). The new genotype (*C. victoriae*) had a population explosion when oat with the *Vb* gene was planted over a large area. When *Vb* oat cultivars were discarded, *C. victoriae* became rare, found only occasionally on senescent grass leaves (Scheffer and Nelson 1967; Scheffer 1989a, 1989b).

The mechanism of toxic action by victorin is still uncertain. There is an early effect on the plasmalemma (within minutes), as shown by increased losses of electrolytes from leaf tissues that previously were infiltrated with water prior to exposure to toxin. Several other criteria indicate very rapid effects on the cell membrane (Samaddar and Scheffer 1968, 1971). Similar rapid responses are induced by some other host-specific toxins (Gardner, Mansour, and Scheffer 1972; Kasai et al. 1975; Scheffer and Livingston 1980). Nevertheless, this is not conclusive evidence for a direct effect on the cell membrane. More recent data show accumulation of labeled toxin in mitochondria in intact tissues, but no accumulation in isolated mitochondria (Wolpert and Macko 1991). However, known mitochondrial poisons such as HmT-toxin (affecting maize) have a delayed effect on electrolyte leakage from tissues (Halloin et al. 1973); victorin has a rapid effect, suggesting a different site of action. Victorin binds to several proteins that occur in both resistant and susceptible tissues (Akimitsu et al. 1992; Wolpert et al. 1994), but the significance is not known.

12.2. Maize Leaf Spot Caused by *C. carbonum* Race 1

The disease caused by *C. carbonum* race 1 (Figure 12.3) was first discovered in experimental plots of plants with the recessive *hm* gene (reviewed by Scheffer 1989a). All susceptible maize inbreds were derived from the open-pollinated cv. Pride of Saline. Later, race 1 was found to be a rare but widely distributed form of *C. carbonum* on senescent leaves of various grasses. Specialized race 1 probably was derived from a less-specialized type (race 0 or race 2) of *C. carbonum,* an opportunistic fungus found on many graminaceous species. There are still other genotypes of *C. carbonum;* some are general invaders of senescent leaves, and others are selectively pathogenic on certain grasses. All appear to be related to *C. sativus* (Scheffer 1991). The selective pathogenicity of *C. carbonum* race 1 is based on its ability to produce the selective toxin (HC-toxin) that affects susceptible (*hm*) maize; resistant (*Hm*) maize is insensitive (Figure 12.4) (Scheffer and Ullstrup 1965).

New races of *C. carbonum* pathogenic to other genotypes of maize continue to appear (Hamid, Ayers, and Hill 1982; Xiao et al. 1990, 1991;

Figure 12.3. Symptoms of natural infection on maize by *Cochliobolus carbonum* race 1. (Picture by R. P. Scheffer)

Scheffer 1991; Welz and Leonard 1993), probably because the fungus is genetically diverse. *C. carbonum* may be evolving toward a gene-for-gene relationship (see Chapter 6, section 1). Introduction of new maize genotypes may lead to still more races of *C. carbonum* and, perhaps, to discovery of new toxins.

The consequence of toxin production by race 1 is clear: An opportunistic pathogen was converted to a host-selective, virulent, and destructive form. This potential problem for agriculture was solved by discarding maize with the recessive allele (*hm/hm*), thanks to astute work by Professor A. J. Ull-strup.

HC-toxin was characterized independently by several groups of researchers. It is a cyclic tetrapeptide with an epoxide group (Figure 12.5) (reviewed by Scheffer and Livingston 1984).

HC-toxin is required for colonization of maize tissues by the producing fungus (Comstock and Scheffer 1973). Purified toxin has some unexpected effects that do not fit old notions that toxin-producers invade only killed

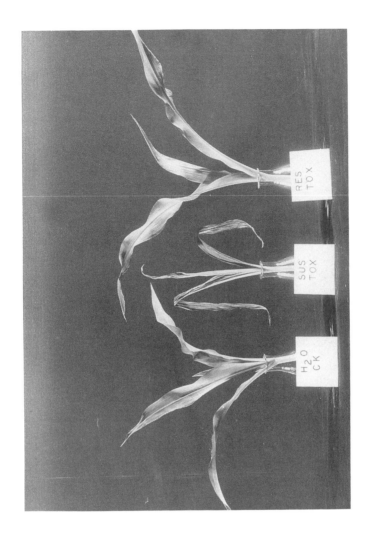

Figure 12.4. Selective effects of HC-toxin on cuttings of susceptible (center) and resistant (right) maize plants. Equal amounts of toxin were added to water in the flasks. Control without toxin is shown on the left. (Scheffer and Ullstrup 1965)

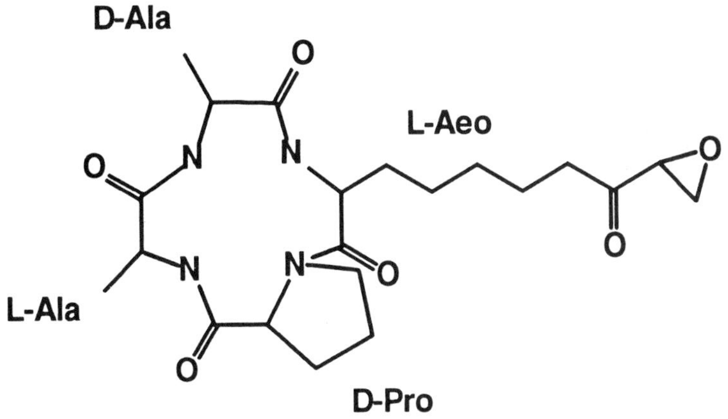

D-Ala

L-Aeo

L-Ala

D-Pro

Figure 12.5. Chemical structure of HC-toxin.

cells (necrotrophs). At very low concentrations, the toxin causes increased growth and synthesis in susceptible seedlings (Kuo, Yoder, and Scheffer 1970). The toxin at slightly higher concentration and with longer exposures causes the same physiological changes in host tissue that are caused by the colonizing fungus; included is increased capacity to fix carbon dioxide in the dark, via increased carboxylation of ribose-5-phosphate (Kuo and Scheffer 1970). Another striking effect is increase in uptake of nitrate and certain other ions. The data suggest that HC-toxin alters plasma membranes (Yoder and Scheffer 1973a, 1973b) but does not cause the drastic derangement that occurs in oat cells soon after exposure to victorin (Samaddar and Scheffer 1968).

The molecular basis of toxin production by *C. carbonum* race 1 has now been clarified. Initial work showed two enzymes involved in toxin bio-synthesis (Walton 1987; Walton and Holden 1988). This work seemed to contradict earlier findings that one Mendelian gene controls toxin production (Scheffer, Nelson, and Ullstrup 1967); however, later work clarified this. A 22-kb chromosomal region, inherited as a single gene (*Tox 2*) encodes a large multifunctional biosynthetase that is not present in toxinless genotypes of *C. carbonum* (Panaccione et al. 1992). Sequencing the 22-kb region revealed an open reading frame (15.7 kb) that encodes a cyclic peptide syn-thetase complex. The 15.7-kb DNA sequence encodes two enzymes (syn-thetases 1 and 2) involved in synthesis of HC-toxin. These two enzymes appear to exist *in vivo* as part of a single unit (Scott-Craig et al. 1993).

The basis of toxin resistance in maize also was determined. The *Hm-1* gene governs tolerance to toxin and resistance to *C. carbonum* race 1. HC-toxin is converted by a reductase in resistant leaves and in cell-free preparations from leaves to a single nontoxic compound (an 8-hydroxy derivative of HC-toxin) (Meeley and Walton 1991). The reductase activity was detected only in extracts of maize that are resistant to *H. carbonum* race 1. At first, the work was confusing because toxin introduced into intact plants via the transpiration stream was inactivated by both resistant and susceptible tissues, as was reported in earlier work (Kuo and Scheffer 1969). Inactivation in the transpiration stream must depend on mechanisms not associated with cellular resistance. The conclusion is that the biochemical basis of resistance to *C. carbonum* race 1 is enzymatic detoxification of HC-toxin (Meeley and Walton 1991; Meeley et al. 1992). In contrast, resistance to *C. heterostrophus* race T is based on lack of a toxin-receptor protein (see Chapter 12, section 3). The HC-toxin inactivating system depends on a nicotinamide dinucleotide phosphate (Jokal and Briggs 1992).

The question remains: What is the mode of action of HC-toxin in the susceptible cell? Recent evidence indicates that the toxin promotes compatibility (susceptibility) between host and pathogen by interfering with histone acetylation, which is implicated in control of chromatin structure, cell-cycle progression, and gene expression. These are reversible processes involved in response to disturbance, including hypothetical resistance reactions. This metabolic change does not lead to rapid killing of cells and is compatible with many previously reported effects of HC-toxin (Brosch et al. 1995).

12.3. Maize Blight Caused by *C. heterostrophus* Race T

A severe epidemic of leaf blight, caused by a new race of *C. heterostrophus* (teleomorph, *Bipolaris maydis*, = *Helminthosporium maydis*), occurred in the United States and elsewhere in 1970. The old form of *C. heterostrophus* (now known as race O) was known since 1923 as cause of a minor leaf disease of maize and teosinte in the southeastern United States and Mexico. Morphologically, *C. heterostrophus* is similar to *C. sacchari* from sugarcane and *C. miyabeanus* from rice. The new race of *C. heterostrophus* (race T) is virulent and host-specialized on Texas-male-sterile (Tms) maize (Smith, Hooker, and Lim 1970), formerly used for economy in production of hybrid seed. The maize crop was genetically uniform for Tms cytoplasm throughout the United States. The 1970 epidemic was the most severe on record, in terms of financial losses for a crop in a single season (Ullstrup 1972).

Race T was first discovered in the Philippine Islands in 1961 and in Iowa in 1968 (Ullstrup 1972). Race T could have been brought from the Philip-

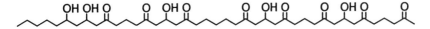

Figure 12.6. Chemical structure of HmT-toxin from *Cochliobolus heterostrophus.*

pines to the United States, but we do not know this; alternatively, it could have originated in the United States from another mutation or other genetic change in race O. The race T gene probably was in the United States for some time, becoming evident only when Tms maize was widely planted and weather conditions over a large area favored disease development. Both races O and T are seed-borne and readily overwinter in crop debris in the northern United States. Thus, it appears unnecessary for race T to spread each year from southern sources, as was proposed in 1970. Race T became rare when Tms-cytoplasm maize was discarded.

There is little doubt that race T originated from race O of *C. hetero-strophus.* First, the two are morphologically identical and are easily mated in laboratory experiments. Two mating types (A and a) occur in both races; race O always had both types in equal proportions. Isolates of race T taken in 1970 were predominantly mating type A, throughout the range of the epidemic. After 1970, mating type a gradually increased in race T, until types A and a were about equal (Leonard 1973, 1977). These findings suggest a single origin of race T from mating type A of race O, probably in the south-central part of the United States corn belt.

The first credible report of a host-specific toxin from race T (HmT-toxin) was that of Hooker et al. (1970). The toxin was characterized by Kono and Daly (1979) as a linear polyketol ($C_{41}H_{68}O_{13}$) (Figure 12.6). Even earlier, HmT-toxin was known to uncouple phosphorylation and oxidation in mitochondria from susceptible but not from resistant maize (Figure 12.7) (Miller and Koeppe 1971). Mitochondria in intact susceptible cells also are affected (Walton et al. 1979). The fungus produces several similar molecules, with lower molecular weights, that have the characteristic toxicity to Tms maize.

Molecular genetic studies on toxins began with the finding that Tms-cytoplasm maize has a 13-kd protein (URF-13 protein) that is not present in normal cytoplasm (toxin-insensitive) maize (Dewey et al. 1988). The URF-13 protein is encoded by the *T-Urf-13* gene (Wise et al. 1987) that was moved to a bacterium (*Escherichia coli*) (Dewey et al. 1988) and to yeast (Huang et al. 1990). Transformed bacteria and yeast are sensitive to HmT-toxin from *C. heterostrophus.* Next, the gene was moved to tobacco, making it sensitive to HmT-toxin. These are the only known toxin-sensitive organisms other than Tms maize. Toxin-sensitive microorganisms are con-

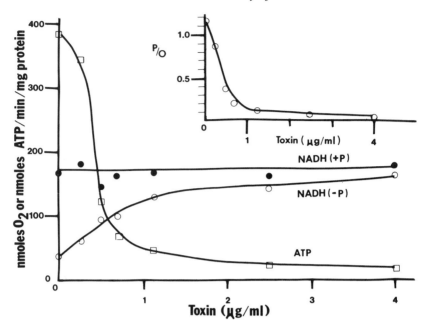

Figure 12.7. Effect of HmT toxin on synthesis of adenosine triphosphate (□) and oxidation of nicotinamide adenine dinucleotide (NADH) by mitochondria from susceptible maize, using phosphorylating (●) and nonphosphorylating (○) conditions. Insert shows effect of toxin on P/O ratios (Bednarski, Izawa, and Scheffer 1977).

venient for toxin bioassays (Ciuffetti, Yoder, and Turgeon 1992). HmT-toxin binds reversibly to the URF-13 protein (Braun, Siedow, and Leavings 1990), thus confirming earlier findings on reversibility of toxic effects on mitochondria (Bednarski, Izawa, and Scheffer 1977).

Phyllosticta maydis produces a toxin with similar selective effects and with a similar structure, but with a lower molecular weight (reviewed by Scheffer 1991). Its selectivity to T-cytoplasm maize suggests either parallel adaptations by fungi of different genera or exchange of genes between different fungi.

12.4. Sugarcane Leaf Spot Caused by *Bipolaris sacchari*

Sugarcane leafspot, often known as eyespot, is caused by *Bipolaris sacchari* (= *Helminthosporium sacchari*). The disease was first observed in 1892 in Java; it became a serious problem after 1920 in Java, Hawaii, Puerto

Rico, and other parts of the world when certain new cultivars (clones) were planted on a large scale (reviewed by Scheffer 1989b).

The most obvious factor favoring *Bipolaris* leafspot was monoculture of genetically uniform sugarcane. Also, sugarcane is an alien plant in many places, now grown throughout the tropics and subtropics. Genetic uniformity in sugarcane is maintained because the crop is propagated vegetatively, starting with selected genotypes. New sugarcane clones had been selected from seedlings by plant breeders, without regard to their reactions to *B. sacchari*. Unfortunately, susceptible clones were planted in Australia, even after the danger was apparent, and the disease again became a problem. It is hoped that plant breeders now have learned that new clones must always be screened for resistance.

Sugarcane is genetically complex, and the inheritance of disease resistance is not known. There are highly resistant, intermediate, and very susceptible cultivars (clones) (Scheffer and Livingston 1980). Other grasses, including Napier (*Pennisetum purpureum*) in Florida and Hawaii, lemongrass (*Cybopogon citratus*) in the Florida Everglades, and pearl millet (*P. glaucum*) are affected by *B. sacchari* or by closely related fungi (Scheffer 1989a); only isolates from sugarcane have had serious study. Widespread planting of sugarcane in new areas could endanger other crops and native flora.

There has been little speculation on the origin of *B. sacchari*. Morphologically, it is similar to several *Bipolaris* (*Cochliobolus*) species, including *C. heterostrophus,* which affects maize, and *C. miyabeanus,* which affects rice. The Cochlioboli (Helminthosporia) that affect graminaceous plants are genetically complex and sometimes interfertile; there are many possibilities for adaptive development.

A host-specific toxin (HS-toxin) from *B. sacchari* was discovered in Hawaii (reviewed by Scheffer 1989a); it was characterized as two galactose units on each side of a sesquiterpene core (Figure 12.8) (Macko et al. 1983). There were earlier reports on the chemistry of HS-toxin, but they were in error. Isomers of the toxin differ in activity against susceptible sugarcane; the most active is toxic at less than 1.0 μM. All isomers are selectively toxic only to those clones that are susceptible to the fungus (Livingston and Scheffer 1984a, 1984b). There also are analogs with fewer galactose units. The analogs protect susceptible tissue from the 4-galactose form (HS-toxin), although one of the 3-galactose analogs is highly toxic to some but not to all susceptible sugarcane clones (Livingston and Scheffer 1984a, 1984b). The galactose analogs are interconvertible in culture (Livingston and Scheffer 1984b; Nakajima and Scheffer 1987). Resistant sugarcane and other plant

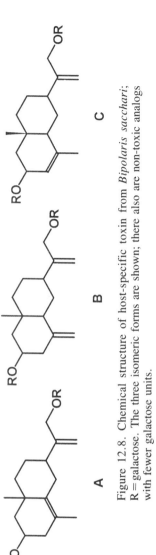

A   B   C

Figure 12.8. Chemical structure of host-specific toxin from *Bipolaris sacchari*; R = galactose. The three isomeric forms are shown; there also are non-toxic analogs with fewer galactose units.

species tolerate all these compounds, making possible the use of HS-toxin to screen for disease resistance.

Resistance to all known host-specific toxins clearly depends on insensitivity or inactivation of the respective toxin, although the mechanisms of toxicity and resistance have not been elucidated in most cases. A mechanism similar to that for HmT-toxin was proposed in 1974 for HS-toxin that affects sugarcane. The hypothesis could be essentially correct, but the supporting data could not be confirmed; the mechanisms for toxicity of HS-toxin are still uncertain (Lesney, Livingston, and Scheffer 1982).

The ecological consequences of HS toxin production by *B. sacchari* are clear enough. Toxin-producing ability led to widespread epidemics and to a vast increase in populations of *B. sacchari*. Even the seasonal incidence of the disease is determined by relative sensitivity of tissues to toxin; sugarcane becomes insensitive at temperatures above 32°C (reviewed by Scheffer 1989a, 1991). This explains why the disease in Florida and Hawaii is severe during the cooler season and tends to disappear during the hottest months. Heat-induced tolerance to toxins is known for several other diseases caused by Helminthosporiae and Alternariae; more detailed studies were conducted with *Periconia circinata* on sorghum (Chapter 8, section 3) (Bronson and Scheffer 1977). Heat-induced resistance also is a factor in some diseases not known to be mediated by toxins.

12.5. Toxin-Mediated Diseases as Models

Toxin-mediated and toxin-unknown diseases are comparable in many ways, including the following: the inception and development of infections as determined by histological methods, physiological responses of host tissue to infection, ecology and epidemiology, evolutionary development of pathogens and their host interactions, genetics of host resistance, and genetics of fungal pathogenicity. These similarities indicate that toxin-mediated diseases are representative models.

The inception of infection and the colonization of plant tissues by known toxin producers and by other pathogens have been examined by histological methods; both light and electron microscopy have been employed. Thorough studies on toxin-mediated cases include diseases caused by *C. victoriae* in oat (Paddock 1953), *C. carbonum* race 1 in maize (Jennings and Ullstrup 1957; Comstock and Scheffer 1973), and *C. heterostrophus* in maize (Contreras and Boothroyd 1975). There are numerous such studies on diseases not mediated by toxins; examples are from the work of Jennings and Ullstrup (1957), Skipp and Deverall (1972), and Mercer, Wood, and Greenwood (1975). All the studies confirm that host–pathogen interactions with the toxin producers are typical of virulent, specialized fungi in general. Later in the colonization process, tissue damage by the toxin producers can spread or become systemic. This does not happen with all toxin producers; for example, *C. carbonum* race 1 causes a defined leafspot (Comstock and Scheffer 1973). Systemic and widespread damage also occurs in some diseases not known to involve toxins – for example, with northern leafblight of maize, caused by *Exserohilum turcicum* (Jennings and Ullstrup 1957).

Fungal penetration of resistant host tissue often results in a hypersensitive response, with death of one or a few host cells. Hypersensitivity occurs when germinating spores of *C. carbonum* race 1 penetrate cell walls of resistant (*Hm*) maize, and when spores of toxinless mutants penetrate either susceptible (*hm/hm*) or resistant (*Hm*) host genotypes. When spores that lack toxin are supplemented with toxin in the infection droplet, hypersensitivity fails to develop. When toxin application is delayed after inoculation, allowing time for possible inhibitors to accumulate, the toxinless fungus resumes colonization (Comstock and Scheffer 1973). These observations appear to rule out phytoalexins (see Chapter 4, section 2) as significant factors in resistance to this disease. Later studies (Cantone and Dunkle 1990) confirmed this work in part but did not reexamine all aspects.

Oat leaves resistant to *C. victoriae* may or may not show hypersensitivity following penetration; the response may be conditioned by supplementary factors (Paddock 1953). These resistance responses of oats and maize to *C. victoriae* and *C. carbonum* are similar to those for many specialized and virulent fungi not known to be toxin producers.

The physiological responses of host tissue to infection by fungi and bacteria have been reported in many papers. Infection is known to cause increases in tissue respiration, dark fixation of CO_2, free amino acids, and phenol or ascorbic acid oxidase activity, plus changes in cellular permeability. These physiological effects are essentially the same for diseases caused by toxin producers and nonproducers. Such changes are induced also by application of host-specific toxin alone. These comparisons are discussed elsewhere in more detail (Scheffer and Yoder 1972; Scheffer 1976).

Genetic controls of fungal pathogenicity to plants are well known; both toxin producers and nonproducers have been examined. The studies with *Cochliobolus* species began with *C. sativus*, a complex species that contains many genotypes with diverse pathogenic potentials (Nelson 1961). Genetic studies were carried further with *Cochliobolus* species that produce host-specific toxins.

Toxin-producing ability in *C. victoriae* is controlled by the same Mendelian gene that determines selective pathogenicity. Toxin-producing isolates of *C. victoriae* were mated with toxinless isolates; the F_1 progeny had a 1:1 ratio of toxin producers and nonproducers. All toxin-producing progeny were selectively pathogenic to oats with the *Vb* (susceptible) gene, and all toxinless isolates were nonpathogenic to oats. Next, isolates of *C. victoriae* that produced toxin were mated with *C. carbonum* race 1, a producer of HC-toxin that affects maize with *hm/hm* genotype. F_1 progeny had a 1:1:1:1 ratio of oat-toxin (victorin) producers, HC-toxin producers, producers of both toxins, and toxinless isolates. This dihybrid ratio shows that production of each toxin is controlled by a single Mendelian gene. Progeny that produced both toxins were pathogenic to both oats and maize of the appropriate genotypes; isolates that produced neither toxin did not infect either host. There were no exceptions (Scheffer, Nelson, and Ullstrup 1967).

A single Mendelian gene controls toxin production by *C. heterostrophus* race T; this same gene also controls selective pathogenicity to maize with Tms cytoplasm (Bronson 1991). The genetics of toxin production and pathogenicity by other producers of host-specific toxins have not been determined. However, hundreds of isolates of *C. sacchari* were tested, and there was a complete correlation between toxin-producing ability and pathogenicity of isolates to sugarcane. Isolates that failed to produce toxin were nonpathogenic to sugarcane. Comparable data are available for production of host-specific toxins and pathogenicity of several *Alternaria* species, for *Periconia circinata,* and for *Phyllosticta maydis.*

Genetic control of host resistance–susceptibility and toxin insensitivity–sensitivity is even better known. In all cases, resistances to these fungi are controlled by the same genes that control insensitivity to the respective host-specific toxin. In oats, the reaction is controlled by one gene locus (*Vb*), with susceptibility being dominant. There are alleles with semidominance; some give intermediate levels of toxin sensitivity and fungal resistance. In maize, resistance to *C. carbonum* race 1 and insensitivity to HC-toxin are controlled by one dominant gene; there is a major gene locus with two or more alleles that give intermediate reactions (Scheffer 1976). Susceptibility to both *C. heterostrophus* race T and to *Phyllosticta maydis,* and to their toxins, are maternally inherited. Host resistance to the toxin-producing Alternariae (Chapter 13) is either dominant or recessive, depending on the host involved. Resistance to *Periconia circinata* is recessive (Scheffer 1989b); resistant plants are insensitive to PC-toxin (Chapter 8, section 3).

Genetic controls of resistance and susceptibility for a vast number of plant diseases that are not known to be mediated by toxins are well documented. The number of cases with dominant or with recessive genes for resistance are approximately equal (Scheffer, unpublished). Resistance controlled by two or more genes also is well known. There are several known cases of maternal inheritance of resistance–susceptibility (Hooker 1974).

Much of our understanding of the diseases mediated by host-specific toxins has come from molecular-genetic and biochemical studies. Such work has led to an understanding of these diseases exceeding that for most others diseases, and it supports the contention that toxin-mediated diseases are excellent models. Conclusive biochemical explanations of disease development, host-specificity, and disease resistance are now known for two diseases mediated by toxins (see Chapter 12, sections 2 and 3). There are biochemical explanations for some of these phenomena in leaf mold of tomato (Chapter 3, section 5, and Chapter 4, section 4), in fire blight (Chapter 3, section 5, and Chapter 8, section 2), in crown gall (Chapter 3, section 3), and in several other bacterial diseases (Chapter 3, section 5, and Chapter 4, section 4). Molecular-genetics studies of toxin-mediated diseases are reviewed in detail elsewhere (Walton and Panaccione 1993).

12.6. General Discussion of Diseases Mediated by Host-Specific Toxins

Host-specific toxins are significant factors in several diseases, and the list of such known toxins is growing slowly. However, it is unlikely that such toxins are involved in all diseases. On the other hand, there must be chemical explanations of virulence, pathogenicity, and host-selectivity. Chemical determinants are not necessarily toxic; there are limited data that suggest such (Xiao et al. 1990). Some toxins appear to have species rather than genotype selectivity; the significance of this has not been clarified (Stermer, Scheffer, and Hart 1984; Takai 1989), but it deserves more study. Compounds involved in disease may act at concentrations well below toxic or detectable levels, and this makes examination more difficult.

Few host-specific toxins are known, as compared to the vast numbers of plant diseases. It is possible that so few are known because of technical difficulties. Significant determinants could be part of cellular surfaces (for example, harpin from *Erwinia amylovora*; see Chapter 8, section 2) and may not be found easily in cell-free preparations. The selective toxins now known may simply be those that are produced in excess or that are stable enough to accumulate. Perhaps special media are needed for some potential toxins to accumulate in culture, as is the case with *P. circinata* toxin (Scheffer 1976). Toxin-sensitive plants presently known may have extreme susceptibility that usually is eliminated from plant populations; in unusual cases, susceptible genotypes were saved only because the gene for susceptibility also gave resistance to other diseases (Pringle and Scheffer 1964). Many disease-susceptible cultivars may actually be intermediates in the suscepti-ble-to-resistant scale or may have a masked reaction to toxin. This is evident with maize leafspot, caused by *C. carbonum* race 1; if research had started with intermediates, it is unlikely that HC-toxin would have been found. Breeding for susceptibility might help in our quest for understanding. It is likely that many other diseases are mediated by host-specific toxins or other compounds that have selective effects.

There are significant studies, at the molecular level, on virulence and avirulence of microbes pathogenic to plants. Researches on crown gall, to-mato leaf mold, fire blight, and several other bacterial diseases are conclu-sive and informative. Still, the most complete biochemical knowledge to date of pathogenicity, disease resistance, and host-selectivity is with dis-eases mediated by host-specific toxins. Explanations of pathogenicity, resis-tance, and host-selectivity based on phytoalexins and extracellular enzymes are still inconclusive (see Chapters 3 and 4).

12.7. Summary

Toxin-mediated plant diseases are similar in many ways to diseases without known toxins. Both have similar host–pathogen relationships; toxin-medi-ated and nontoxin diseases and their pathogens also have comparable ori-gins, evolution, ecology/epidemiology, and the same genetic systems con-trolling pathogenicity and resistance. The only obvious difference is knowledge of mechanisms in the toxin-mediated cases.

Certain species or genotypes in at least six fungal genera are known to produce low-mol-weight, host-specific toxins. At least seventeen such toxins are now known, and there are limited data for others. Some toxin-producing genera also have species or genotypes that are not known to produce selec-tive toxins, yet they incite diseases similar to those mediated by toxins. Still other forms of these and other fungal genera are either saprophytes or oppor-tunistic pathogens that affect senescent tissue of many host species. The rational conclusion is that the toxin producers evolved from opportunistic, nonspecialized pathogens and that the toxin producers are good models for study of a larger group of plant pathogens and the diseases they cause. Toxin-mediated diseases are now among the best understood of plant dis-eases in regard to biochemical mechanisms of disease development, disease resistance, and host-selectivity.

Toxic proteins are known to be involved in several diseases (see Chapter 3, section 1, and Chapter 8, section 2); at least two of these have host-specificity.

13

Monoculture:
Alternaria Diseases with Toxins

Eight or more pathotypes or races of *Alternaria alternata* are known to excrete compounds that are selectively toxic to suscept plants. Most pathogenic species of *Alternaria* affect dicotyledonous plants, in contrast to the Cochlioboli, most of which affect grasses (monocotyledons). Otherwise, there are many parallels between diseases caused by fungi in the two genera.

Pathogenicity and host-selectivity of these fungi is based on genetic adaptations that result in toxin production. Genetic change can occur via parasexuality and gene segregation or through transfer of genes from other races or species. Pathogenic Alternariae have one experimental disadvantage, compared with the Cochlioboli: Sexual stages are not known, making genetic analyses more difficult. However, the possibilities of nonsexual genetic interactions *in vivo* are good, as indicated by reports that several different forms of *Alternaria* were found in a single lesion on Japanese pear leaves (Adachi and Tsuge 1994). Isolates of *A. alternata* that produce both AK- and AM- (apple-affecting) toxins have been found (see Scheffer 1992). This suggests genetic recombinations via parasexuality.

13.1. Black Spot of Pear in Japan

The black spot disease of Japanese pear cv. Nijisseiki, caused by *A. alternata* f. sp. *kikuchiana* (= *A. kikuchiana*), was the first disease known to be mediated by a host-specific toxin. The Nijisseiki story began in 1888, when a thirteen-year-old boy in Chiba, Japan, found a pear seedling growing in a garbage dump; this was the origin of the delicious Nijisseiki (Twentieth Century) pear. It was soon widely planted, occupying more than half the area planted to pear in Japan. Then *kokuhan-byō* (black spot disease) (Figure 13.1) appeared; the causal agent, a virulent fungus, affects only Nijisseiki

Figure 13.1. Black spot symptoms caused by *Alternaria alternata* f. sp. *kikuchiana* on Japanese pear, cv. Nijisseiki. (Picture from S. Nishimura)

and a few similar pear cultivars (Nishimura and Kohmoto 1983; Scheffer 1992). Black spot was well known by 1917 in all areas of Japan and Korea where Nijisseiki pears were grown (Tanaka 1933). Since then, the Nijisseiki pear has been maintained at great cost by the use of fungicides.

A host-specific toxin (AK-toxin) that affects only the Nijisseiki pear was discovered by Tanaka (1933). The report stimulated intense and sustained research, and AK-toxin has become one of the models for work on host-specific toxins. Its chemical structure was finally elucidated by Nakashima, Ueno, and Fukami (1982), who reported two closely related forms of a complex molecule that contains an epoxide group, one N, and an aromatic group (Figure 13.2). Most pear cultivars and other species are resistant to

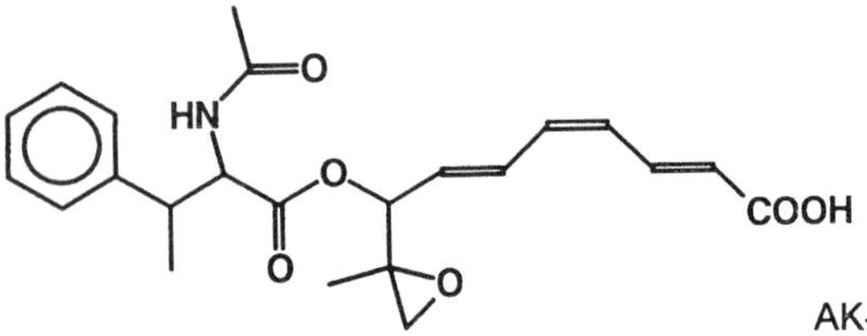

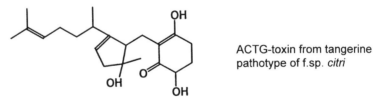

AK-toxin 1 from f.sp. *kikuchiana*

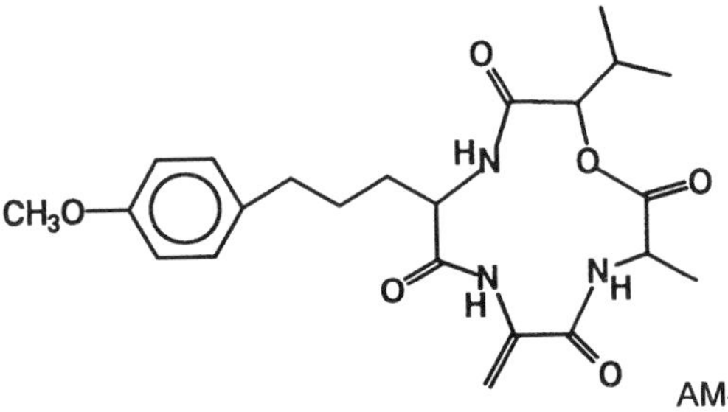

AM-toxin 1 from f.sp. *mali*

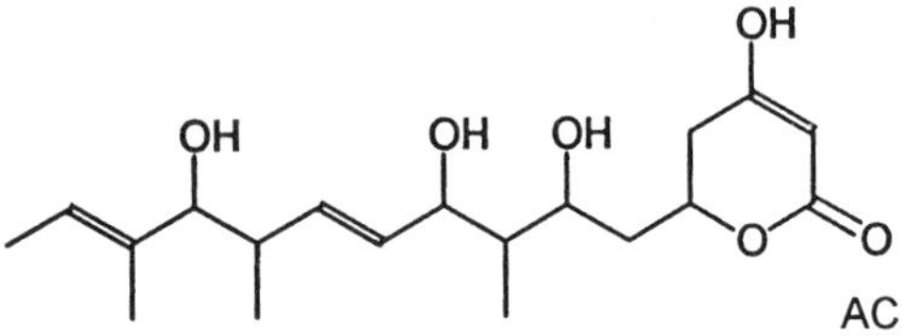

ACR-toxin from rough lemon pathotype of f.sp. *citri*

ACTG-toxin from tangerine pathotype of f.sp. *citri*

Figure 13.2. Chemical structures of host-specific toxins from several f. sp. of *Alternaria alternata*.

the fungus and insensitive to AK-toxin, which causes necrosis in tissues of susceptible pears at 10^{-8}M (Nishimura and Kohmoto 1983). Susceptibility is controlled by one dominant gene.

The origin of black spot disease became a topic of speculation following its sudden appearance. The rational hypothesis is a mutation or other genetic change in the fungus, resulting in production of host-specific toxin. Such metabolites can give the producer an ecological advantage, leading to explosions of pathogen populations and to new disease problems (Nishimura and Kohmoto 1983).

Adaptability of *A. alternata* is evident in development of tolerance to toxicants. Polyoxin, an antibiotic fungicide, was used for many years to control black spot. Then, in 1971, strains of the fungus appeared that were tolerant of polyoxin (Nishimura, Kohmoto, and Udagawa 1976); this was the first example of development of tolerance to a fungicide in Japan. Tolerant strains of the pear pathogen became common, and polyoxin became useless for control of black spot. Tolerant strains were still present ten years after polyoxin was discarded (Udagawa et al. 1983).

AK-toxin has rapid and drastic effects on cell membranes, causing losses of electrolytes and organic materials from susceptible cells within minutes after exposure (Nishimura and Kohmoto 1983). However, a direct effect on the plasmalemma has not been demonstrated. Electrolyte losses also are evident in tissues colonized by the fungus. Several other toxic compounds from the fungus did not induce such leakage, suggesting that they are not significant mediators in disease.

Sensitivity of pear tissues to AK-toxin and susceptibility to the fungus are affected by heat. There is a much lower incidence of new infections by the fungus during the heat of summer, as compared with spring; this is correlated with sensitivity to AK-toxin. Laboratory experiments showed that sensitivity of pear leaves is less at elevated temperatures; two seconds in water at 55°C or sixteen hours in air at 35° caused tissues to become insensitive to toxin. Heat treatments at these temperatures caused no obvious damage to the leaf (Otani, Nishimura, and Kohmoto 1974). New leaves are much more sensitive than old leaves to toxin, and this also is correlated with seasonal incidence of disease (Nishimura et al. 1978). These findings are further confirmations that AK-toxin has a significant role in disease development.

Still other observations and data confirm the central role of AK-toxin. Spores of the fungus release toxin on germination, and toxin suppresses both resistance and the immediate responses of tissues to disturbance (Hayami et al. 1982). Both toxin-producing and toxin-minus isolates of the fungus will penetrate both susceptible and resistant plant tissues; only the toxic form develops further in metabolically active tissues, and only in susceptible host genotypes. The toxin-minus form will colonize senescent or stressed tissue. Release of toxin by spores and suppression of tissue responses are significant factors in understanding how the black spot fungus causes disease.

13.2. Alternaria Blotch of Apple

A serious new disease of apple was discovered in 1956 in the Iwate prefecture of Japan. The disease was caused by a virulent and host-specific form of *A. alternata* f. sp. *mali* (= *A. mali*). Blotch first became a problem on apple cultivars Indo and Delicious (Sawamura and Yanase 1963) that were recently introduced in Japan; the disease soon spread throughout Japan to cultivars intermediate in resistance and, finally, to some cultivars previously thought to be resistant. Alternaria blotch is now the most important apple disease in Japan. Later, this toxin-producing fungus was found in North Carolina (Filajdić and Sutton 1991). Opportunistic forms of *A. mali* had been present in both countries for many years prior to the sudden appearance of the cultivar-specific and virulent race. The most rational hypothesis for both Japan and the United States is mutation or other genetic change to a toxin-producing form of the fungus. There are no genetic studies on toxin production in the Alternariae, but the ability of similar fungi to produce host-specific toxins is known to be under the control of single Mendelian genes.

Selective toxicity of culture filtrates from the blotch fungus was demonstrated soon after the disease appeared (Sawamura and Yanase 1963). At least three closely related compounds with selective toxicity are now known (Ueno et al. 1975a, 1975b); they are cyclic peptide analogs (Figure 13.2) that can differentiate between susceptible, intermediate, and resistant cultivars of apple. The major form (AM-toxin I) induces necrosis in susceptible apple tissues at 10^{-8}M; intermediate apple types are affected at 10^{-4}M (see Scheffer 1992), and resistant cultivars tolerate 10^{-4}M. AM-toxin also causes veinal necrosis in Nijisseiki but not in Chojuro pears; however, parasitic fitness of the apple fungus to pear was less than that of AK-toxin producers (Maeno et al. 1984).

Apple cultivars vary in resistance to the fungus and to AM-toxin. The following categories in the susceptible-to-resistant scale are evident: (a) very susceptible (cvs. Indo, Orei, and Red Delicious), (b) intermediate (cvs. Jonathan, Golden Delicious, McIntosh, and Mutsu), and (c) resistant (cvs. Gala and Splendor). Relative sensitivity to toxin and resistance or susceptibility to the fungus are correlated. Three or more genes are thought to be involved in the different levels of resistance. Some cultivars of Japanese pear are susceptible to the apple race of the fungus and sensitive to AM-toxin, whereas other pear cultivars are resistant to both apple and pear races of *Alternaria* and are insensitive to both toxins (Maeno et al. 1984; Scheffer 1992). Nijisseiki pear also is sensitive to the AF-toxin from an *Alternaria* that affects strawberry.

The primary site of action by AM-toxin appears to be the plasma membrane (Kohmoto et al. 1982), but the evidence is circumstantial rather than direct. There are some puzzling findings regarding tissue sensitivity. Nongreen tissues, especially flower petals, are less sensitive than are green tissues; this is true also for tentoxin, a nonspecific but related compound from an opportunistic form of *Alternaria* (see Chapter 13, section 8). Thus, AM-toxin may have a second site of action, in the chloropast. Other puzzling aspects involve differences in tissue sensitivities among cultivars. The following tissue sensitivity categories have been observed: (a) cultivars with only green tissues that are sensitive, (b) cultivars with both green and nongreen tissues that are sensitive, and (c) cultivars that are insensitive in both green and nongreen parts. Cultivars in these categories are susceptible, moderately resistant, and resistant (respectively) to the fungus. Multigenic control of resistance may be involved (Kohmoto, Nishimura, and Otani 1982).

Another puzzling effect is that of light on action of AM-toxin. Five hours or more of darkness are needed for full expression of toxicity by AM-toxin; continuous light exposure following toxin treatment results in loss of sensitivity. However, light does not affect invasion of tissues by the fungus or losses of electrolytes after infection or after toxin exposure; it inhibits only lesion development. These observed effects could be related to a site of action in the chloroplast (Kohmoto and Otani 1991).

Pathogenic potential of toxin-producing and opportunistic (nontoxic) isolates of *A. mali* were compared (Sawamura and Yanase 1963). The toxin producers colonized and caused necrotic spots on young leaves and fruit. Opportunistic isolates did not infect intact young leaves, but they invaded points of injury on both apple and pear leaves and caused rot when inserted into ripe apple fruit. Old leaves were much more resistant to infection by the toxin producers than were young leaves. These observations support the hypothesis that toxin producers are variants of an opportunistic form of *A. alternata* (=*A. mali*). Generally speaking, opportunistic pathogens colonize only senescent, stressed, or injured tissues.

13.3. Black Spot of Strawberry

Strawberries are popular in Japan and have been bred intensively. Around 1970, a superior new cultivar, Morioka-16, was introduced; it soon became popular and was widely planted. By 1975, a new disease, known as black spot, was present in northern Japan; the cause was a previously unknown form of *Alternaria* that affected only the Morioka-16 strawberry (reviewed by Scheffer 1992). Nishimura and associates identified the fungus as a race or pathotype of *A. alternata* that produces a toxin specific to the Morioka-16 strawberry. However, inoculation experiments showed that the fungus also infected Nijisseiki pears and two apple cultivars (Indo and Delicious). The strawberry black spot fungus has been found to date only in Japan and Korea.

Toxin from cultures and from germinating spores was specifically active only against Morioka-16 strawberry, Nijisseiki pear, and the susceptible apples. Fungal isolates that lost pathogenicity also lost toxin-producing ability. Purified toxin was lethal to strawberry leaves at 0.1 ug/ml but was labile. The producing fungus is thought to be a new toxin-producing variant of *A. alternata* (Maekawa et al. 1984; Yamamoto et al. 1984).

Three related forms of AF-toxin have been described (AF-toxins I, II, and III). All are esters of epoxydecatrieonic acid (Figure 13.2), and there are stereoisomers of each. These structures are similar to those of AK-toxin and ACT-toxin. An isolate of *A. alternata* was found that produced three toxins and was pathogenic to apple, pear, and strawberry (reviewed by Scheffer 1992). This isolate could be a product of parasexuality.

There are many complexities in biological activities of several host-specific toxins from the Alternariae, and they appear to be correlated with slight differences in chemical structures. AF-toxin I causes necrosis in leaves of Morioka-16 strawberry and Nijisseiki pear; AF-toxin II affects pear but not strawberry; and AF-toxin III is very toxic to Morioka-16 strawberry and somewhat toxic to pear (Maekawa et al. 1984). In contrast, AK-toxins are highly selective against pear, with no effect on strawberry (Otani et al. 1985). In inoculation experiments, these toxin producers infected all hosts that were sensitive to their toxins. However, the strawberry *Alternaria* was seldom isolated from pears in orchards, possibly because other factors affect parasitic fitness (Hayashi and Nishimura 1989).

Data on spore germination and possible interaction of toxin variants are of interest. Spores release AF-toxins during germination; up to 0.08 pg of AF-toxin I was produced by a spore in six hours, along with a detectable amount of AF-toxin II. This is important because it prevents the host cell from recognizing the fungus as "non-self" (Hayashi et al. 1990). The AF-toxins are thought to have their primary sites of action in the cell membrane. AF-toxin II, to which strawberry is not sensitive, gave some protection of strawberry tissues against AF-toxin I, which suggests competition for receptor sites (Namiki et al. 1986).

Spores of several isolates of *A. alternata* were reported to release a high-mol-weight metabolite(s) that causes strawberry leaf tissues to resist infection. Treatment with a solution that contained both the resistance inducer, which has not been characterized, and a host-specific toxin resulted in infection; the appropriate toxin appears to suppress resistance (giving compatibility) even to pathogens that do not ordinarily affect strawberry (Yamamoto et al. 1984). This report needs confirmation and further study because susceptibility to plant diseases, in general, appears to be correlated with suppression of general reactions to disturbance.

Susceptibility to strawberry *Alternaria* and sensitivity to its toxin are controlled by the same semidominant gene with two alleles (Yamamoto et al. 1985). Thus, AF-toxins can be used to screen for disease resistance, using intact seedlings. There is also the possibility of use in cell culture work.

13.4. *Alternaria* Diseases of Citrus

An opportunistic form of *A. alternata* affects senescent tissues of many citrus species. This form (*A. alternata* f. sp. *citri*) appears to have spawned several virulent and host-selective races or pathotypes. The first such specialized pathotype was found on Emperor mandarin (*Citrus reticulata*) in Australia; the same race appeared later in Florida on Dancy tangerine, a cultivar closely related to Emperor mandarin. Filtrates from cultures of both the Australia and Florida fungi proved to be selectively toxic to tissues of the hosts (Kohmoto, Scheffer, and Whiteside 1979; Scheffer 1992). The toxic compounds have been isolated and characterized (Kohmoto and Otani 1991; Kohmoto et al. 1993).

Several different host-specific compounds are produced by the tangerine pathotype of *A. alternata* f. sp. *citri*. Two groups are now recognized; they are designated ACTG-toxins and ACT-toxins I and II (Kohmoto & Otani 1991). The ACT-toxins are methyldecatrienoic acids (Figure 13.2) (Otani and Kohmoto 1992) that affect sensitive citrus tissues at 10 ng/ml; they are similar chemically to AK-toxin, affecting Nijisseiki pear, and AF-toxins, affecting strawberry. The ACTG-toxins (A, B, and a related harmless form, C) differ from the ACT-group in chemical structure (Figure 13.2).

A second race or pathotype of *A. alternata* f. sp. *citri* was found in Florida causing leafspots on rough lemon that was used for rootstocks. The disease was shown in 1937 to be caused by a specialized strain or race of *A. citri* (= *A. alternata* f. sp. *citri*). This host-specialized and virulent race affects young leaves of rough lemon and Rangpur lime, but not tangerine and other citrus, in contrast to the nonspecialized form of the fungus that

Figure 13.3. Selective effects of host-specific toxin from *Alternaria alternata* f. sp. *citri* (tangerine pathotype) on leaves of host (Dancy tangerine, center) and nonhost (rough lemon, left) plants. Control without toxin is shown on the right. A small droplet of toxin solution was placed at the center of the blackened area. (Kohmoto, Scheffer, and Whiteside 1979)

affects senescent leaves of many citrus species. Next came the discovery that culture filtrates from the rough-lemon fungus are selectively toxic to rough-lemon leaves, not to tangerine. In contrast, filtrates from the tangerine fungus are toxic to Dancy tangerine, not to rough lemon (Figure 13.3). All these isolates are morphologically identical. The story was carried further by Kohmoto, Scheffer, and Whiteside (1979), who isolated the host-specific toxins of each pathotype and demonstrated selectivity of purified toxins.

Chemical characterization of toxins specific to rough lemon was reported in 1985. The major toxin is a 19-carbon polyalcohol with a 2-dehydropyron ring (Figure 13.2); this compound is highly active (ED_{50}, 18–36 ng/ml) and is very selective, affecting only rough lemon and Rangpur lime, among sixty-five citrus species tested. Toxin-sensitive citrus species are the only known hosts of the fungus (Gardner, Kono, and Chandler 1986; Kohmoto and Otani 1991). Rough-lemon toxin is much more selective than are the tangerine toxins. Four minor lemon toxins, chemically similar to the major toxin, were found. All are mitochondrial poisons, comparable in this respect to another long chain compound, the host-specific toxin from *Cochliobolus heterostrophus* race T (Chapter 12, section 3) (Scheffer and Livingston 1984). Light suppresses the effects of rough-lemon toxin on susceptible leaves (Akimitsu et al. 1994).

The rough-lemon pathotype appears to be a virulent and host-specialized race of *A. alternata* f. sp. *citri.* Genetic analyses and DNA homologies have not been done, but there are many parallels with similar species of Co-chlioboli. Tentative conclusion is that acquisition of genes for production of host-selective toxins converts opportunistic pathogens to virulent and host-specific forms.

Complex effects of the several toxins from *A. alternata* f. sp. *citri* are confusing. One form of the tangerine toxin (ACT-toxin Ib) is toxic to the Nijisseiki pear, not previously known to be affected by the tangerine race in the field. ACT-toxin IIb, the 5″-deoxy derivative cf toxin Ib, is much less toxic than is Ib to Dancy tangerine, but toxin IIb is ten times more toxic than Ib to Nijisseiki pear (Kohmoto and Otani 1991); IIb is as active against pear as is AK-toxin. However, AK-toxin does not affect the Emperor mandarin, as do both groups of tangerine toxins. One isolate of *A. alternata* from citrus was found to be pathogenic to both Nijisseiki pear and Emperor mandarin and to produce ACT-toxins. The Nijisseiki pear is now known to be sensitive to several different host-specific toxins. Sensitivities of strawberry, tangerine, and several other hosts to toxins are much more specific and restricted.

There are data and speculations on synthesis, sites of action, and consequences of toxin production by tangerine, pear, and strawberry pathotypes of *A. alternata.* The ACT-, AK-, and AF-toxins probably are produced via a common biochemical pathway. Variations in toxin molecules appear to be important in host-selectivity; slight changes in structure can result in toxicity to new hosts. ACT-, AK-, and AF-toxins are chemically related and may be similar in modes of action; plasma membranes appear to be primary sites (Kohmoto and Otani 1991). ACT-toxin IIb is released from toxin-producing isolates during spore germination. Addition of toxin IIb to the site of infection allowed a toxin-minus isolate to colonize susceptible citrus (Kohmoto et al. 1993).

13.5. Alternaria Stem Canker of Tomato

This disease was first found about 1970 in fresh market tomato plants near San Diego, California. The causal agent was found to be a new pathotype

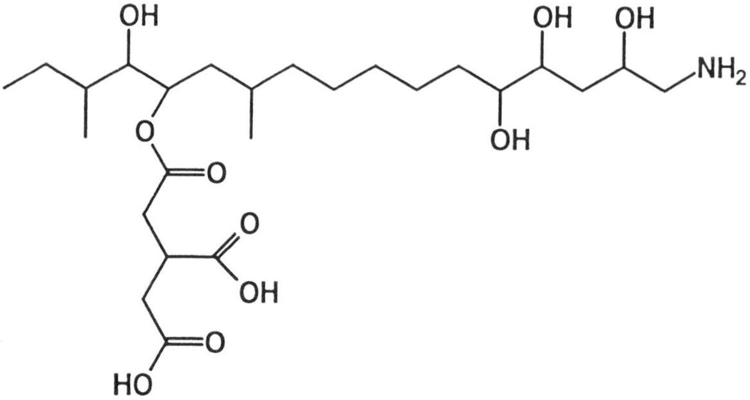

Figure 13.4. Chemical structure of the major form of host-specific toxin from *Alternaria alternata* f. sp. *lycopersici*.

(*A. alternata* f. sp. *lycopersici*), which attacks only a few cultivars of tomato, notably Earlypac-7, which was widely planted in southern California. Most cultivars are resistant, and Earlypac was soon replaced by resistant types. The fungus produces a host-specific toxin with the same plant specificity as the fungus itself (reviewed by Scheffer 1992). The stem canker disease next appeared in 1977 in Japan on a popular greenhouse-grown tomato (cv. First). The California and the Japanese isolates of the fungus had the same morphology and produced the same toxin (AL-toxin) (Kohmoto et al. 1982). Origins in California and Japan may have been from two different genetic events, possibly mutations. The new fungus, which spread rapidly in the genetically uniform crop, became a serious problem in both areas.

Genetic analyses showed that susceptibility to the fungus and sensitivity to its toxin depend on the same gene. Resistance is semidominant, with two alleles at the resistance locus; thus, there are intermediate levels of resistance. Susceptibility to the fungus was not separated from sensitivity to the toxin in eight inbred generations (reviewed by Scheffer 1991).

There are two related forms of AL-toxin; they are esters of 1-amino-11, 15 dimethylheptadeca-2, 4, 5, 13, 15-pentol (Figure 13.4). Each ester has two isomers, all found in cultures and in infected tissue; all forms are equally toxic (at 25 ng/ml) and equally selective (Kohmoto, Nishimura, and Otani 1982; Siler and Gilchrist 1983; Gilchrist and Harada 1989). No other races or species of *Alternaria* produce the toxin, although related compounds from *Fusarium* are known. There have been studies on sites of action in tomato, but to date there are no definitive conclusions (Gilchrist and

Harada 1989). Several other *Alternaria* toxins disrupt the plasma membrane, but this may not be the case with AF-toxin. A mitochondrial site seems probable (Kohmoto and Otani 1991).

The tomato stem canker disease fits the usual pattern for new diseases; there frequently are sudden outbreaks following introduction of genetically altered crops.

13.6. Brown Spot of Tobacco

Brown spot disease was discovered in 1924 on leaves of flue-cured tobacco in Florida. Leaves of tobacco of this type are left on the plant to mature before harvest. The tobacco-infecting race of *A. alternata* (= *A. longipes*) in the United States lacks virulence on young leaves; the fungus first infects lower leaves that are senescent and then moves up the stalk as leaves mature, as is typical of opportunistic pathogens (see Scheffer 1992). The *Alternaria* disease of tobacco in the United States is damaging only because it reduces the value of the harvested leaf.

A new race of *Alternaria* on tobacco was found in Japan about 1980; this race produces a host-specific toxin (AT-toxin) that was purified, but its structure has not been determined to date. The toxin affects only plants of *Nicotiana tabacum*. The primary site of action appears to be in the mitochondrion (Kohmoto and Otani 1991).

Literature on the brown spot disease is confusing. Japanese workers used a new and virulent race of the fungus, whereas older work in the United States was with a generalized, opportunistic form of the fungus. Japanese isolates infect young leaves; the United States isolates attack only senescent leaves. There are no reports of host-specific toxins from the tobacco pathogen in the United States.

13.7. Other *Alternaria* Diseases with Toxins

A host-specific toxin has now been found for *A. tenuissima*, which infects certain cultivars of pigeon pea (*Cajanus cajan*). The toxin, not yet characterized, causes rapid changes in mitochondria in susceptible, but not in resistant, cultivars (Nutsugah et al. 1993, 1994).

A. brassicae was reported to produce a toxin with host-species rather than genotype specificity. *Brassica nigra* and *B. campestris* (rapeseed, or canola) are susceptible to the fungus and sensitive to the toxin. *B. juncea, B. hirta*, and *B. napus* are intermediate in both resistance to the fungus and sensitivity to the toxin; *B. rapa* and other *Brassica* species are resistant and low in toxin sensitivity. There still are reservations about this toxin because of low toxicity (toxic at 15–30 mg/ml) and because resistant *B. rapa* tolerated only ten-fold-higher concentrations than did susceptible *B. nigra*. The toxic compound is a low-mol-weight peptide with the same structure as destruction B, the product of a fungus that attacks insects (reviewed by Scheffer 1992).

A possible toxin with species rather than genotype specificity is produced by *A. eichhorniae*, a destructive pathogen of water hyacinth in India. Culture filtrates were selectively toxic to water hyacinth (*Eichhornia crassipea*) and *Monochoria vaginalis,* both hosts of the fungus; filtrates did not affect a number of nonhost plants. The work was confirmed in part, and a species-specific toxin was isolated (reviewed by Scheffer 1992).

Recently, alternaric acid from *A. solani,* a potato and tomato pathogen, was shown to have limited host-specificity. The data support a role for alternaric acid as a disease determinant, but the results are still inconclusive (reviewed by Scheffer 1992).

Host-specific toxins from these several pathotypes of *Alternaria* suggest that all virulent and host-specific forms should be examined for toxins. The most promising for experimentation are those that have both resistant and susceptible host genotypes, especially those with single-gene control of host reaction. Many *Alternaria* pathotypes fit these criteria.

Wide differences in pathogenic potentials in species, races, and genotypes of *Alternaria,* ranging from saprophytes to virulent and host-specific pathogens, are believed to reflect an evolving situation. Studies with several host-specific Alternariae have demonstrated the chemical bases of virulence and selectivity. Host specialization can lead to genetic isolation and other adaptations; this, in turn, can lead to speciation.

13.8. Toxin Production by a Nonspecialized Alternariae

Certain isolates of *Alternaria alternata* (=*A. tenuis*) cause rot in seeds and seedlings of various plant species. Under certain conditions, such infections result in striking chlorosis in young plants. This effect is most apparent in cotton and citrus seedlings, which have received most of the study. The chlorosis is induced by tentoxin, a cyclic tetrapeptide released by the fungus. Most isolates of *A. alternata* that produce tentoxin are low-grade or opportunistic pathogens that affect seeds and seedlings. The fungus is present on and in seeds; chlorosis becomes evident when seedlings emerge. However, seedlings appear to become chlorotic only at a critical stage in development; if they survive, they carry yellow spots or bands on the leaves, but tissues that develop later are not affected. There is some host-specificity in chlorosis development; for example, cucumbers are yellowed, but cabbage is not. Isolated tentoxin also affects only certain host species, at certain stages in development. Sensitivity to tentoxin is inherited maternally. Tentoxin is produced also by some isolates of the specialized forms of *A. alternata* that affect apple and citrus, but the significance of tentoxin in these diseases is unknown (reviewed by Scheffer 1992).

Tentoxin interferes, directly or indirectly, with many metabolic processes. It decouples normal plastid development, leading to chlorosis. A direct site of action is on coupling factor 1 of photosynthesis, making tentoxin a specific enzyme inhibitor. Tentoxin clearly is a determinant in these rather nonspecialized diseases; it is highly toxic, has selective effects at certain sites, and has some species selectivity (reviewed by Scheffer 1992).

Many other toxic metabolites from *Alternaria* species are known, but few of them have been evaluated adequately for roles in disease.

13.9. Summary

Research on diseases caused by specialized *Alternaria* species has increased our understanding of the biochemistry of disease development and resistance. It also has contributed to concepts of microevolution in pathogens and to a basic understanding of the epidemiology and ecology of plant disease.

Form species, races, and genotypes of *A. alternata* vary from saprophytes to opportunistic pathogens to virulent host-specialized pathogens, suggesting evolutionary lines. Several specialized pathotypes have been studied by Japanese scientists; pathogenicity and host-specificity in these are based on release of highly active compounds (fully characterized) that selectively affect only hosts of the producing fungus. Each *Alternaria* pathotype attacks a different host; isolates from Japanese pear, apple, Dancy tangerine, rough lemon, strawberry, tomato, tobacco, and, possibly, *Brassica* species and water hyacinth produce host-specific toxins.

The host-specific toxins from *Alternaria* differ in chemical structure, relative toxicity, plants affected, and mechanisms of action. Target sites in the cell appear to be plasmalemma and mitochondria. The *Alternaria* models are remarkably similar to those discussed for *Cochliobolus* diseases; the Alternariae and Cochlioboli diseases are now among the best understood of plant diseases.

14

Diseases Amplified
by Changes in Agriculture

There is a lack of appreciation that even minor changes in crop or forest management can have significant effects on the severity or incidence of plant diseases. Changes in the following procedures can affect disease: method of propagation or harvest, seed treatment, fire protection or lack of it in forests, thinning in managed forests, mechanization, rotations, fertilization, choice of cultivars, and others. Some significant examples are discussed.

14.1. Tristeza of Citrus

Tristeza is the outstanding example of how a change in agricultural technique can lead to serious disease problems. The technique was grafting sweet orange (*Citrus sinensis*), our most popular citrus type, to roots of sour orange (*C. aurantium*). The purpose of grafting was to solve an earlier threat from Phytophothora root rot; sour orange is resistant. Disaster occurred when tristeza virus was moved, with trade, into new centers of citrus production where grafted plants were widely used.

Species and modern cultivars of citrus are closely related and are subject to many of the same diseases. Wild citrus species are found in southeastern Asia, northern Australia, and the nearby islands of the Indies. Citrus species in southeastern Asia were domesticated at a very early time. Genetic analyses indicate that the many modern species probably originated as natural hybrids between three species: citron (*Citrus medica*), mandarin (*C. reticulata*), and pumello (*C. grandis*) (Bar-Joseph, Marcus, and Lee 1989). Desirable hybrids or species were soon spread over the world. Southeastern Asia probably also was the endemic home of the tristeza virus (Wallace, Oberholzer, and Hofmeyer 1956), which had no more than a mild effect on native citrus species in Asia.

210

The major production centers of citrus are now in the United States, Brazil, Spain, Israel, and South Africa; all are areas where *Citrus* is an exotic or introduced plant. Sweet orange plantings in these areas were mostly on sour orange roots; the plantations were established before tristeza became known.

The very destructive tristeza, or "sad disease," now occurs worldwide wherever citrus is grown. Oranges, lemons, limes, and other citrus types were moved originally by seed to all parts of the tropics and subtropics. Tristeza virus is not seed-borne; hence it was left behind in subtropical China and the East Indies. Beginning in 1890, and especially since 1930, a thriving trade in citrus plants and budwood developed; the virus was carried on plants of desirable cultivars into most citrus-growing areas that previously had been free of it. The results were disastrous, as we might now expect. Worldwide, tristeza probably destroyed 50 to 60 million trees in sixty years; further losses occurred from decreased production by surviving trees. Argentina, Brazil, California, Spain, and South Africa were especially hard hit (Bar-Joseph, Marcus, and Lee 1989).

The most severe expression of tristeza occurs in plants with sweet orange (*C. sinensis*) scions (shoots or buds) grafted to sour orange rootstocks. The problem with sour orange roots was first noticed in 1915 in South Africa, where the cause was thought to be incompatibility between sweet and sour orange tissues. The hypothesis was soon discarded.

The first destructive outbreak of tristeza occurred in 1930 in Argentina, on newly established plantations of sweet orange (Bar-Joseph, Marcus, and Lee 1989). The outbreak followed import of nursery trees from South Africa. The disease, cause then still unknown, also was observed between 1924 and 1940 in Java, Uruguay, Brazil, Paraguay, California, and Australia. A form of tristeza probably was in Florida by 1914, certainly by 1942, but disease in Florida was mild until much later. There was at least one known example of infected budwood that came to Florida from South Africa (Wallace, Oberholzer, and Hofmeyer 1956). Tristeza was the descriptive name coined in Brazil; in California, the disease was known as "quick decline." Proposed causes included soil acidity, nutrient deficiencies, soil toxins, and nematodes.

Finally, scientists at the Citrus Research Station in Riverside, California, reported that quick decline was transmitted by bud grafts, thus implicating a virus (Fawcett & Wallace 1946). An aphid vector was discovered in Brazil, also evidence of viral causation (Bennett and Costa 1949). Aphid vectors were soon found wherever tristeza occurred.

Symptoms of tristeza can vary from mild to severe, depending on strain of the virus, host cultivar or species, and temperature. The most severe manifestation, quick decline, is mainly a disease of susceptible citrus grafted to sour orange rootstocks. Symptoms of quick decline include phloem necrosis at the scion union, sudden wilting, yellowing, leaf desiccation, and pitting beneath the bark. Infected seedlings also show vein clearing and flecking in leaves, plus stunting. Phloem necrosis probably is a result of hypersensitivity of sour orange tissues. This notion is based on the fact that sour orange plants on their own roots are difficult to inoculate; the combination plant (sweet orange on sour orange roots) is inoculated through the sweet orange part. Symptoms can be suppressed in seedlings at high temperatures; the virus often is present in symptomless plants of various citrus types (Roistacher et al. 1974). Sweet orange and mandarin plants usually escape severe damage when grown on their own roots (McClean and Van der Plank 1955).

The tristeza virus is a closterovirus, a slender flexous structure about 2,000 nm long, containing coat protein (23,000 ^{m}r) and a 20-kb RNA core (Lee et al. 1988). Virus is located in phloem cells. The best source for tristeza virus isolation and purification is bark and fruit (Tsuchizaki, Sasaki, and Saito 1978).

Purified virus was used to develop reliable assays for detection of tristeza *in planta* and to identify virus strains. New assays include the use of polyclonal antibodies and immunoabsorbance (Vela et al. 1986). Viral RNA was cloned, and libraries of cDNA were prepared in Florida and Israel. These modern assay methods made possible the detection of tristeza virus in symptomless plants (Bar-Joseph et al. 1979; Rosner, Lee, and Bar-Joseph 1986), and the identification of viral variants in citrus populations (Moreno and Muñoz 1990). Virulent strains of tristeza, identified by nucleic acid technology, were finally detected in Florida, an area previously infested only with a mild form.

The oriental black citrus aphid (*Toxoptera citricida*) is the most efficient vector of tristeza virus (Rocha-Peña et al. 1995). This insect occurs in South America, South Africa, Asia, and Australia but is unknown in the United States, where the melon aphid (*Aphis gossypii*) is the main vector (Grant, Klotz, and Wallace 1953). An aphid can carry the virus for at least twenty-four hours after feeding on an infected plant but usually loses the virus by forty-eight hours. Aphid transfer efficiency varies, depending on viral strain. The insect can transmit more than one strain simultaneously, and a single infected tree can contain more than one viral strain (Raccah, Loebenstein, and Singer 1980). The melon aphid is capable of transmitting many different tristeza virus strains from various parts of the world. The recent high rate of new infections in Florida, including those from virulent strains, was associated with increased populations and spread of the vector (Yokomi et al. 1989). The virus also can be spread by pruning instruments (Garnsey, Gonsalves, and Purcifull 1977).

Spread of tristeza over the world is well documented. The disease was in South Africa by 1915 but probably was present before 1900; presumably, it came from southeast Asia to South Africa. Over the years, citrus plants were moved many times into new areas from South Africa, thus spreading tristeza. There was a sudden outbreak of tristeza in the San Gabriel Valley of California in 1930; virus for this incident probably was brought from South Africa on citrus plants for experimental purposes. Virus-infected grapefruit (cv. Cicily) from South Africa was growing in Riverside, California, since 1924; there is circumstantial evidence for much earlier occurrence of tristeza there. Scattered trees with tristeza were found in Florida in 1953, but the virus probably was there by 1914 (Wallace, Oberholzer, and Hofmeyer 1956). Plants and budwood of some citrus cultivars can carry the virus with no evident symptoms, which explains some of the rapid spread of tristeza.

There is convincing circumstantial evidence for an Oriental origin of tristeza. The Meyer dwarf lemon, a popular dooryard plant, was brought to California from China in 1908. Meyer lemon plants in China always have the virus, often without symptoms; tristeza probably first came to California in Meyer lemon plants. Presence of the virus may account for early reports that cv. Meyer and sour orange were incompatible in California (Wallace, Oberholzer, and Hofmeyer 1956). In Texas, satsuma oranges growing near tristeza-infected Meyer lemons had tristeza by 1954; other oranges, grapefruit, tangelo, and kumquat growing nearby were tristeza-free, but the disease was spreading slowly. There were similar situations in other places. All the earliest known infections appear to trace back to imports from Asia, mostly from China (Wallace, Oberholzer, and Hofmeyer 1956). Later movements were from South Africa.

The earliest research on tristeza control included tests of various citrus types as rootstocks. This research began when the U.S. Department of Agriculture sent a researcher to Brazil to study tristeza as a potential threat to the citrus industry in Florida (Bennett and Costa 1949). Many cultivars of grapefruit, pummelo, shaddock, mandarin, acid lemons and limes, tangelos, and others were tested as rootstocks, especially for sweet orange, in a cooperative, worldwide program. Roots of each type were classified as either tolerant or intolerant of tristeza. Tolerant rootstock types included most mandarins and their hybrids, trifoliate orange, sweet orange, and rough lemon; sweet orange grew well on these rootstocks, even in the presence of the virus (Grant, Klotz, and Wallace 1953). Even at present, the most important controls of tristeza are the use of tolerant rootstocks plus the use of budwood certified as free of the virus.

The early work from Brazil also showed that seedlings of some citrus types growing on their own roots have severe symptoms of tristeza; other types, including sweet orange, have mild or no symptoms. The early research revealed that supposedly different citrus diseases, in different citrus types, are caused by the same tristeza virus and that mild strains of the virus can give some protection against virulent strains. This cross-protection has been used in Brazil and elsewhere for tristeza control (Gonsalves and Garnsey 1989; VanVuuren, Collins, and da Graca 1993).

Many strains of the tristeza virus are now known. Strains differ in virulence, from mild to very severe. Strains also differ in transmissibility by several aphid vectors (Martinez and Wallace 1964; Bar-Joseph and Loebenstein 1973; Raccah, Loebenstein, and Singer 1980). Protection against virulent strains by prior inoculation with mild strains indicates relatedness, a notion confirmed by various serological tests, immunoabsorbance assays, and nucleic acid and protein analyses (Dodds et al. 1987; Guerri, Moreno, and Lee 1990). However, cross-protection is not always effective; the virulent VT strain spread in Israel and Florida, even in the presence of mild strains of tristeza virus (Bar-Joseph 1978). Selective vectoring of new strains by aphids may be involved.

A new and more virulent tristeza strain finally appeared in Florida; the new strain differed from the old one in composition of viral coat proteins. Orchards were surveyed by modern analytical methods to determine spread of the virulent strain (Lee et al. 1988) and to detect severe virus strains among populations of mild strains (Rosner, Lee, and Bar-Joseph 1986). RNA analyses have been used in Spain to detect severe tristeza virus strains (Moreno, Guerri, and Muñoz 1990).

Summary

Production of sweet oranges depends on use of other citrus types as rootstock, to combat a root disease. Use of such combination plants leads to severe losses from tristeza, a viral disease. Tristeza is unusually destructive when sweet oranges are grown on sour orange roots. Apparently, sour orange tissue is hypersensitive to the virus; tissues at graft unions die, killing the whole plant. Other rootstocks are now replacing sour orange for sweet orange production. Tristeza is controlled by use of resistant rootstocks, plus programs for certification of tristeza-free budwood.

Tristeza apparently originated in subtropical China and/or the tropics of southeastern Asia, the ancestral home of cultivated citrus. Citrus species were first taken as seed to all warm parts of the world, leaving tristeza behind because the virus is not seed-borne. The great centers of citrus production are now in areas where citrus is not endemic. Tristeza virus spread into these areas when transportation became rapid and economical, with movement of seedlings and budwood of the most desirable cultivars. The spread of tristeza over the world is well documented.

Tristeza virus is a closterovirus that is confined to phloem tissue. There are many variants or strains, but all are vectored by aphids. Techniques of molecular biology, including serological procedures, nucleic acid analysis, and protein analysis, are used to detect tristeza in symptomless plants and to identify variants.

14.2. Fusarium Wilt of Banana

Fusarium wilt, or "Panama disease," of banana (*Musa acuminata*) (Figure 14.1) became one of the world's most serious plant diseases, with significant economic and social effects in the producing countries. The causal fungus (*Fusarium oxysporum* f. sp. *cubense*) became widely distributed over the tropics before the disease was recognized and before bananas became an export item. The early distributions occurred with occasional international movement of planting materials and with local trade of propagation stock by small farmers. Outbreaks of the disease often went unnoticed in the small, gardenlike plantings of subsistence cultivators. Their scattered gardens, usually with bananas mixed with other species, slowed the local spread of this soil-borne disease; nearby banana gardens often remained free of the disease for a long time (Stover 1962b).

The situation changed when bananas became an item in world trade, beginning late in the last century and expanding rapidly after 1900. Great plantations were developed, and Fusarium wilt became a major problem. The crop is propagated vegetatively, thus is genetically uniform. For years, the banana cv. Gros Michel, a very susceptible type (Stover 1962b), was the mainstay of the export industry. Fusarium wilt of banana is a prime example of a soil-borne disease that spread worldwide with horticultural trade. The method of propagation and the large plantings of a single genotype are the two key factors that led to destructive effects.

Some knowledge of the anatomy of the banana plant is needed to understand how Fusarium wilt became a problem. The plant is a giant herb, consisting of roots, a large rhizome below ground, a trunklike pseudostem consisting of leaf sheaths, and large leaves at the top. A true stem develops within the pseudostem and terminates with a fruit stalk (Figure 14.1). The commer-

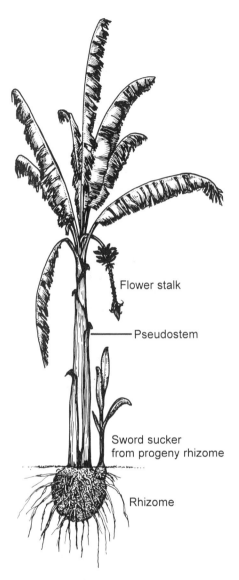

Flower stalk

Pseudostem

Sword sucker
from progeny rhizome

Rhizome

Figure 14.1. Banana plant with Fusarium wilt, showing the gross anatomy of the plant. New rhizomes and "suckers" are formed on the old rhizome, and these are used to propagate the crop. The fungus is disseminated with such propagating material. (Drawing by Marlene Cameron)

cially acceptable cultivars are sterile triploids with no seed; the plant is propagated vegetatively from "sword suckers," which develop from buds on the rhizome, or from 1.8- to 3.6-kg pieces of the rhizome that contain at least one bud. Each stem bears fruit in twelve to fifteen months, followed by death of the stem. Its place is taken by another stem from a bud on the perennial rhizome. Only one sword sucker is allowed to grow per rhizome, or "mat."

Bananas are endemic in the moist tropical regions of southeastern Asia (Simmonds and Shepherd 1955; Simmonds 1966). There, the fruit was utilized as food for thousands of years, perhaps since the Pliocene, in part because bananas were easy to propagate by primitive farmers. Alexander found cultivated bananas in India by 327 B.C., and early records show that early Greeks, Romans, Chinese, and Arabs knew the plant and its virtues. There are stone carvings of bananas from ancient Assyria and Egypt. Bananas were growing in East Africa by the time of Christ; Arab traders carried the plant to West Africa, where early Portuguese explorers observed it by A.D. 1500. The plant was moved eastward to the Pacific islands at an even earlier date. Portuguese sailors took banana to the Canary Islands, where it became important. A Spanish friar is credited with moving banana to Santo Domingo and Panama in the New World, soon after 1500 (Simmonds 1966; Carefoot and Sprott 1967).

Most of the edible wild or primitive bananas are diploid strains of *M. acuminata*. They are still present in southeast Asia, where the natives consume the fruit, inner sheaths of pseudostems, and male flower buds. Edible triploid bananas probably originated in Malaysia. Hybridization with *M. balbisiana* may have occurred on the fringes of the range of *M. acuminata*. This natural hybridization, along with formation of triploids and tetraploids, probably was the final step in evolution of cultivated bananas. The first preserved records of cultivation were from India (Simmonds 1966).

Most modern cultivars of bananas are triploids that do not exist in the wild; a few are diploids (22 chromosomes) or tetraploids. Cultivars are classified on the basis of ploidy: for example, there are AAA types (triploids of *M. acuminata*), AAB types (triploids, with chromosomes from *M. acuminata* and from *M. balbisiana*), ABB types, BB types, and so forth. Most bananas that are eaten as fresh fruit are AAA types (examples, cvs. Gros Michel and Cavendish), whereas most cultivars used for cooking are ABB or AAB (example, the widely used Bluggoe plantain). The plantains are more starchy, more drought resistant, and more resistant to certain diseases than are the AAA types (Simmonds and Shepherd 1955).

Agricultural practices used for bananas vary from region to region; knowledge of these practices will help to understand Fusarium wilt. In Central America, large plantations are on the well-drained and fertile alluvial soils of flood plains. Formerly, the *Fusarium*-susceptible types (cv. Gros Michel) were moved frequently to new disease-free fields. Presently, large plantations usually are of the resistant cv. Cavendish, which lasts for many years, allowing the use of fixed irrigation systems and other permanent infrastructures. Many small farmers still expect to plant new fields every five years (Stover 1972 and 1986). The propagation method gives an efficient means of disease spread. An investment of several thousand dollars per hectare is required to establish new fields, which sometimes are flood-fallowed to destroy *Fusarium* in the soil. Production costs are increased further because sprays with light oils are required to control the sigatoka disease (*Cercospora* leafspot; see Chapter 9, section 3). In Taiwan, bananas are cultivated almost as an annual crop, in rotation with sugarcane or paddy rice.

Banana is the most important tropical fruit and probably the most consumed of all fruit; total world production in recent years has been estimated at 68 or more million metric tons (Ploetz et al. 1990). Half of this production is thought to be eaten fresh and half is cooked; bananas and cassava are two major sources of dietary starch in the tropics, and populations in many countries depend on these crops. Uganda, at one time thought to be the world's largest producer, grew 15 percent of the world's bananas, practically all consumed locally (Simmonds 1966). Brazil probably is now the largest producer, but Brazil exports few bananas (Ploetz et al. 1990). The starchy, or cooking, bananas (plantains) are roasted, fried, baked, or made into chips. Bananas are a major export item, thus are a significant part of the economics of many countries. Any production problem can lead to trouble and is taken seriously.

World trade in bananas began in 1804, when a few bunches of fresh bananas were brought from the Caribbean area to New York. A small unorganized trade soon developed, with transport by clipper ships. An early import company, started after the U.S. Civil War, soon failed because of unreliable sources and poor transport. A successful production and transport business was organized about 1885; this was the Boston Fruit Company, which brought bananas from Costa Rica to the United States. Faster, steam-driven ships contributed to success, and the firm's farming operations soon spread to other countries in Central America. The company merged, over time, with many small competitors, finally with the large Cuyamel Fruit Company, to form United Fruit. By 1929, the United Fruit company was a colossus, with a monopoly on the North American trade, and a major exporter to Europe. It soon became a dominant factor in the governments and economies of countries in the American tropics (Carefoot and Sprott 1967; McCann 1976).

At one time, United Fruit (later called United Brands) was among the world's largest corporations. Clearly, it was the world's largest agricultural enterprise, holding in reserve or in use vast tracts of the best land for bananas. New lands became necessary because Fusarium wilt forced moves to virgin areas; thus, the wilt fungus was spread to new, noninfested areas. Small growers were soon eliminated, in part because of large capital investments required for the export trade and for disease control. Some countries gave up on banana exports. The plus side was an increase in the standard of living among farm workers in some countries and the development of a substantial middle class, especially in Costa Rica (Carefoot and Sprott 1967; McCann 1976). United Fruit's monopoly finally was broken; at present, there are three large corporations that control only 30 percent of the export trade. Export bananas are grown on large, modern farms in Central America, Colombia, Ecuador, and Mindanao province of the Philippines. Small-to-medium export farms still function in the Caribbean Islands, Taiwan, Cameroon, Ivory Coast, and Somalia (Stover 1986).

Fusarium wilt probably was endemic in southeastern Asia and was later spread over the world's tropics. Some *Musa* species native to southeastern Asia are resistant (Vakili 1965; Buddenhagen 1990). This fits the contention that the wilt *Fusarium* is endemic in southeastern Asia; otherwise, we know little about the origin of Fusarium wilt. We cannot rule out the possibility that Fusarium wilt also evolved in the western hemisphere, from opportunistic forms of *Fusarium* that are found worldwide (Ploetz 1990; Ploetz et al. 1990) or from races infecting related native species in tropical America. In either case, it is clear that the fungus has been distributed widely by man. The *Fusarium* that causes banana wilt was first isolated by Erwin F. Smith in 1910, from Cuban materials. Final proof of pathogenicity was not published until 1919 (Brandes 1919; Thurston 1984).

Major losses occurred following adoption of banana cv. Gros Michel, the mainstay of the export industry until about 1960. However, the fungus was widely distributed before Gros Michel became the major cultivar. Early descriptions of banana troubles indicate that cv. Silk was infected in Australia by 1876, at least thirty-five years before Gros Michel was taken to Australia. Silk was widely planted in the Caribbean area by 1750; the indications are that Silk had Fusarium wilt before the highly susceptible Gros Michel was planted there in the early 1800s (Stover 1962b).

From 1890 to 1910, *F. oxysporum* f. sp. *cubense* was spread further with infected rhizomes of cv. Gros Michel, before the nature of the disease was known. There are recorded examples of this spread. Infected rhizomes were taken from Costa Rica to Hawaii, thus infecting a new area; the disease is unknown elsewhere in the mid-Pacific islands (Ploetz et al. 1990). The United Fruit Company supplied rhizomes for planting in Surinam in 1906, and Fusarium wilt appeared the next year; by 1910, every field was infected, and many planters were ruined. More recently, race 4 of the fungus was moved with rhizomes from Taiwan to the Philippines (Stover 1986).

Heavy losses occurred in the Caribbean area, beginning about 1914. The first devastating outbreaks were in Panama (hence the name Panama disease) in 1904, and, by 1935, the fungus was in the soil of practically all banana-producing areas of the western hemisphere (Carefoot and Sprott 1967). The fungus was carried over long distances by rhizomes for propagation and was carried locally by drainage water, tools, rhizomes, and agricultural workers. Once in the soil, *Fusarium* remains for many years as dormant structures (chlamydospores) and as an opportunistic inhabitant on the roots of nonhost species, including certain grasses (Waite and Dunlop 1953; Thurston 1984; Su, Hwang, and Ko 1986).

Fusarium wilt became a major limitation in banana cultivation in the Caribbean Islands, Central America, and South America. Until recently, it had minor effects in Africa, in part because some local cultivars were resistant and, possibly, because most bananas were grown in small plots by subsistence farmers. At present, the disease causes severe losses for village farmers in Africa and elsewhere (Buddenhagen 1990). Losses in large plantations in the western hemisphere declined slowly when resistant Cavendish cultivars were accepted for export, beginning about 1960. However, Fusarium wilt is still a serious threat because the fungus can evolve new races.

Fusarium wilt has been most serious in four cultivars: the widely distributed cvs. Silk and Gros Michel, cv. Chalamuco in Puerto Rico, and cv. Mangana in Cuba, Jamaica, and Central America. Spread of the disease is a matter of record; it appeared in Cuba in 1910, Jamaica in 1911, Australia (definitely confirmed) in 1912, and throughout the American tropics, except Colombia, by 1919. Plantains for cooking were seriously affected in Puerto Rico by 1910, causing much hardship (Brandes 1919). The disease was first reported in East Africa in the early 1950s and is now widespread there on the new cultivars; older land cultivars, used for cooking and beer, are resistant (Stover 1990). Overall, records of spread indicate that Fusarium wilt was of minor concern until the great plantations were established for the export trade, beginning early in the twentieth century. All banana production in the western hemisphere was threatened by 1955. United Fruit was forced to abandon the susceptible cv. Gros Michel and turn to resistant Cavendish cultivars, which required changes in growing and shipping procedures.

Three triploid types of edible bananas are very susceptible to Fusarium wilt; these include AAA, AAB, and ABB cultivars. Some AA clones and some wild strains of *M. acuminata* also are susceptible; the latter probably were the source of susceptibility genes in triploid bananas. Cv. Gros Michel (AAA) was of such superior quality that it soon became the major export fruit, grown worldwide except in eastern and southern Africa. However, it is little grown today because of Fusarium wilt. The plantain cv. Bluggoe was once resistant, but it is now in serious trouble in the Caribbean region and in Central America after a new race of the wilt fungus appeared (Stover 1990). Another susceptible species is *M. textilis,* a plant grown for fiber. *Heliconia caribea* and *H. mariae,* closely related to the banana, were reported to have a Fusarium wilt in the jungles of Panama; other *Heliconia* species are resistant. Some of the *Fusarium* isolates from *Heliconia* are pathogenic to cultivated bananas. Resistance is found in some AA, AAA (notably Cavendish types), AAB, ABB (notably certain selections of cv. Bluggoe), and in at least one ABBB clone (Stover 1962b, 1972; Ploetz et al. 1990).

Symptoms of Fusarium wilt show first as pale yellow patches at the base of a lower petiole, appearing two to five months after infections occur through the rhizome or roots. Initial symptoms in cv. Gros Michel bananas are followed either by striking chlorosis that progresses up the plant (the yellowing syndrome) or by collapse and fall of leaves without chlorosis. Plants with either syndrome have internal browning in the xylem, and leaves eventually turn brown and die. The symptoms are similar to those of the moko disease, caused by a systemic bacterium (*Pseudomonas solanacearum*). Both Fusarium wilt and moko are spread with propagating materials; in both cases, symptoms in new shoots may not be evident for a long time after infection (Stover 1962b).

The causal fungus of banana wilt is morphologically identical to other forms of *F. oxysporum.* The species has saprophytic forms, opportunistic forms, and form species that are highly specialized to certain hosts; important specialized forms include f. sp. *lycopersici* (infects tomato), *conglutinans* (cabbage), *cubense* (banana), *vasinfectum* (cotton), and at least fifty others. No sexual or perfect stage is known for *F. oxysporum,* but other known Fusaria are Ascomycetes. The wilt fungi are filamentous, producing yeastlike microconidia in infected xylem vessels, crescent-shaped macroconidia in decayed tissues, and thick-walled chlamydospores, which can remain dormant in soil for years (Beckman 1987).

At least four races of the banana wilt fungus are known. Race 1 is distributed all over the tropics, attacking several banana cultivars including Silk and Gros Michel, some clones of *M. acuminata* (AA), *M. balbisiana* (BB), and *M. textilis* (Stover 1962, 1990; Vakili 1965; Ploetz et al. 1990). Race 2 was first discovered on plantain cv. Bluggoe (ABB); race 2 also infects *M. schizocarpa,* but not the major banana cultivars (Gros Michel, Silk, Cavendish, and Lacatan), *M. balbisiana,* and some clones of *M. textilis* (Stover 1962; Ploetz et al. 1990). Race 3 infects *Heliconia* species in tropical America but has little or no effect on cultivars Gros Michel and Bluggoe (Waite 1963). Race 3 is now rare or has disappeared (Ploetz and Correll 1988).

Race 4, first described in Taiwan (Sun, Su, and Ko 1973), has been found in subtropical areas in the eastern hemisphere (Canary Islands, Philippines, Taiwan, Australia, and South Africa) (Ploetz and Correll 1988). Race 4 is pathogenic to banana cultivars Cavendish, Gros Michel, Silk, and Bluggoe, and to *M. balbisiana* (Thurston 1984). The Cavendish cultivars are now the mainstay of the export trade, making race 4 a serious potential threat. However, race 4, to date, has thrived only in the subtropics. It or other races could still spawn new races that have race 4 pathogenicity and that thrive in the tropics. Much more research is needed on the race situation; there are enough confusing reports to indicate that other races exist. Sound banana-breeding programs will require more information on races (Buddenhagen 1990).

F. oxysporum has no known sexual stage; variations and new races can arise either by mutations or by heterokaryosis and the parasexual cycle. These conditions have had much study in other genera of fungi. In general, the fusaria are homokaryotic; they become multinucleate or heterokaryotic by vegetative fusion. A diploid (2 N) is formed sooner or later after vegetative fusion occurs, and this is followed by meiosis with genetic segregation (Leslie 1990). Vegetative fusion does not occur at random within or between isolates of *F. oxysporum;* instead, the process is limited by vegetative incompatibility. Isolates in the same group are compatible, and their hyphae can fuse; isolates from different compatibility groups do not fuse. This probably is a function of specific proteins (Ploetz and Shepard 1988; Ploetz 1990; Ploetz et al. 1990). Many compatibility groups have been found (Puhalla 1985).

Vegetative compatibility can be used to confirm origins and trace movements of races. New races of *F. oxysporum* f. sp. *cubense* contain only one compatibility group, for the most part; many old races have isolates in two or more such groups. The situation has been most clear with *F. oxysporum* from crucifers (Leslie 1990). Saprophytic populations of *F. oxysporum* generally have several compatibility groups, whereas pathogenic forms tend to have isolates with only one or two groups. Pathogenic forms or races that have a single vegetative compatibility group are likely to be clones of common origin, even when obtained from different areas. Isolates with the same pathogenicity but with vegetative incompatibility probably differ in origin and are not clones (Ploetz 1990; Ploetz et al. 1990).

If these assumptions are true, then clues are available as to whether a given isolate belongs to an old race or was spawned recently from an opportunistic form of *F. oxysporum*. The possibilities were demonstrated in southern Florida, where race 1 appeared suddenly without new plant introductions (Ploetz and Shepard 1988). Race 1 in Florida belongs to a compatibility group that is unknown elsewhere in the world; thus, Florida race 1 probably is a new clone, most likely of local origin (Ploetz et al. 1990). These ideas were discussed in detail by Leslie (1990), and possible explanatory models were given by Kistler and Momol (1990).

Precise molecular techniques are now available that should clarify our understanding of the origins, diversity, phylogeny, and spread of races. Population biology data now seem to support diverse geographical origins of *F. oxysporum* affecting banana (Ploetz 1990). Genetic markers useful in such studies include virulence, vegetative compatibility, and *nit* mutants that have resistance to chlorate. There are preliminary data from analyses of plasmids, mitochondrial and nuclear DNA (Kistler and Momol 1990), karotype variability (Miao 1990), and other parameters. So far, the data are not sufficient for firm conclusions.

Several environmental conditions affect development of Fusarium wilt of banana. All the Fusarium wilts are favored by relatively high soil and air temperatures, usually with most rapid development at about 28°C. Certain infested acid loam soils will produce bananas for only three to five years, whereas some alkaline clay loams are productive for as long as twenty-five years in the presence of the pathogen. There is no adequate explanation of the "resistant" soils; microbiological factors may be involved (Stover 1962b). Such soils have been utilized to cope with the disease; Gros Michel bananas are still being grown in limited areas in resistant soils (Ploetz 1990). Low pH, low Ca, and low total salts in soil favor rapid disease development, as does poor drainage and fluctuating water supply. Roots either grow to the inoculum, or the inoculum is carried in soil to roots; thus spread is favored by tillage. Spread from plant to plant can occur when roots of infected and noninfected plants are together (Stover 1962b).

There are many histological and physiological studies of banana wilt. Infection occurs through secondary roots (Stover 1962), perhaps through root caps, as is known for other Fusarium wilts (Smith and Walker 1930; Bishop and Cooper 1983). The fungus becomes systemic in the xylem, moving to the rhizome and the pseudostem. Beckman, Halmos, and Mace (1962) placed *Fusarium* spores in xylem vessels of resistant and susceptible roots; in resistant roots, the parenchyma cells adjacent to vessels responded quickly, forming xylem occlusions that prevented further spread of the fungus. Spores germinated in the xylem of susceptible roots, formed mycelium, and sporulated. Gel plugs formed quickly, but did not become discolored and usually dissolved. New spores were formed and distributed with transpirational flow. Only later did a firm xylem blockage occur in susceptible plants. Thus, the susceptible plant responds slowly, and the fungus passes on before the host blocks it. However, the blocks in susceptible plants eventually lead to wilting of the plant for lack of water (Beckman 1990).

Similar reactions occur in susceptible and resistant plants of other species, in response to their specialized Fusarium wilt fungi (Beckman 1987). The rapid cellular reactions in resistant plants result in local necrosis in the xylem. The process appears to be essentially the same as occurs in many localized infections of plants, with a rapid hypersensitive response in the resistant tissues and a delayed response in susceptible tissues. This may be part of the phenomenon of recognition of "self" and "non-self" (Beckman 1990).

There are many studies on the roles of toxins and extracellular enzymes from the pathogen and their possible effects on development of wilt diseases. Fusaric acid and lycomarasmin are produced in culture by the banana and several other Fusaria, but, to date, significant roles in disease development have not been established (Kuo and Scheffer 1964; Scheffer 1983). This is true also for hydrolytic enzymes from the Fusaria. However, circumstantial evidence does indicate that unknown toxins have a role, but not as the direct cause of wilt (Scheffer 1983).

The primary control of Fusarium wilt of banana is use of resistant cultivars; presently, the Cavendish cultivars are resistant except in certain areas of the subtropics where race 4 is present. Large planters in lowlands and flood plains have successfully disinfested fields by flood fallowing, although the procedure is not cost-effective in general. Fields are diked in such a way that flooding occurs to a depth of at least 30 cm for six months. This procedure eliminates the wilt fungus in the soil, probably for lack of oxygen, and bananas can be grown for several years (Stover 1962).

It is important to use disease-free planting materials for all fields that are free of the fungus. Such planting materials have been developed in Taiwan; meristem cultures were established, and plantlets from the cultures were grown and multiplied for planting stocks (Su, Hwang, and Ko 1986; Krikorian 1990). This system has increasing use in Taiwan and limited use in other areas. Rotation of bananas with flooded paddy rice, which eliminates the wilt fungus, also is used in Taiwan.

Breeding for resistance to Fusarium wilt of banana has had little attention since the export industry started using Cavendish bananas. No new cultivars of commercial significance were released from 1955 to 1986 (Stover 1986). New and aggressive breeding programs are needed for the several types of bananas, including those that are susceptible to race 4 (Buddenhagen 1990). It is hoped that prospects for breeding and for basic studies will be improved with development of the International Network for Improvement of Banana and Plantain (INIBAP) (Stover 1986).

Summary

Fusarium wilt of banana was spread over the tropics with movement of planting materials. The disease became a serious problem when large plantations with monocultures were developed for the export trade. The fungus is thought to be endemic, along with bananas, in southeastern Asia; there, the disease was of no consequence because bananas were grown by subsistence farmers in small plots, along with other crops.

The cultivated banana lacks seed and is propagated vegetatively by rhizomes and sword suckers; thus, cultivars are clones, with genetic uniformity. Rhizomes and symptomless suckers can carry the pathogen and spread the disease, both locally and to other geographical areas. When Fusarium wilt became widespread and damaging, export producers were forced to discard the highly desirable, but susceptible, cv. Gros Michel and to replace it with the resistant Cavendish cultivars. However, new races of the wilt fungus appear, including race 4, which attacks the Cavendish bananas. New races are a worldwide threat to the banana industry.

Fusarium wilt and its control are affected by many factors, including agricultural practices, the export trade, anatomy of the banana plant, propagation methods, ploidy of cultivars, fungal races, and fungal compatibility groups. These factors are discussed briefly, as is work on disease physiology, genetics of pathogen, molecular biology of *Fusarium,* and control.

14.3. Fusiform Rust of Pine in the Southeastern United States

Fusiform rust is an example of plant disease caused by an endemic pathogen that became destructive following changes in forest management. The two major factors involved were protection from fire and large-scale planting of pines in monoculture.

The southeast is the second most important area in the United States for production of wood. Pines predominate in the area's forest industry; there are now large monoculture plantations of two species (*Pinus taeda* and *P. eliottii*). The primeval forest contained a mixture of species, which buffered against disease. Originally, the forest included four major pines, with partially overlapping ranges: (1) loblolly (*Pinus taeda*), (2) slash (*P. eliottii* var. *eliottii*), (3) longleaf (*P. palustris*), and (4) shortleaf (*P. echinada*) (Dinus 1974).

Two forms of a gall-producing rust (*Cronartium quercuum* f. sp. *fusiforme* and f. sp. *quercuum*) (=*C. fusiforme* and *C. quercuum*) affect loblolly and slash pines. These rust fungi, of no consequence in the native forest, became major limiting factors in plantation forestry. Fusiform rust is now one of North America's most serious plant diseases.

The native southeastern pine forest was a fire subclimax. Fires have swept the forest area for thousands of years; otherwise, pines would have been succeeded by hardwoods (Dinus 1974). The longleaf pine, which is the most fire resistant, was favored in the fire subclimax. Fusiform rust was present but relatively uncommon in this fire forest; oaks, the essential alternate hosts, were less common than today because they are more sensitive than pines to ground fires. When fire protection became more effective, oaks became more numerous (Dinus 1974; Powers, Schmidt, and Snow 1981).

Pines, which grow rapidly in the southeast, were planted as cotton production decreased; large timber companies became involved. Loblolly pine is the most desirable species for both lumber and pulpwood, and it was planted over millions of hectares, possibly, at one time, at the rate of 600,000 ha per year. Slash pine also was widely planted. Both species are very susceptible to fusiform rust. In some areas, the longleaf pine, a rust-resistant species, was replaced by loblolly; longleaf was not planted because it grows more slowly than loblolly and is susceptible to brown spot needle blight, caused by *Scirrhia acicola,* a fungus. Plantations of loblolly and slash pines have now been extended well beyond their natural ranges. Loblolly is, by far, the most used, in plantings from the Carolinas to southwestern Louisiana (Dinus 1974; Powers, Schmidt, and Snow 1981).

Fusiform rust was inconsequential before 1900, even though the southeastern United States has a rust-inducive climate. The disease became serious for several specific reasons, all concerned with forest management. First, effective fire prevention favored oaks, the essential alternate host of the rust. Next, vast areas were planted to loblolly and slash pines, both highly susceptible. The two planted species replaced longleaf and shortleaf pines, both resistant. Other factors in rust epidemics included reforestation with infected seedlings from nurseries and fertilization, which predisposes seedlings to rust infection (Manion 1981). The average level of rust-infected trees in young plantations of loblolly pine is 60 percent, and this percentage seems to be increasing (Snow, Dinus, and Walkinshaw 1976; Powers, Schmidt, and Snow 1981). There is some sentiment for replacing loblolly with longleaf pines in some areas because more rapid growth and resistance to needle blight can be achieved by selection.

Cronartium quercuum (= *C. fusiforme*) was first described in 1898 (Powers, Schmidt, and Snow 1981). There is speculation that the rust fungus either originated or survived during glacial periods in Mexico or Central America (Tainter and Anderson 1993); at present, the fungus is found only in the southeastern United States. *C. quercuum* causes galls and cankers, which decrease tree value and often lead to breakage. Seedlings and young trees frequently are killed.

The pine rust fungus is a typical long-cycle rust; two years are required for completion of the cycle. Pine needles and soft stems are infected in spring by basidiospores from telia on oaks. The fungus grows from needles into the stem, causing spindle-shaped (fusiform) galls in four to six months. Rust overwinters in these perennial galls, producing spermagonia (involved in sexual fertilization) from October to December, usually during the second year. Blisterlike aecia are formed on galls in spring, liberating powdery yellow aeciospores that infect expanding oak leaves. Uredia and urediospores are formed on the undersurface of oak leaves within a few days; urediospores reinfect oak. Telia and teliospores develop next; these produce basidiospores that infect pine, completing the cycle. The fungus often becomes dormant or dies at temperatures above 29°C (Sinclair, Lyon, and Johnson 1987).

Many species of *Pinus* proved to be susceptible to fusiform rust in inoculation tests; of dozens tested, only *P. resinosa* produced no galls. The major *Pinus* species in the western United States are susceptible, although fusiform rust does not occur there. Also, many European, Asian, and Caribbean species are susceptible (Raddi and Powers 1982; Tainter and Anderson 1993). Pines in these areas probably evolved in the absence of *C. quercuum;* in contrast, some pines native to the southeastern United States (*P. echinata, P. palustris,* and *P. glabra*) are relatively resistant (Powers 1975), suggesting a long period of coexistence with rust (Tainter and Anderson 1993). The susceptibility of pines in the western United States, Europe, and Asia makes fusiform rust a potential threat in any area that has pines, oaks, and a rust-inducive climate; this area includes parts of the western United States and Europe. Resistant selections of loblolly and slash pines are now being planted in the southeastern United States, and there is evidence that resistance can be increased by breeding (Powers and Duncan 1976). However, pine selections that are resistant to one race are susceptible to another (Sinclair, Lyon, and Johnson 1987). The fungus is adaptable, able to form new races.

C. fusiforme and *C. quercuum* formerly were classified as individual species. At present, they are considered to be forms of the same species (*C. quercuum*) because they are identical in morphology (Powers, Schmidt, and Snow 1981). Also, serological comparisons showed that f. sp. *moniliforme* and f. sp. *quercuum* are strongly cross-reactive, only differing in cross-adsorption analyses by one or very few antigens. The f. sp. *quercuum* differed from *C. ribicola* (Gooding & Powers 1965). Furthermore, esterase isozyme analyses gave supporting data, with identical results for aeciospores from sand and Virginia pines. The pattern differed from that for spores from loblolly, jack, and shortleaf pines; the latter three differed from one another (Powers, Lin, and Hubbes 1989; Powers et al. 1991). These data suggest that the rust from sand and Virginia pines should be classified as a new form species (*Virginianae*) (Powers et al. 1991). Esterase assays showed no differences in isolates within each form species (Powers, Lin, and Hubbes 1989).

Comparison of DNA fragments (DNA "fingerprinting") is conclusive in detection of genetic differences. The technique has been tested with pycniospores from galls and with fungal isolates derived from single spores of the fusiform rust fungus. The DNA fragments differed for each of ten different fungal lines and for pycniospores from several sources. The phenotypes of spores from four galls differed, and some galls contained more than one fungal type (Doudrick et al. 1993). The method should be applicable for race determinations.

C. quercuum f. sp. *quercuum* generally causes globose galls, whereas f. sp. *fusiforme* causes spindle-shaped galls. However, this can vary; it is not a good marker for identification. A better identification is based on the fact that black oak (*Quercus velulina*) is resistant to f. sp. *fusiforme,* but susceptible to f. sp. *quercuum* (Dwinell 1971). Otherwise, f. sp. *quercuum* and f. sp. *fusiforme* have similar oak hosts.

Oaks are essential alternate hosts for *C. quercuum;* without oak, there is no disease on pine. Several oak species are suscepts; water oak (*Q. nigra*) and willow oak (*Q. phellos*) are common in the pine areas and are very susceptible. Other susceptible oak species include blackjack (*Q. marilandica*), bluejack or turkey (*Q. incana*), southern red (*Q. falcata*), and northern red (*Q. rubra*) (Dwinell 1974). Some western oaks, including the California black (*Q. kelloggii*), are very susceptible to fusiform rust (Dwinell and Powers 1974). Scarlet (*Q. coccinea*) and black (*Q. velutina*) oaks are hypersensitive-resistant to f. sp. *fusiforme* (Dwinell 1971). In contrast to pine, oaks are seldom damaged seriously. The white oak group, in general, is resistant.

C. quercuum f. sp. *fusiforme* clearly has a number of races, as is the case with wheat and other rusts (Chapter 11, section 1). The first solid indication of races in fusiform rust came from observations on disease in geographical family lines of slash pine (Snow, Dinus, and Kais 1975). One pine line proved to have resistance to many rust collections, a second line appeared to be immune, and other lines were susceptible. Later work (Snow, Dinus, and Walkinshaw 1976) showed that rust collected from resistant slash pines gave more infection in resistant seedlings than did rust from the general population of pines, an indication of fungal specialization (races).

Findings on slash pine races were soon confirmed and expanded with loblolly pine (Powers, Matthews, and Dwinell 1977). Inoculum was taken from many galls on trees from seven states of the southeastern United States; seedlings of three loblolly family lines (resistant, intermediate, and susceptible) were inoculated. There were variations in virulence of the rust from different states and from within states. No immune loblolly family lines were found; the most resistant line was susceptible to inoculum from certain galls. Thus, the fungus varies greatly in pathogenicity, and this must be considered in tree-improvement schemes. A wide genetic base in pines is very important, but it may not be available.

Better data came from use of half-sibling families of loblolly pine that varied in resistance. Pine seedlings were inoculated with five different, highly virulent rust selections from different areas, and the percentage of seedlings that became galled (that is, infection levels) was determined. Pine families differed in susceptibility; of the five, only one was resistant to all five selections of the fungus. Two families were intermediate but had adequate resistance to at least one fungal selection. Also, there were variations in fungal lines selected from a single gall; the variability was as great as that reported for rust collections from different states (Powers and Matthews 1979). Certain random fungal isolates from susceptible pines were as virulent on resistant pines as were isolates that came exclusively from resistant pines (Walkinshaw and Bey 1981). These data must be considered in the possible development of resistant seed orchards for specific areas, as was proposed. Obviously there is a narrow known base for resistance and great variability in the pathogenic potential of *C. quercuum*.

Our understanding of races of *C. quercuum* f. sp. *fusiforme* was refined still further by use of single aeciospores propagated on red oak seedlings. Three half-sibling families of loblolly pine (resistant, intermediate, and susceptible) were inoculated with the basidiospore progeny from ten fungal isolates. There were differences in virulence levels in the fungal lines, but the resistant pine family had no more than a low level of infection by any of them (Powers 1980). This work was carried further by the use of loblolly pine seedlings inoculated with fungal lines derived from single spores (Kuhlman 1990). Several of the lines were virulent on each of three pine families that had been selected for resistance; one such pine family had been widely planted. Some of the virulent fungal isolates came from areas other than the source area of the pine families.

The experiments described seem complex, but such approaches are necessary for a disease such as pine rust. There are two inherent problems. First, there is no way to clearly define individual races because there are no defined, nonsegregating genotypes of the host to use for testing. This is not easily attained with trees. The second problem is the difficulty in growing the fungus in culture, thus complicating experimentation. Nevertheless, it is now clear that both form species of *C. quercuum* have many races that differ in virulence and pathogenicity. The race problem can be clarified by the use of DNA fingerprinting (Douderick et al. 1993).

Pine-progeny families from open-pollinated trees in areas with high disease incidence are, in general, more resistant than are families from low-disease areas. For example, inoculated seedlings from low-disease areas had 85 percent disease incidence, whereas inoculated seedlings from high-disease areas had 61 percent disease incidence (Goddard, Schmidt, and Vande Linde 1975). The difference is not impressive, but there are confirmatory findings (Powers and Matthews 1980; Schmidt et al. 1986). Also, seedlings from each source area are, in general, more susceptible to the fungus from the same area than to fungus from other areas (Powers and Matthews 1980). Forest managers are now using bulk lots of seed, from specific areas, that are known to have some resistance. One of the best sources has been Livingston Parish (county), Louisiana; seeds from there have produced plants that were reasonably resistant over all of the southeastern United States (Wells and Wakeley 1966; Powers, Schmidt, and Snow 1981). Rust resistance can be increased by hybridization (Powers and Duncan 1976).

Fusiform rust is most severe in the most fertile soils where both oak and pine grow well. However, the major factor associated with a high incidence of disease is the amount of inoculum for pine. Thus the more oak in an area, the more rust on pine (Hollis and Schmidt 1977). Pines are infected by basidiospores at temperatures from 8° to 28°C if moisture is present (Kuhlman and Pepper 1994).

The major thrust for control has been the use of tolerant pines for planting. The only other direct control measure has been the use of fungicides in seedling nurseries. This method of control has been of limited value, in part because it is not practical to spray young trees once they have been transplanted to the field. Disease incidence in seedling nurseries is minimized by proper site selection, eradication of nearby oaks, and other management practices (Powers, Schmidt, and Snow 1981).

Little is known regarding mechanisms of resistance. Histologically, there seem to be three types: (1) The fungus does not penetrate the plant surface; (2) tissues are colonized without gall formation; and (3) hypersensitive responses, with the fungus confined to penetrated cells. The hypersensitive cells are darkened, and fungal hyphae become nonfunctional in them (Jewell and Speirs 1976; Miller et al. 1976). Tannin deposits are high in infection sites, but this does not appear to be the basis of resistance because tannins also are high in the galls that occur on susceptible pine (Walkinshaw 1978); high tannin levels may be only a result of disturbance. Some families of slash pine that were selected for resistance develop stem-girdling galls, with a swelling above a constriction zone, causing severe disruption; such stems usually die quickly (Walkinshaw and Roland 1990), and this may suppress further disease development. Typical galls contain a tenfold-higher-than-normal level of cytokinin, a plant growth hormone (Rowan 1970). Cytokinin may be a causal factor in the overgrowth, as with crown gall (see Chapter 3, section 3).

Summary

Fusiform rust of pine (*Cronartium quercuum*) was endemic, but not destructive, in the native fire-subclimax forest of the southeastern United States. The disease became severe during the twentieth century, with the advent of forest management. Two major factors were involved: (1) Fire protection became efficient, and (2) rust-susceptible pines were planted over large areas. These practices favored fusiform rust. Fire protection favored growth of oak, the essential alternate host, creating more inoculum for pines. Large monoculture plantations favored spread of fusiform rust. There are other parts of the northern hemisphere that have both susceptible pines and oaks but still do not have fusiform rust; the disease is a potential threat in all such areas.

Some control is achieved by the use of resistant pine families, selected in areas of high disease incidence. However, races of the fungus vary in virulence and pathogenicity to pine selections, and control is always uncertain. Rust races are not well defined, but several form species of *C. quercuum* can be identified by isozyme and serological analyses. DNA technology could be useful for identification of races.

14.4. Other Examples Involving Changes in Agriculture

Many changes in agriculture and forestry, both major and minor, have led to changes in the incidence of disease. These include changes in rotations, irrigation, planting times, harvest methods, fertilization, and size of fields. A few examples that have encouraged damaging diseases are summarized.

Several plant diseases became damaging following the use of minimum tillage for soil conservation. With minimum tillage, the crop is seeded directly into the soil, through the crop residue of the previous year. Many fungal and bacterial pathogens survive from season to season in such debris and go readily to new seedlings. Deep plowing destroys most such pathogens, especially if used along with a two-or-more-year rotation with another crop. Examples of maize diseases encouraged by minimum tillage include anthracnose, *Cercospora* leafspot, *Cochliobolus* blights, *Phyllosticta* leafspot, and *Physoderma* brown spot (Sumner, Doupnik, and Boosalis 1981). Bacterial blight and anthracnose are encouraged in beans, as is tan spot in wheat (Schuh 1990).

Maize anthracnose, caused by *Colletotrichum graminicola,* was not a problem before about 1970; it became increasingly important with increased use of minimum tillage (Nicholson 1992). Another factor in severity of the disease was increased use of certain highly susceptible inbred lines for producing hybrid seed. Infection causes leaf blight and root rot of young seedlings, plus stalk, ear, and kernel rots in mature plants. Young seedlings are infected by inoculum from infested debris. More mature plants have few new infections, but at the time of anthesis (preceding senescence), the leaves, ears, and stalks again become very susceptible (Warren and Nicholson 1975; Warren and Shepherd 1976; Lipps 1983).

Cercospora, or gray leafspot of maize, became a problem for the same reasons that led to an increase in anthracnose (Payne, Duncan, and Adkins 1987). Tan spot of wheat, caused by *Pyrenophoa tritici-repentis,* also increased in prevalence with minimum tillage (Schuh 1990).

Root rot of forest trees in some areas became destructive when management favored thinning of stands on a large scale. *Heterobasidion annosum* (= *Fomes annosus*) causes root and butt rots of many species, especially conifers. It is a minor problem in natural stands, occasionally causing root and butt rots in mature trees; in contrast, young trees in plantations often are killed. The fungus spreads via basidiospores and with root grafts between diseased and healthy trees (Powers and Verall 1962; Hodges 1969; Ross 1975). The disease becomes serious in plantations when they are thinned; stumps created by thinning or harvesting are quickly colonized by airborne spores. The fungus then spreads to adjacent undisturbed trees via root grafts, creating a spreading site that eventually stops, for unknown reasons. Treatment of stumps with borax, urea, or sodium nitrate is effective in control (Driver 1963). There also is an effective biological control (Chapter 6, section 4), achieved by inoculating fresh stumps with a competitive saprophyte (*Phlebiopsis gigantea,* a fungus) (Sinclair, Lyon, and Johnson 1984). Another disease similar to *annosus* rot is caused by *Phallinus pini.*

Pitch canker of pine, caused by *Fusarium moniliforme* var. *subglutinans* (= *F. subglutinans*) was discovered in the southern Appalachian region of the United States in the mid-1940s. The disease was of no significance for the next thirty years. Pitch canker is now a major disease in plantation pines from Virginia south to Florida and west to Texas; natural stands are seldom affected. Many *Pinus* and other species are suscepts; the disease is of serious concern in pitch, Virginia, slash, and shortleaf pines, often killing 25 percent or more of young plantation trees and deforming many more. The disease is known as pitch canker because lesions become resin soaked.

The pitch canker pathogen requires wounds for entry. It became serious in seedling nurseries, seed orchards, and plantations; all are recent management developments in southeastern forestry. Trees in such situations are damaged frequently by branch pruning, mowing, and cone harvesting. Insect bites serve as points of entry, and insect populations often are higher in plantations than in natural forests. Wind, ice, and hail also cause wounds. Monoculture is a factor. Stem cankers can become perennial and may girdle main trunks, killing the tree. However, twig blighting, resulting from small branch girdles, is the most common symptom (Sinclair, Lyon, and Johnson 1987).

Some diseases are encouraged and others are discouraged by mechanization of agriculture. *Ceratocystis* canker of stone fruit trees increased greatly with mechanization in orchards. Infections occur through injuries in the bark; these injuries increased greatly when mechanical harvesters were employed in the almond, prune, peach, and apricot orchards of California and elsewhere (Barnes 1964). Power sprayers and aircraft application of pesticides have decreased the severity of some diseases. Sprinkler irrigation often favors leaf diseases (Curl and Weaver 1958) such as anthracnose. Irrigation also favors certain virus diseases by facilitating their spread each year between successive crops and by more survival of weed hosts during dry seasons (Thresh 1982).

Increased use of fertilizers can lead to increased populations of virus vectors (see Chapter 15, section 1) and can predispose crops such as rice to disease. When large fields replace small plots, many viruses spread more readily because vector populations can build up, leading to long-distance spread. Other diseases are less severe in large than in small fields because there are fewer borders with weed hosts and other crops that harbor viruses; there is an example with cucumber mosaic in Florida. Crop sequences without rotation can lead to buildup of certain pathogens that are easily passed from one crop generation to the next (Thresh 1982).

Bacterial blight of rice, caused by *Xanthomonas oryzae* pv. *oryzae,* is a systemic infection that blocks the xylem. It was a minor disease prior to the development of modern cultivars that resist lodging; their use led to increased use of nitrogen fertilizers, which favored blight. The disease is now a major problem in Asia, and it is of increasing importance in Africa and South America (Mew 1987; Mew et al. 1993).

Virus diseases of chickpea became major problems in California when planting time was changed from summer to winter. This change led to increased populations of aphid vectors, which, in turn, led to increased infection by six different viruses, including cucumber mosaic and legume yellows viruses (Bosque-Perez and Buddenhagen 1990).

The bacterial pathogen *Pseudomonas solanacearum* is of special interest; further study could give insights into the origin and evolution of bacterial pathogens. The bacterium has been isolated from virgin soils (that is, nonagricultural) in Florida, Indonesia, and Central America; thus, it was widely distributed, probably before cultivation started. *P. solanacearum* may have evolved from a soil saprophyte at various sites; related saprophytic bacteria are common and widely distributed. *P. solanacearum* often loses pathogenicity in culture, although survival ability of these saprophytic forms is unknown (Buddenhagen and Kelman 1964).

P. solanacearum, widespread in warm areas, is best known as the cause of serious wilt diseases of tobacco and banana; it also infects tomato, potato, other Solanaceous species, and peanuts, with destructive effects. The disease appeared in various places at about the same time, always when cultivation of susceptible crops was intensified; there is little doubt that both monoculture and intensive cultivation are involved. The disease became increasingly severe at each site with succeeding crops.

There were many early reports of plant diseases in the tropics that later were attributed to *P. solanacearum.* A wilt disease of tobacco was reported in Indonesia in 1898 but was known to farmers there by 1864. Other early reports for tobacco were from Sumatra (Indonesia) and Japan (1881). The first credible report of pathogenicity in *P. solanacearum* came from Erwin F. Smith in 1890, in reference to a disease of tomato and potato in Mississippi (United States). Soon, the disease was reported in the Carolinas and Florida.

A destructive disease of tobacco was reported in 1903 from Granville County, North Carolina, where the finest of all flue-cured tobacco was intensively cultivated; farmers in the area had known the disease since 1881. Granville wilt had severe social and economic consequences in the bright-leaf tobacco areas of the southeastern United States; there were refugees from rural areas of North Carolina. *P. solanacearum* is now known in Brazil, the Philippines, Australia, Central America, the Caribbean area. Portugal, South Africa, and China (Kelman 1953; He, Sequeira, and Kelman 1983).

P. solanacearum has many means of dispersal: seedling transplants from infested areas, rhizomes of banana used in propagation, potato seed tubers, insects, and pruning tools. There are many weed hosts that allow survival (Hayward 1991). Thus, it has been impossible to determine areas of origin. However, the pathogen and/or its hypothetical saprophytic predecessor are widespread in the tropics.

Once soil is infested, control of *P. solanacearum* is difficult; there have been some successes by the use of resistance genes, but only in tobacco and peanuts (Hayward 1991). Three races are well known, and there are various strains (genotypes) in each race. Race 1 affects primarily Solanaceous species. Race 2 is specialized to banana and *Heliconia,* a related plant. The major host of race 3 is potato. The banana race is distinctly different from and more specialized than the others; triploid bananas are not affected by races 1 and 3 (Buddenhagen and Kelman 1964). Data for *P. solanacearum* from China indicate divergent evolution in that area; two additional races and many biotypes were found (He, Sequeira, and Kelman 1983).

Molecular-genetics methods are being used with *P. solanacearum,* but the results are still inconclusive. The roles of bacterial *avr, hrp,* and exoenzyme genes in disease development have been examined; mutants lacking endo-polygalacturonase and endogluconase were still pathogenic, and *hrp* mutants remained infectious. The role of exopolysaccharidases was not clarified (Boucher, Gough, and Arlat 1992).

The hoja blanca disease of rice in Venezuela became serious with the use of irrigation, which allowed for two rice crops per year. Hoja blanca is a widely distributed virus disease that is spread by leafhoppers. In Venezuela, vector populations were favored by the higher humidity in fields, and this led to increased spread of the disease from weed-grass reservoirs that were 100 percent infected (Ten Houten 1974).

Pear decline was absent in the western United States until 1940. The disease became very damaging in the Pacific Coast region from 1952 to 1964. The reason was the increased use of Oriental pear rootstocks, which are resistant to other diseases but are very susceptible to pear decline, an insect-vectored virus disease (Ten Houten 1974). Pear decline is discussed as an alien plant exposed to a native pathogen in Chapter 8, section 5.

Changes in cultivars often have unexpected results. When banana production was changed from cv. Gros Michel to Cavendish, a nematode disease became a problem because the latter is very susceptible. The "green revolution," based on the use of new cultivars, gave new disease problems that required further breeding for correction (Ten Houten 1974).

There are many more examples of diseases that increased with changes in agricultural practice, but the ones discussed are sufficient to make the point.

15

Anthropogenic Reintroduction Each Year

Many annual crops are invaded each year by pathogens that move from reservoir plants, either crop or weed. The most destructive examples occur in crops with intensive cultivation that covers large areas; this coverage provides source plants for massive inoculum and host plants for targets. Both local and long-distance distributions can occur, depending on the disease and circumstances.

Reservoirs for some pathogens are in weeds, often perennials, that occur in and around cultivated fields. Weeds, in turn, sometimes are heavily infected because of their proximity to susceptible crops, a situation that favors dissemination to weeds. Some crop diseases exist only because of inoculum reservoirs in weeds or native flora. In other cases, double-cropping each year or even rotations of crops that are susceptible to the same pathogens can provide the continuity needed for heavy infections. These are the various anthropogenic inputs that contribute to some diseases.

The most dramatic examples of annual reintroduction of pathogens are with yellow dwarf of cereals and downy mildew of tobacco. The annual input of pathogens into crops, from distant reservoir plants, is a contributing factor for many other diseases, notably stem rust of wheat (Chapter 11, section 1).

15.1. Yellow Dwarf Diseases of Cereals

The virus that causes dwarfing and discoloration of cereals is a significant example of pathogens that move each year into large areas of cultivation. The virus is not seed or mechanically transmitted, and it does not survive from season to season in some areas; thus, the farmer starts each year with a virus-free crop. The only means of spread into these virgin fields is via virus-carrying aphids that survive locally on other plant species or that move in each year with the wind from distant sources. The relative importance of local or distant sources of the virus depends on the location. A knowledge of aphids as vectors of plant disease is the key to understanding the epidemiology of yellow dwarf diseases of cereals.

The cereal dwarf diseases are major problems in small grains. Worldwide, they probably are the most damaging of all virus diseases in cereals. Losses vary from year to year, but loss of 10 percent of the crop is common; losses during severe years can go to 50 percent in some local areas (Bruehl 1961). The literature on yellow dwarf and the causal virus is enormous (Bruehl 1961; Irwin and Thresh 1990; D'Arcy and Burnett 1995).

Major symptoms of yellow dwarf are discolorations (yellowing or reddening) and stunting, with drastic decreases in numbers of tillers and heads. These conditions lead to decreases in yield, the extent depending on time of initial infection of a large number of plants in a field. The symptoms are not distinguishable from those caused by malnutrition, insect attacks, unfavorable temperatures, or various noxious chemicals. Nondistinctive symptoms were responsible, in part, for the fact that the true nature of the disease was not recognized for many years. Eventually, all other causes were ruled out, and the virus relationship was demonstrated. Cereal dwarf diseases are still hard to diagnose, especially in autumn and in pasture or weed grasses. Epidemics that start in fields of young seedlings have the most striking symptoms and lead to the greatest losses; later infections are less severe. Reliable detection of the virus in fields depends on the use of monoclonal antibody assays (D'Arcy, Hewings, and Eastman 1992).

The causal agent is properly known as the barley yellow dwarf virus, in accord with custom in plant pathology; viruses are named in regard to the first-discovered host. The virus causes a common and very serious disease in oats, often called the oat red leaf disease; oat fields in the mid–United States rarely escape. Wheat is another important host, with widespread losses. In addition, the virus has many other hosts in the *Poaceae* (*Gramineae*), affecting more than one hundred species of grasses. Hosts include sorghum, maize, rye, rice, triticale, fescues, Johnson grass, bluegrass, orchard grass, brome, wild barley and oats, and canary grass (Oswald and Houston 1953b). Some of these hosts, including maize and tall fescue, harbor the virus but usually show no symptoms; they can serve as reservoirs of the virus, which subsequently spreads to other crops. Weed hosts and volunteer cereals are important epidemiological factors.

The origin of barley yellow dwarf virus is unknown. It belongs to a large group known as *Luteoviruses* (= yellowing); included are beet yellows, carrot leafroll, potato leafroll, soybean dwarf, bean leafroll, and tobacco necrotic leaf viruses. Data from serological and nucleic-acid hybridization tests, plus host range data, show relationships between the *Luteoviruses* (Martin et al. 1990). Minor changes in the chemical composition of any of these viruses might have created the barley yellow dwarf virus. Perhaps there was a common ancestor of all the *Luteoviruses,* and it or its variants were present in the native vegetation of some areas long before the virus moved to cultivated crops. Mild or missing symptoms in some infected host grasses suggest a balanced host–parasite relationship. In any case, the barley virus was in cultivated crops long before it was identified.

Many early reports of epidemics of unknown cause probably can be attributed to the barley yellow dwarf virus. There were such epidemics on oats and other cereals in the United States from New England to Georgia by 1890. Oat redleaf was serious in New York in 1937, and there were many other bad years (Bruehl 1961).

The cause of the mysterious cereal diseases finally was found to be an aphid-transmitted virus (Oswald and Houston 1953a). Intensive studies were stimulated by serious outbreaks of discoloration and dwarfing in barley fields in California. Aphids from diseased barley plants were collected and caged with healthy barley seedlings; typical symptoms appeared in the previously healthy seedlings in two to three weeks. This was the origin of the name "barley yellow dwarf." Similar results were soon reported by researchers in Minnesota, Arkansas, Mississippi, Iowa, and elsewhere. Following the lead of the California group, cereal diseases caused by the barley yellow dwarf virus were soon found throughout cereal-growing regions of the United States (including Alaska), Canada, Mexico, Europe, Australia, Asia, New Zealand, and elsewhere (Bruehl 1961).

Reports after 1960 suggest explosive worldwide increases in populations of both aphids and cereal dwarf diseases. The larger aphid population appears to be correlated with worldwide increases in the use of fertilizers. The added nitrogen is known to increase reproduction of aphids and to stimulate conversion to the winged form (Baranyovits 1973). The increased use of irrigation, as well as other changes in agriculture, also may have contributed to increases in aphid numbers and to the incidence of cereal dwarf diseases (Bruehl 1961; Brown, Wyatt, and Hazelwood 1984). New cereal cultivars introduced since 1960 have higher yields than old types, but they require heavy use of fertilizers.

The cereal dwarf diseases have a complex epidemiology that requires an understanding of the aphid vectors and their movements. There are more than twenty species of aphid vectors that differ in migratory habits and in relative ability to carry the virus. Also, there are many distinct variants of the virus that differ somewhat in hosts and that affect efficiency of vectors. Both the virus and the vectors are worldwide in distribution, and populations of each will vary with weather conditions. The viruses are totally dependent on aphids, which differ in behavior. Once infected, an aphid carries the virus and can spread it for life, even through several molts, although the virus is not transmitted through the eggs (Irwin and Thresh 1990). In addition, there are many ubiquitous host species, some that are perennial reservoirs for the virus. There are, of course, the usual complexities of environmental effects on the host–pathogen relationship.

The barley yellow dwarf virus has several characteristics that affect its spread. The virus is restricted to the phloem of host plants; the phloem is the tissue utilized by feeding aphids. A given virus strain does not protect a plant or a vector from infection by other strains; plants often carry several (Allen 1957; Rochow 1970). It soon became evident that many variants of barley yellow dwarf virus exist and that aphid species are selective as to the strain that each will vector (Allen 1957). Virus strains have been described mostly on the basis of selective vectors.

The virus strains are named with acronyms to designate their major vector species. For examples: Strain MAV is vectored by the aphid *Macrosiphum avenae* (since renamed *Sitobion avenae*); RPV by *Rhopolosiphum padi* (oat-cherry aphid); RMV by *R. maidis* (corn leaf aphid); and SGV by *Schizaphis graminum* (English grain aphid). The most prevalent strain in America is PAV, perhaps because there are many vectors that carry many strains of the virus; the most effective vector of strain PAV is the oat-cherry aphid (*R. padi*). PAV is extremely variable, so researchers often characterize the variants as PAVlike. Nucleic acid analyses of the virus were used to evaluate and confirm the differences in the viral stains (Gildow, Ballinger, and Rochow 1983; Fattouh et al. 1990). However, the classification based on major vector species may break down as more vectors are examined. The ribonucleic acid constituent of barley yellow dwarf virus was first described by Brakke and Rochow (1974).

Aphid movements are complex, both locally and long range, and knowledge of this is necessary to understand the cereal dwarf epidemics. Individual aphids, both winged and nonwinged forms, move often between plants in a field (Holms 1988); this movement accounts for enlargements of focal points of infection. The same aphid forms, often carrying the virus, are carried by wind to nearby fields. Some winged aphids are transported via wind for great distances. For example, aphids survive wind transport from northern Europe to the Arctic island of Spitzbergen, a distance of 800 miles (Elton 1925). Long-distance movements are very important for the occurrence of cereal dwarf epidemics in some parts of the world.

Major epidemics in mid–North America often begin from long distance movements of viruliferous aphids. Cereals are grown from Mexico northward through mid–United States to southern Canada, with range and pasture grasses interspersed. This movement makes a broad and almost continuous area of plant hosts for several species of migratory aphids that overwinter only in the south. Both viruliferous aphids and virus survive in Texas from season to season in infected weed reservoirs. During some years, the insects build up to enormous populations in Texas and Oklahoma, with totals depending on weather conditions. A mild winter with light rain, combined with a cool, damp spring, results in high aphid populations and a bad virus year farther north. The cereal crops and weeds mature in Texas and begin to dry in late spring; the resident aphids then begin a northward movement to younger grain fields. The corn leaf aphid and the greenbug (an aphid, *Toxoptera graminum*) are major migratory species. Both are carriers of the virus, and neither overwinters north of Arkansas or southern Illinois.

Winged forms of aphids from the huge southern population become windborne, and they can move 50 to 200 miles nonstop. The aphids land and feed after each northward movement. This is the pattern of disease transmission and virus reproduction. The next northward step is taken by the offspring of a previous generation of vectors. Eventually, viruliferous aphids reach southern Canada, where they do not survive winter but do start destructive epidemics in cereal crops. The aphids are carried northward by the same air movements (Figure 11.4) that carry rust spores, which result in the great epidemics of stem rust of wheat (Chapter 11, section 1).

How do we know that all this occurs? Aphids carrying strains of the virus previously unknown in local areas but common farther south were collected. Insects collected from aphid "showers" were shown conclusively to carry the virus. Symptom development in grain crops occurs simultaneously over the 50–200 mile "steps," two to three weeks (the usual incubation time) after the "shower" occurred. Long-range transport of aphids, in steps from Texas and Oklahoma to Minnesota, was documented (Hodson and Cook 1960), as was movement to Wisconsin (Medler and Smith 1960) and Iowa (Wallin and Loonan 1971). The great movements of rust spores are documented even better (Nagarajan and Singh 1990). Some aphid species can survive winter in these northern states and can start virus outbreaks. However, the damaging and widespread epidemics in mid–United States appear to come with the invaders from the south. Outbreaks from local sources in mid–United States usually are later to start, and plants infected when older are not so severely damaged.

The situations differ in other parts of the world. In some places, the epidemics start from viruses that survive locally from one growing season to another in aphids and in infected weed hosts. For example, irrigated maize supports both aphids and virus over the summer in the Pacific Northwest in the United States; maize is a symptomless host. Cereal grains are planted in fall, and aphids move from maturing maize to new seedlings. These aphids often carry the barley yellow dwarf virus; for example, 61 percent of aphids (*R. padi*) in young grain fields in Washington state were viruliferous in 1980. In these dry areas, sparse weeds in summer will support few or no aphids, but irrigation allows both maize and lush weeds to grow. Thus maize is a major carry-over reservoir for the virus in eastern Washington (Bruehl 1961; Brown, Wyatt, and Hazelwood 1984). The virus in maize probably comes each year from wheat and other small grain crops harvested soon after maize is planted. Local virus reservoirs in grasses, mild winters, and both fall and spring planting of cereals are usual in the Pacific coastal area of the United States; there, local sources of the virus and vectors are major factors in epidemics. Local weather conditions determine whether or not a given year will have serious outbreaks of aphids and cereal yellow dwarf.

Winter is the critical season for survival of both virus and vectors in some other regions. Many aphid species do not survive winter in southern Canada. Other species, such as *R. padi,* survive on a woody host (*Prunus*) but are not viruliferous until they feed on infected grasses (Irwin and Thresh 1990). Local reservoirs for virus survival are not always evident. Small grains are seeded in fall in southern Ontario, Canada; locally derived aphids carry virus to new grain seedlings, and virus survives there until spring. In the northern United States, unusually mild winters allow some viruliferous vectors to survive until spring.

The critical season for virus survival in the Mediterranean region is summer. Maize is the oversummer reservoir in Italy, and aphids move virus to newly seeded small grains in fall (Coceano and Peressini 1989).

Grass species are major reservoirs in some areas (Fargette, Lister, and Hood 1982). Tall fescue grass is a major reservoir in Missouri and Arkansas; 60 percent of fescue samples taken in Missouri carried the virus (Grafton et al. 1982; Mahmood et al. 1993). Tall fescue is a forage grass in the area, and it is planted along highways; there were 800,000 ha in Arkansas that were commonly infected with barley yellow dwarf virus but were lacking symptoms. Fescue also is a favorite host of the oat-cherry aphid, a good vector of the prevalent PAV strain of the virus.

Effective control measures for the cereal dwarf diseases are not available at this time. Some oat cultivars are more tolerant than others (Jedlinski, Rochow, and Brown 1977), and their use has been encouraged. A few genes that give some resistance are known, but they have disadvantages. Yd_2 and Yd_3 genes from wheatgrass give resistance to some strains of the virus, but undesirable traits are linked with these genes (Baltenberger, Ohm, and Foster 1987). So far, genes from highly resistant grasses have not been exploited; this should be possible with recombinant DNA techniques. Breeding for resistance to aphid vectors is another possibility. Pesticides for vector control have given poor results (Irwin and Thresh 1990), perhaps because a single aphid feeding for a relatively short time can transmit the disease (Oswald and Houston 1953). With age, small grain plants become increasingly resistant; thus, early planting would seem to be advantageous. However, early plantings have a greater chance of exposure to aphids that carry the virus and to insects such as the Hessian fly.

Resistance to yellow dwarf was found in barley from Ethiopia; more than 20 percent of accessions from there to the United States had some degree of resistance. Although barley did not originate in Ethiopia, it has been grown there for thousands of years; native land races are well adapted to the yellow dwarf virus (Harlan 1977).

Summary

Viral-induced dwarf diseases of cereals cause economically significant losses, known only by a few plant scientists. The causal agent, barley yellow dwarf virus, was named from the first identified host. This virus would not exist without aphid vectors; there are no other means of transmission. Major crops affected are oats, barley, wheat, and rice; there also are more than one hundred host species in the family Graminae (grasses). Maize and several common grasses are symptomless carriers that serve as reservoirs from which the virus is carried by aphids to newly planted cereals. This worldwide malady has a complex and difficult epidemiology. There are many strains of the virus that differ in effects on host species, and there are more than twenty species of aphid vectors with selective abilities to transmit different virus strains. Aphids differ in migratory patterns; they carry the virus over great distances via prevailing winds. Aphids also spread the virus locally. Available controls generally are ineffective, although cereal cultivars vary in tolerance to cereal dwarf virus.

15.2. Blue Mold of Tobacco

Blue mold of tobacco is reintroduced each year in some areas by spores carried with prevailing winds; otherwise, the disease could disappear from such areas. The reintroductions are possible because there are concentrations of susceptible crops in both source and target areas. Two other factors that can affect disease incidence are movement of diseased plant materials by commerce, and monoculture. Blue mold is associated with intense cultivation.

Blue mold (*Peronospora tabacina* = *P. hyoscyani*) is a downy mildew that belongs to the phylum Oomycota, order Peronosporales. Fungi of this type were formerly known as Phycomycetes. There are many species of downy mildews, all host-specific obligate parasites. Downy mildew genera include *Peronospora, Pseudoperonospora, Sclerospora, Pseudosclerospora, Bremia,* and *Plasmopora* (Spencer 1981). Other crops that have important downy mildew diseases are hop, grape, *Brassica* species, cucurbits, gramineous plants, onion, soybean, sunflower, and beet. Of these, downy mildews of grape, tobacco, and small grains are best known. *Plasmopara viticola* on the grape, important in the history of plant pathology, has been a major problem in Europe (Millardet 1885; Viennot-Bourgin 1981) (Chapter 7, section 4).

Peronospora tabacina, in nature, occurs only on plants of the genus *Nicotiana,* although some related genera of the *Solanaceae* have been infected in inoculation experiments. There are several suggested centers of origin for *Nicotiana,* all in temperate regions: Australia, the western United States and adjacent Mexico, and Argentina (Clayton and Stevenson 1943; Wolf 1947; Johnson 1989). Several wild tobaccos are endemic in Australia, including *N. debneyi, N. rustica, N. goodspeedii,* and *N. suaveolens.* The Australian species are, in general, resistant to blue mold, but they can host the fungus. Five native American species occur in the Rio Grande valley of Texas and Mexico: *N. repanda, N. glauca, N. loniflora, N. trigonophylla,* and *N. plumbagnifolia* (Wolf 1947). All the North American species are susceptible to blue mold. *N. loniflora* is thought to be endemic in Argentina, and several *Nicotiana* species grow wild in California and other western states of the United States. No native *Nicotiana* species occurred in the United States east of the Mississippi River.

The origin of cultivated tobacco (*N. tabacum*) is unknown; it does not exist as a wild plant and appears to have been created by people. The eastern region of the United States is now the major area of tobacco production; little or no tobacco is grown commercially in the western United States. The major areas of production in the western hemisphere are in south Georgia/north Florida, the Carolinas/Virginia (flue-cured or bright-leaf type), Kentucky (burley type), Connecticut (shade-grown cigar wrapper), Ontario (flue-cured), and Cuba (cigar tobacco). The first users of tobacco, the Amerindians, smoked leaves of *N. rustica,* a blue mold–resistant species (Clayton and Stevenson 1943).

Origin of the blue mold fungus is still uncertain. Some scientists consider it to be endemic in all native areas of the genus *Nicotiana*. However, Australia appears to be the most likely place of origin; *P. tabacina* was first found there, and the expected tolerance to *P. tabacina* occurs in wild Australian populations of *Nicotiana*. The fungus was found at an early date in the Rio Grande valley of North America, where the native species are susceptible. Wild plants of *Nicotiana* were widely scattered before agriculture was developed in the region, and disease did not spread readily. Cultivation favored the wild *Nicotiana* species as weeds, and *N. repanda* became common (Wolf 1947; Renfro and Shankara-Bhat 1981). Thus, susceptibility of the Rio Grande species does not rule out the area as a possible endemic home of the blue mold fungus. *P. tabacina* was found in the western United States as early as 1895, on native *Nicotiana* species (Johnson 1989). Very early reports of *P. tabacina* in South America are questionable; credible reports place it in South America in 1939. *P. tabacina* has not been found in India, China, New Guinea, New Zealand, or Africa south of the Sahara (Weltzien 1981; Johnson 1989).

Tobacco became a major crop along the eastern seaboard of the present United States almost from the beginning of European colonization, but blue mold was not seen there from 1620 to 1921. Perhaps the fungus came from Texas with the wind (Valleau 1947); if so, how did it skip tobacco in Louisiana, along the way? A Texas origin is rational only because the disease appeared in Georgia and Florida soon after weed tobacco and blue mold populations increased in Texas; it appeared also soon after tobacco production was greatly increased in the southeastern United States. Others speculated that blue mold came to Florida from the Caribbean area; however, the nearest and largest center of tobacco cultivation there is Cuba, and blue mold was unknown in Cuba until 1957. Perhaps the fungus came from Australia as latent mycelium in seeds (Viennot-Bourgin 1987). The major question about an Australian origin arose because blue mold appeared in the eastern United States but not at that time in New Zealand, South Africa, Europe, and Asia (Clayton and Stevenson 1943).

Blue mold in Australia appeared soon after tobacco cultivation began in 1840, possibly as early as 1850. The disease was first described on cultivated tobacco in Queensland and, by 1900, was the major tobacco disease in eastern Australia. There is little doubt that blue mold on *N. tabacum* in Australia came from native species of *Nicotiana* (Johnson 1989).

A small outbreak of blue mold in Georgia (United States), in 1921, marked the first appearance of *P. tabacina* in eastern North America (Lucas 1980; Weltzien 1981). This infestation apparently was eradicated, or at least disappeared, until 1931, when the disease reappeared in the same area and soon spread to the north, through the Carolinas and Virginia to Maryland (Clayton and Stevenson 1943). Blue mold has been in this area ever since, although it was a minor disease from 1955 to 1970. The disease was found in Cuba in 1957, then disappeared until 1979 (Lucas 1980). Blue mold now affects tobacco production in North America everywhere that tobacco is grown, from Florida north to Connecticut and Ontario. Most blue mold prior to 1979 was in tobacco seedbeds; then suddenly, in 1979, blue mold became a major disease in fields of burley, flue-cured, and shade-grown tobacco. This continental epidemic was thought to be caused by a new and virulent race of *P. tabacina;* computer analyses indicated the Caribbean area as the source of the new race.

In the United States, blue mold now moves north each year, from Florida and southern Georgia to Connecticut (Lucas 1980; Aylor 1986). Its movement is correlated with air flows that carry spores in short and long steps, up to 700 km each (Davis 1987). The yearly movement is possible, in part, because the major tobacco cultivars used in the eastern United States and Ontario are susceptible to blue mold. There appear to be similar patterns of seasonal movement in Europe and Australia.

P. tabacina was assumed, for a long time, to overwinter as oospores, often found in decaying tobacco tissue. This assumption appears to have been based on behavior of other downy mildews. There is no conclusive evidence that blue mold overwinters as oospores in areas with cold winters (Main and Davis 1989); oospore survival in the United States is not known to occur north of Georgia (Valleau 1947). Thus, oospores are now discounted as a significant factor in epidemiology. The fungus survives the winter in Florida and southern Georgia as systemic infections in dead stalks and in roots of tobacco plants that live through the winter. Infected volunteer plants and, possibly, oospores as well, also survive there. The fungus in systemically infected plants sporulates in spring, and sporangia are carried by wind to new tobacco seedbeds.

The major epidemiological factor for blue mold is spores carried northward each spring with prevailing winds (Davis 1987; Main and Davis 1989). Movement to the north also can occur with seedling plants from the Deep South that are sold in northern areas (McKeen 1989). Some observers think that blue mold spores can be blown in steps from Texas to Kentucky

(Nesmith 1984), but there is little or no evidence for such, either direct or circumstantial (McGrath and Miller 1958).

Blue mold was taken from Australia to Britain in 1959 by a chemical company, to be used to test fungicides in glasshouses (Klinkowski 1962; Delon and Schiltz 1989). That same year, blue mold was taken from England to the Netherlands on tobacco plants for virology research. The disease spread quickly via wind-borne spores to tobacco fields in the Netherlands, Belgium, and Germany (Populer 1981). The effect on tobacco in Europe was devastating. The scourge continued to spread; within four years, it reached North Africa and the Near East. By 1967, blue mold was across Asia into Cambodia (Democratic Kampuchea), traveling at a rate of more than 1,000 km per year (Gregory 1973).

P. tabacina now spreads north each year over Europe, with spores carried by the wind. A blue mold warning system was started in Europe in 1961. The cooperative Center for Scientific Research Relative to Tobacco (CORESTA) monitors its annual spread from south to north and warns growers of blue mold danger. The blue mold starting point each year is the Mediterranean area, where tobacco plants survive the mild winters. Also, the Mediterranean countries now have *N. glauca* as an introduced weed, and blue mold can survive the winter in it. There is still the possibility that the fungus sometimes survives farther north as systemic infections in old tobacco stalks left in the field. However, oospores do not seem to have a role in blue mold epidemiology in Europe and elsewhere (Delon and Schiltz 1989).

Symptoms of blue mold infection are variable, depending on weather conditions, age of plants when infected, and stage of disease development. Early symptoms on plants in the seedbed are pale green spots on the upper leaf surface, with a fuzzy blue fungal growth on the lower surface (hence the name blue mold). Tissues soon collapse and turn brown, giving a scalded or frost-injury appearance to the seedbed. Infected plants recover under dry conditions but frequently are killed if wet conditions continue. Leaves of plants in the field have necrotic spots or are killed when conditions favor *P. tabacina*. Blue mold in individual fields usually starts on a few infected plants; wind-blown spores from initial infections enlarge the infested area (Waggoner and Taylor 1955).

Blue mold causes damage during seasons that are cooler and wetter than normal; thus, blue mold outbreaks are erratic. In the eastern United States, there was a widespread epidemic in 1931–2, from Georgia to Maryland and west to Louisiana. The greatest seedbed epidemic was in the Carolinas in 1949 (Lucas 1980). Another extremely serious epidemic occurred in 1979, a cool and wet year, from Cuba north to Ontario; only Wisconsin was spared. Previously, Ontario seldom had the problem, but in 1979 more than 30 percent of the crop was destroyed. The Ontario disaster was attributed to importation of infected seedlings from Florida; this allowed a very early start for the disease (McKeen 1989).

The year 1979 also was exceptional in that it marked the beginning of blue mold as a field as well as a seedbed problem. A new race of *P. tabacina* probably was involved. Blue mold destroyed 90 percent of the crop in Cuba in 1979; the disease also was severe in Central America and elsewhere in the Caribbean zone. Since 1979, blue mold has declined in eastern North America, due, in part, to drier conditions and preventive programs, including a warning system and the use of fungicides (Nesmith 1984).

The U.S. Department of Agriculture established a blue mold warning system, in cooperation with state agricultural experiment stations, in 1945. Reports on the presence of blue mold and predictions of epidemics were sent to all tobacco areas. This system was the model for the European system (CORESTA). Unfortunately, activity by the United States program had declined with a lull in blue mold outbreaks, and it was not in effect in 1979. A new system was started immediately, modeled on, but improved from, the old system. When blue mold danger increases, growers are advised to increase control measures, including the use of fungicides (Nesmith 1984).

Blue mold control has always included the use of protective fungicides. A systemic fungicide (metalaxyl) is now used in seedbeds. However, the fungus can develop resistance to metalaxyl (Johnson 1989), first noticed in Mexico, the Caribbean area, and North Carolina (Bruck, Gooding, and Main 1982; Rufty 1989).

Breeding for resistance to blue mold has had limited success (Matthews 1981). There was early breeding work in Australia, the United States, and Europe (Schiltz 1981; Johnson 1989; Rufty 1989). Clayton, an early researcher, hybridized *N. tabacum* with *N. debneyi* but had trouble with sterility in progeny. The efforts finally were successful, resulting in resistant lines of burley, flue-cured, and cigar-wrapper tobaccos. The *Nicotiana* species of Australia have at least one major dominant gene for resistance, but it does not perform well in *N. tabacum;* the cultivars (including those from Clayton) that carry this gene have had limited use, in part because the resistance is greatly affected by nongenetic factors (Rufty 1989). Resistance is poorly understood, and its biochemical basis is unknown. All six *Nicotiana* species from Australia carry resistance or immunity of some kind; all native American species (ten or more) are susceptible to blue mold (Rufty 1989).

What is needed is horizontal or multigene resistance, which should give some resistance to all races of the fungus.

Races of *P. tabacina* were first found in Australia. Race ATP1 is the wild race of Australia; resistance to it was found in *N. goodspeedii* and was transferred to *N. tabacum*. Race ATP2, affecting plants with this resistance, appeared in 1966, and all cultivars with the *N. goodspeedii* gene became susceptible in the seedling stage. Race APT3 was found in previously resistant *N. velutina* in Australia (McKeen 1989). There are virulent but unnamed races in Europe. The 1979 epidemic in eastern North America is believed to have been caused by a new race, but this has not been examined carefully.

Summary

Blue mold of tobacco, caused by *Peronospora tabacina,* was discovered in Australia on native species of *Nicotiana;* it became a problem there on cultivated tobacco (*N. tabacum*) in 1891. *P. tabacina* may also have been endemic elsewhere. A major tobacco area is the eastern United States; *P. tabacina* was unknown there until 1921. The source of *P. tabacina* in eastern North America is unknown. The fungus was taken in 1959 from Australia to Britain and to continental Europe with plant materials; since then, it has spread over Europe and into Asia.

P. tabacina does not survive in areas with low winter temperatures; it does survive in live plants in regions with mild winters. Each year the fungus moves north with prevailing winds, into major tobacco-producing areas of the United States, Canada, and Europe. The system requires adequate populations of susceptible crops in source and target areas. The disease is erratic, depending on weather conditions; this tendency led to the development of warning systems. Breeding for resistance was limited because of fungal adaptability and other factors. Fungicides are used for control.

15.3. Other Diseases with Annual Reintroduction

Seasonal invasions of many annual crops occur each year; alternative hosts of the pathogens often are involved. The point is illustrated by several virus diseases that are not seed-borne. Cucumber mosaic offers an excellent example of local spread each year from weeds to crops. The virus infects many hosts, some that show no symptoms. Weeds can be either the overwintering or the oversummering hosts, depending on geographical location. The same weeds can be hosts of aphid vectors that carry the virus to crop plants (Tomlinson et al. 1970; Duffus 1971). Another good example is beet western yellows; the causal virus survives winter in the western United States in weed crucifers and other weed species. Viruliferous aphids move the virus from weeds to sugarbeets in spring (Duffus 1971).

A related topic of interest is the origin of new virus diseases. These often appear when crops are moved to new agricultural areas. There are well-documented examples, with the virus later identified in native plants of the region; maize streak is a good case in point. When maize was taken to Africa, the streak disease appeared; its causal virus was found in at least twenty-two species of native grasses. Another example is cacao swollen shoot, a disease that first appeared when the crop was moved to Africa; the causal virus is endemic in forest trees there (Duffus 1971).

16

Abiotic Diseases:
Damage from Air Pollution

Plants often are damaged by adverse environmental conditions, without the intervention of parasites. Environmentally induced damages clearly are within accepted definitions of disease, in plants as in animals. The same kinds of abiotic disorders affect both plants and animals; in plants, causal factors can be nutritional (low or high N, P, K, and minor elements), low or high temperatures, excesses and deficiencies of water, improper oxygen relations, excess heavy metals, and atmospheric pollution. Notable is the absence of vitamin and amino acid deficiencies; plants make their own.

Salinization of soils, a problem in areas with intense irrigation, is of concern. Available water in such areas often is high in dissolved mineral salts, which remain in the soil after evaporation; eventually, plant-damaging concentrations accumulate. Good agricultural soils have been damaged or destroyed in parts of the western United States, western China, Egypt, and elsewhere. The only available remedy is to flush the soil with water, but affected areas seldom have this option. Salinization of soils is increasing.

The abiotic disorders constitute a vast subject that will not be covered here. Instead, the discussion will be restricted to a more current and much more controversial topic: death of forest trees in various parts of the world and possible involvement of air pollution in some of it (MacKenzie and El-Ashry 1990). Good overall discussions of nonparasitic or abiotic diseases can be found elsewhere (Walker 1969; Agrios 1988).

There are some unequivocal examples of forest destruction by air pollution. One example is the destruction of vegetation in the copper basin of eastern Tennessee, at a place called Ducktown. Many square kilometers of the forest around Ducktown were killed, beginning about 150 years ago, by sulfate fumes from copper smelters (Miller and McBride 1975). The smelters were abandoned approximately 100 years ago, but much of the region still lacks vegetation. This is the mystery of Ducktown, in an area where abandoned farms usually grow a dense crop of weeds, then become covered with young pines within 15 years. A harvestable forest can grow on abandoned fields, with another generation of trees underway, all by natural regeneration within 60 years. Yet this did not occur at Ducktown in more than 100 years. Why? Unfortunately, there have been no thorough studies. Erosion of the denuded area has been severe, but this does not account for lack of vegetation in the valleys and on the relatively level upland spots that can be found.

Another clear example of forest destruction by air pollutants is at Sudbury, Ontario, Canada. Sudbury is the site of nickel mining and smelting; again, the release of sulfur oxide fumes is involved. Recently, high smokestacks have been built, and local vegetation seems to be recovering. There are similar sites elsewhere in the world (California, Montana, British Columbia) (Miller and McBride 1975). Extensive forest destruction, similar to that at Ducktown, was observed in Tasmania (Figure 16.1) and in central China (Scheffer, unpublished).

Photooxidants were known for many years to damage vegetation in southern California, site of the first air-pollution studies. Hydrocarbon and nitrogen oxides react photochemically in the atmosphere, producing ozone (O_3), peroxyacetyl nitrate (PAN), and related compounds. Both ozone and PAN are toxic to plants, and damaging concentrations are common in California, around Mexico City, and in other areas. There is no question that photooxidants cause forest and crop damage. The major sources of hydrocarbons and nitrogen oxides are automobile exhausts and power plants. The decline of vegetation in some areas may result from the combined effects of photooxidants and acid rain.

Acid rain is a long-time suspect for damage to vegetation, but its role has been difficult to prove (Mohnen 1988; Murach and Ulrich 1988; Tomlinson 1990). Acid rain is derived from SO_2 and NO_x in the air; their sources are automobile exhausts and industrial plants. The role of acid rain in forest damage has been controversial, perhaps because indirect or secondary effects are involved. Some data on this problem will be presented. Tall stacks transfer the pollution problem from a local area, as at Ducktown, to distant sites, especially those at higher altitudes.

The decline and death of Norway spruce (*Picea abies*) in central Europe has been a cause of alarm and controversy. Obviously, forest death (*Waldsterben*) is not a general occurrence all over Europe, as sometimes was implied in the past; there are even claims that *Waldsterben* is a fantasy (Skelley and Innes 1994). But forest death does occur in localized areas, and dramatically. There may be more than one cause of forest problems in Europe;

Figure 16.1. SO$_2$ emissions from copper smelters have destroyed the forest over a large area in Tasmania. (Picture by J. H. Hart)

perhaps some of the prevalent disagreements are the result of researchers assuming that their particular study area has general application.

Around 1980, there was general alarm that all European forests were in a state of decline. However, this was never conclusively documented; some of the impressions may have been based on normal cyclic events in the forests. On the other hand, there are areas that are severely affected; in general, these are at higher elevations in the scattered mountain ranges of central and eastern Europe, north and east of the Alps. Dramatic examples are evident in Poland, Germany, and the Giant and Isar ranges in Czechoslovakia (now Czech Republic) (Figure 16.2), where death has spread over thousands of hectares of the spruce forest. The problem is obvious, but less dramatic, in the Fichtelgeberge, Solling, Hils and Harz mountains of Germany, and in the Göttinger Wald (Ulrich and Pankrath 1983). There are small pockets of destruction in the Black Forest, but one must know where to look.

Forest stand die-off can be seen in localized areas in many parts of the world. I have observed such in Europe, Hawaii, Uganda, the Andes Mountains of South America, mountains in central China, California, Japan, and Australia. The causes of such die-offs usually are unknown; in some cases, pathogenic microorganisms are involved. For example, death of the Jarrah forest in Australia occurred after man disturbed the forest, bringing virulent strains of a fungus (*Phytophthora cinnamomi*) into the area. A different situation has led to death of ohia (*Metrosidesoa polymorpha*) trees in Hawaii; this phenomenon appears to be a natural one (not caused by human intervention), occurring in groves that have matured in unison after they were established in unison on old lava flows (Mueller-Dombois 1992). A less striking example, cause still unknown, is death of small areas (known as pocket decline) of red pines (*Pinus resinosa*) in planted forests in the Great Lakes region of the United States. Some of the destruction of planted Monterey pine (*P. radiata*) in South America, New Zealand, and South Africa has been attributed to *Dothistroma* and other blights (Chapter 8, section 4), fungal diseases that kill twigs and branches.

Many causes have been proposed for forest decline and destruction in Europe. The following are a few of the more rational causes: fungi, bacteria, viruses, cyclic decline, air pollution and acid rain (both direct and indirect effects), climate and weather changes, repeated harvests that have depleted the soil over the years, and insects. Most of these have been ruled out as widespread causal factors. Europe's forests are now largely planted by humans, mostly to spruce; hence monoculture must be considered, at least as a contributing factor.

The first sign of trouble in the European spruce forest is yellowing and

Figure 16.2. Forest destruction in the Giant Mountains of Czechoslovakia, 1987. Dead trees in the foreground had been removed. (Picture by R. P. Scheffer)

browning of older leaves, followed by defoliation. Normally, each spruce leaf is held for six to eight years. As disease progresses, finally only the youngest or terminal leaves are left. At this stage, the tree is definitely in trouble. At the same time, tree growth is slowed, as shown by sizes of annual rings. Foresters found evidence that growth was affected over large areas, possibly over half the forests of Europe; they were unable to correlate this with weather conditions. This started a general public alarm. However, more recent observations suggest that the claims were greatly exaggerated.

Germany was especially alarmed. Research institutes were established, with the best modern equipment and buildings, including phytotrons. Millions of deutschmarks were spent. The equipment allowed researchers to gas young trees and crop plants with various air pollutants, at different concentrations and exposure times, at temperatures from below freezing to 40°C. The resulting data showed that some plants, notably some cultivars of tobacco, are very sensitive to ozone. However, spruce was tolerant of ozone at concentrations that usually occur in many parts of Europe; the tree also tolerated airborne SO_2 and NO_x. But this was only a beginning; later work uncovered long-term factors, overlooked in the early experiments.

Significant studies of the forest problem in Europe were initiated by Professor Bernhard Ulrich, at Göttingen University (Ulrich 1985, 1986). Ulrich has summarized voluminous data on amounts of pollutants released into the air in Europe. Each year, more than 3 million tons each of SO_2 and NO_x have been spewed into the air over West Germany (Figure 16.3). The totals for East Germany and Czechoslovakia are thought to be much higher; East Germany is said to be the most polluted country in the world. Releases of SO_2 and NO_x in the United States are at least tenfold greater than in Germany, but are over a much larger area.

Ulrich has estimated the total pollution released since 1860, based on the amounts of coal burned each year; combustion of coal is the prime source of SO_2. There are good correlations between coal consumption and acid deposition in fields and forests. Pollutants come down with rain and fog and as dry deposits of particles. An important point to consider is that H_2SO_4 and HNO_3, formed from SO_2 and NO_x in moist air, are deposited in the soil, where they displace nutrient atoms from soil particles. Nutrient elements are lost by leaching, and the process is cumulative over time. The forest collects more acids than do open fields, and spruce collects more than beech (Table 16.1) (Ulrich 1985). This is because much of the acid input comes from fogs, and the uneven canopy of a forest harvests more fog than does a field. Spruce is evergreen; hence it collects more acid fog in winter than does deciduous beech.

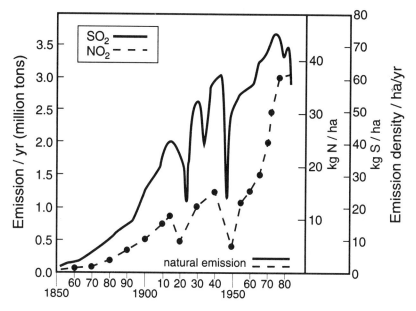

Figure 16.3. SO_2 and NO_x released into the air over West Germany, 1850–1990, estimated from coal consumption (data from Ulrich 1985). Low emissions (1920 and 1945–50) reflect economic activity. Levels from nonanthropogenic sources are indicated at lower right. Values for recent years correlate with direct measurements of surface deposits; nitrate deposits from 20 to 50 kg per hectare often were recorded.

Table 16.1. *Pollution deposits in Solling Mountains of Germany (Kg/ha/year, 1969–85). This is an excerpt from data of Ulrich, 1985.*

Site	H+	Sulfur (SO_4)	Cl
Open fields	0.8	23.2	16.7
Beech forest	2.0	50.0	32.5
Spruce	3.8	83.1	38.6

The accumulation of acids in forests of central Europe = 50 to 300 k mol H+/ha. This could be countered by 1000–6000 kg Ca or 2.75–7.50 tons of ground limestone.

Professor Ulrich has studied soil acidification and aluminum toxicity for many years. His published data show total acid deposits in a spruce forest from 1860 to date. Early data were calculated from the amount of coal burned, whereas later data were from direct measurements of deposits in forests and fields; the calculated and direct measurements were closely correlated. Ulrich also has measured acidity, N, Ca, Mg, and other elemental changes that have occurred in soils. More acid is introduced into the soil each year than is lost, which leads to a gradual accumulation over time, with displacement by H^+ of nutrient elements from soil particles. Ulrich's estimates of sulfur and nitrogen deposition have been confirmed elsewhere; for example, 23 kg of S and 13 kg of N per hectare per year were recorded for a coniferous forest in southern Sweden (Grennfelt, Bengston, and Skarby 1980).

Nitrogen in forest soils has increased dramatically in recent years; this increase is correlated with increased automobile traffic in Europe's cities and highways. There is now excess N input into the forest, causing late-season succulent growth of trees and leading to susceptibility to winter injury, frequently observed in Europe. Total N input per year in some areas is estimated to be five times more than the forest needs for optimum growth. As the soil pH falls below 4.5, some of the aluminum that is present in all soils becomes soluble, and Al ions in solution are toxic to plant roots (Godbold, Fritz, and Hutterman 1988). Even prior to root death, nutrient imbalances become evident, leading to nutrient deficiencies, especially for Ca, Mn, and Mg (Shortle and Smith 1988).

There are considerable data in support of the general concept outlined. Seepage water, that which percolates through the soil, was shown to have high levels of soluble Al and low Ca:Al ratios (Figure 16.4). This is conclusive proof that the buffering capacity of the soil was overwhelmed. High levels of sulfates and nitrates in seepage water is further evidence that acidity is high around roots. Seepage water that contains Al moves to streams, then to lakes, killing freshwater fish; this has occurred in some areas of the northeastern United States, eastern Canada, and Scandinavia. Mg and Mn are lost from soil particles, and these also appear in seepage water. Concentrations of soluble Al in soil water around roots of dying trees often are well above the level known to be toxic to plants (Godbold, Fritz, and Hutterman 1988).

Skeptics of Ulrich's conclusions maintain that soils are well buffered and resist changes from acids in rain. This is true, but Ulrich's data show that buffering is overwhelmed. The skeptics seem to overlook that soil acidification from rain, fog, and solid particles is cumulative and that the accumula-

Seepage Water

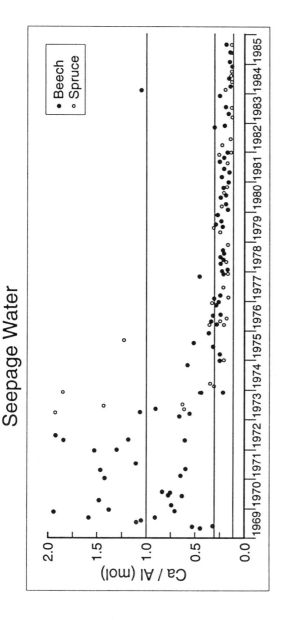

Figure 16.4. Ca:Al ratios in seepage water from spruce and beech stands in the Solling Mountains of Germany (redrawn from data of Murach and Ulrich 1988). Ca:Al ratios below 1.0 are considered hazardous; lower ratios shown here cause Al toxicity to roots. Al levels as high as 18 mg per liter were recorded, with 20 mg or more of sulfates. Mg:Al ratios were similar.

tions of years may be involved. We must consider also the fact that damage has occurred first on noncalcareous soils that are poorly buffered; such soils are formed on granite, sandstone, and basalt, not on limestone-derived soils. But even calcareous soils can be overwhelmed in time by acids (Ulrich 1985). Eventually, water tables can be acidified; this would be disastrous if it occurs on a large scale.

Recent analyses based on ^{15}N and ^{18}O in water leached from soil have revealed a new aspect of soil acidification. Nitrate from combustion has more ^{15}N (heavy N) and ^{18}O (heavy O) than that in nitrates normally found in the soil. This provides a method to trace sources and has confirmed that air pollution adds much nitrogen to soils. Seepage water from a healthy spruce stand had 16 to 20 percent of its nitrate in the form of ^{15}N; in contrast, seepage from a declining stand had 60 to 100 percent of its nitrate as ^{15}N. This result was unexpected. The nitrogen anion moves through the soil in solution, pulling cations along; Mg and Ca are the most common cations in soil, and these are lost with nitrates in seepage water. Soils and trees soon become deficient in Mg and Ca, as has been recorded many times in Europe. As cations are removed from soil particles, they are replaced by protons; soil acidity increases, leading to soluble Al and decreasing buffering capacity. This spiraling effect may continue until the soil system collapses. Natural weathering does not allow soil to recover rapidly from Ca and Mg losses (Durka et al. 1994; Hedlin 1994; Raloff 1995).

Still other observations and data fit the hypothesis that soil acidification leads to forest damage. Healthy young tree seedlings often flourish for a while under dead and dying trees (Figure 16.5). Roots of these seedlings are in the organic layer of topsoil; soluble Al is tied up by organic materials, allowing the seedlings to thrive. Eventually, the roots penetrate to mineral soil and are damaged, leaving the trees with shallow root systems and stubby roots. Such trees are easily uprooted; this has occurred over large areas during windstorms (Ulrich 1985). Even without wind, affected trees topple easily when they grow larger. The hypothesis also is supported by the fact that tree damage in the acidified areas can be reversed by heavy liming to counter the acid. However, liming is not a practical control in forests, for logistical reasons.

Spruce is not the only species affected in Europe. The silver fir often is first to go. Deciduous trees, in general, are last to be damaged; they also harvest less acid from fogs (Ulrich 1985), and some are relatively tolerant of Al in solution. However, at severe sites in Europe, even the beech has been affected.

There is little doubt that soil acidification is damaging trees at certain

Figure 16.5. Forest destruction on Mount Mitchell, North Carolina (United States), in 1986. Trees were Fraser fir, with some red spruce. (Picture by R. P. Scheffer)

high-elevation sites in parts of Europe. The only question concerns the general or widespread occurrence of damage caused by soil acidification.

Forest dieback is occurring at elevations greater than 1,400 m in the Appalachian Mountains of the eastern United States (Johnson 1992). At these elevations, Fraser fir and red spruce predominate, and the trees are good at collecting acids from the frequent fogs. Also, many soils in these mountain areas are poorly buffered, formed from rocks that lack Ca. A good example of forest destruction is on Mount Mitchell, North Carolina, where death of trees started near the top and is gradually spreading down the slopes (Figure 16.5). There are similar examples for both spruce and fir at many sites in the Appalachian chain.

A study site was established in 1986 on Mount Mitchell, and voluminous data were collected; however, the study site was largely abandoned in 1992, before the studies were conclusive. Data from Mount Mitchell show occasional levels of ozone and photooxidants high enough to damage vegetation. Also, acid clouds and acid rains are common (Bruck, Robarge, and McDaniel 1989). Preliminary data indicate a possible soil-acidification factor. The sources of pollution are many miles to the west and northwest, from coal-burning plants with high stacks in the Tennessee Valley and in Ohio, carried long distances by prevailing winds. However, the cause of tree death on Mount Mitchell appears to be complex; the fir trees have been stressed by infestations of woolly aphids, an alien species from Europe. Forest damage in the southern Appalachians continues to spread. By 1993, it was apparent on Clingman's Dome in the Smoky Mountain National Park, where spruce and fir trees are dying over a large area at high elevations.

The northern Appalachians and the Adirondack Mountains also receive much air pollution, and forest damage is evident (Eager and Adams 1992). Photooxidants are lower there than they are in the south, but acid rains and acid clouds are frequent (Cronan and Schofield 1979). Some mountain lakes there and in eastern Canada have been acidified, resulting in death of fish and other water fauna. Also, there are pockets of forest damage, cause unknown, in the coastal areas of Maine and the Maritime Provinces of Canada. Pollutants come from the great cities to the south; acidic clouds and fogs are common and long lasting. There are cases of direct damage to foliage, and indirect damage from soil acidification is a possibility. Maple trees are declining at some sites in Ontario, Quebec, and New England; again, the causes are unknown.

A rational hypothesis is that the delay in recovery of vegetation at Ducktown, as described previously, is because the soil was acidified by sulfuric acid, overwhelming the buffering capacity. Work elsewhere has shown that once acidification occurs, many years will pass before natural weathering of the soil and parent materials restores the soil to a condition favorable for tree growth (Ulrich and Pankrath 1983). Could this be the fate of damaged forest areas in Europe and in parts of Appalachia? If so, we can expect the former forest lands to be no more than sedge and grass steppes for many years.

We have the technology to remove sulfur from power plant emissions; "scrubbers" for tall smokestacks are effective. Of course, electricity will be more expensive, but, overall, this may be cheaper than doing nothing. Also, there are new laws for air protection; enforcement of the laws should be helpful.

Summary

Data now available prove that air pollution is killing trees in certain local areas of Europe, North America, and elsewhere. Direct effects of polluting photooxidants are the problem in some areas, and they may be a contributing factor in other areas. Stresses on plants resulting from combined effects of photooxidants, soil acidification, drought, insects, and other factors may be involved. The cause of forest death at some European sites appears to be airborne deposition of SO_2 and NO_x, which causes changes in soil chemistry.

Death of trees at several European sites was correlated with three conditions: (1) mineral deficiencies, especially of Mg, Mn, and Ca; (2) winter kill, from excess nitrogen; and (3) aluminum toxicity. Soil acidification from atmospheric pollutants is basic to all these conditions. Similar mechanisms may occur in Appalachia in the United States, where conditions are similar to those in Europe.

17
Prospectus

Revolutionary developments in plant pathology, and indeed in all of biology, are under way. Many recent advances were made possible by discoveries and development in at least three areas of technology: (1) Development and general availability of instruments for purification and characterization of natural products: High-performance liquid chromatography, mass spectrometry, and nuclear magnetic resonance techniques have made possible the rapid characterization of complex natural products such as the host-specific toxins; (2) development of molecular biology, leading to new insights in genetics and to conclusive evidence for modes of action in biological systems; and (3) computer technology, which has been involved in almost every advance in knowledge of plant diseases since about 1980, and no doubt will be of increasing importance in the future. The advances in biochemical and genetic knowledge are contributing to, and are being integrated by, modern concepts in ecology, as illustrated by various examples discussed in this book. New understanding is coming at an increasing rate.

So far, molecular genetics and DNA technology have been most useful for contributions to understanding the biochemistry of plant disease development. For examples, we now understand how some toxins mediate plant disease development and the basis of resistance in these cases (see Chapters 12 and 13). The new technology has given an understanding of the role of a bacterial protein (harpin) in development of the fire blight disease (Chapter 8, section 2). Many more such insights are under development or are expected in the future.

Molecular biology techniques have potential uses in taxonomic and evolutionary studies of pathogens. Origins, taxonomic relationships, lineages, and movement of races can be traced by the use of DNA "fingerprints." There has been limited use of this technology with *Phytophthora infestans,* the cause of potato late blight (Chapter 9, section 1), and preliminary studies have been reported for several other plant pathogens. Such approaches will undoubtedly be expanded in the future, leading to better understanding of the biology of disease.

Many practical uses of DNA technology for plant disease control are apparent, but, to date, the applications have been limited. The technology greatly enlarges the possibilities of incorporating genes for resistance into crop plants, speeds the breeding process, opens new sources for disease resistance, and creates novel possibilities for entirely new kinds of resistance. An example of the latter was Beachy's (1990) report that genes for synthesis of viral coat protein can be introduced into the plant genome and that plants containing such protein are resistant to specific viral infections.

Useful computer-driven predictions of plant disease epidemics have been developed (see Chapter 8, section 2); more and better systems of prediction no doubt will be available in the future.

There have been great expectations for the development of biological controls, but they have been slow to come. There are successes for special disease cases (see Chapter 6, section 4); we can expect more of these, but the theoretical potential of biocontrol may never be realized. On the other hand, the wide use of biocontrol as a supplementary procedure is not too much to expect. Many management procedures in crop production amount to biological controls, and we can expect these to be refined and extended. There will be a continued need for fungicides.

Biocontrol often is considered to include the use of resistance genes, our major plant disease control method. Reliance on resistance genes is expected to continue, supported, in part, by the new methods of "genetic engineering." But final solutions must never be expected; plant pathogens are far too shifty.

New plant diseases appear each year, and the trend is expected to continue or increase in rate, driven by the extension of agriculture into new areas. The development of agriculture in new areas leads to contact of crop plants with new pathogens that previously existed only in native vegetation. For example, cultivation is being expanded into valleys and hillsides at high elevations in the tropics; temperate-climate vegetables and fruits are grown. These crops are being exposed to new pathogens, some of which may spread to other areas. This situation parallels those in medicine and veterinary practice, as worldwide interactions with previously isolated populations of people and animals are increased. Pathogens such as HIV, Ebola virus, Lassa virus, and the Legionnaires' bacterium appear suddenly. Increased understanding, plus integration of information and organizations, will be needed to cope with new problems.

The destruction of vegetation by air pollution is a continuing problem, even though there are remedies for much of it. Still another cause of alarm is the increase in soil salinization that is occurring in dry areas with heavy irrigation. Soils in many such areas are becoming useless for agriculture. The only known remedy is to flush salts from soil, but there is insufficient water for this option in many affected areas.

Glossary: Technical Terms
Used in the Text

Alternate host: One of two host species required by certain plant pathogenic fungi to complete the life cycle. Do not confuse with *alternative host*, which is not required for the life cycle.

Anamorph: The asexual or imperfect stage in the life cycle of pleomorphic fungi. See *Teleomorph*.

Anastomosis: Vegetative fusion of fungal hyphae. Transfer of genes can occur after fusion of cells with different genotypes. Part of the *parasexual cycle*.

Anthropogenic: Created by people.

Ascomycete: A fungus that produces sexual spores in an ascus, a saclike cell.

Auxin: A plant growth hormone (indole-3-acetic acid) involved in cell enlargement.

Basidiomycete: A fungus that produces sexual spores on a cell known as a *basidium*.

Biotroph: An organism that lives and reproduces only on another living organism; an obligate parasite.

Blight: Rapid kill of leaves, flowers, and stem.

Canker: On a stem, a necrotic or sunken area surrounded by living tissue.

Compatibility: Herein, used to describe cells of different genotypes that can exist together. *Incompatible* cells are rejected or killed.

Conidium: A fungal spore that is produced asexually. Formed on a *conidiophore*.

Conjugation: Sexual reproduction involving fusion of morphologically similar gametes. Common in bacteria.

Cross-protection: Protection of plant tissues by one viral strain against another strain, usually more virulent.

Cytokinin: A plant growth hormone that regulates cell division. Cytokinins contain N^6-substituted adenine; example, zeatin.

Disease: A malfunction of cells and tissues, resulting from continuous irritation. There are *biotic diseases,* caused by living pathogens, and *abiotic diseases,* caused by adverse environmental factors. The *disease cycle* is the chain of events in disease development, including effects on host and stages in development of pathogen.

DNA: Deoxyribonucleic acid. DNA is in chromosomes and is the key molecule of genes that determines heredity by regulation of protein synthesis. cDNA is double stranded; mDNA is single stranded; T-DNA encodes for tumor production. See *RNA.*

Elicitor: A molecule that induces responses in a biological system. Elicitor, as used herein, is a pathogen-produced molecule that induces production of phytoalexin by the host plant.

Endemic: Peculiar to a particular geographic area; that is, native rather than introduced; also used to denote constant presence in a region, as an *endemic pathogen.*

Epidemic: Widespread occurrence or rapid increase in disease. There are *pandemic diseases,* which involve large areas and several seasons. *Epidemiology* is the study of epidemics.

Facultative parasite: A saprophyte that, under some conditions, can live as a parasite. See *Pathogen (opportunistic).*

Form species (f. sp.): A species subgroup. In plant pathology, all form species have identical morphologies but are specialized to different host species.

Fungicide: A compound toxic to fungi. There are *protective fungicides,* which prevent initial infection, and *systemic fungicides,* which become distributed throughout the tissues and can be curative.

Gall: An overgrowth with excessive local proliferation of cells without tissue differentiation.

Gene-for-gene (hypothesis): In host–parasite relationships, correspondence of each gene for resistance in the host with a specific gene for virulence in the pathogen.

Habituation (of cell cultures): An adaptive change in cells, allowing them to grow on a minimal medium that will not support normal cells.

Helminthosporia: An informal, collective term referring to the fungi formerly placed in the genus *Helminthosporium.* Includes asexual genera *Bipolaris, Exserohilum, Dreschlera, Curvularia,* and *Helminthosporuim,* and sexually reproducing genera *Cochliobolus, Pyrenophora,* and *Setosphaeria.*

Heteroecious: Requiring two different species of hosts for completion of life cycle. Some rust fungi are heteroecious.

Heterokaryon: A fungal cell with two genetically different nuclei.

Host range: The spectrum of plant species or genotypes that can be hosts of a parasite.

Host-specific toxin (also host-selective toxin): A pathogen-produced substance or compound that is toxic only to hosts of that pathogen.

Hypersensitive: A term used in plant pathology to denote excessive sensitivity and rapid kill of cells after exposure to certain microbes. The *hypersensitive reaction* is thought to give a type of disease resistance.

Hypovirulence: Reduced virulence, often the result of acquisition of double-stranded RNA or virus by a fungal pathogen.

Immunity: Complete disease resistance, with no reaction of plant tissue to potential pathogen.

Infection: The establishment of a pathogen in or on its host. Includes penetration of cell walls and colonization of tissues.

Infestation: The contamination of nonliving objects or areas by various organisms, usually with reference to microbial pathogens.

Inoculum: The propagule or part (such as a spore) involved in transmitting disease. *To inoculate* is to transfer a pathogen to a host.

Isozyme: One of several forms of a specific enzyme. *Isozyme analysis* is the determination of all forms of a specific enzyme produced by an organism.

Land race: In reference to crops: a primitive cultivar that contains a mixture of genotypes, usually propagated by subsistence farmers.

Leaf spot: A leaf lesion that is self-limiting in size, in contrast to blight or spreading lesion.

Lesion: On plants, a localized area of dead or discolored tissue.

Meiosis: The division of cell nuclei with reduction of chromosome number from *diploid* (2N) to *haploid* (N).

Mendelian gene: A gene unit that controls a given character, inherited according to Mendel's principles; not necessarily the same as a molecular gene.

Mitosis: The division of a cell nucleus that involves chromosome duplication without change in chromosome numbers, as in vegetative cell division.

Necrosis: The death and discoloration of cells and tissues. A *necrotroph* is an organism that lives on dead tissue, in contrast to a *biotroph,* which requires live host tissue.

Oospore: A resting spore produced by union of morphologically different gametangia, as with *Oomycetes.*

Opine: An amino acid derivative produced in crown gall (plant) cells. Opines can be utilized only by the gall-causing bacterium. *Octopine, nopaline,* and *agropine* are opines.

Parasexualism: The system that allows gene recombination in fungi with two genetically different nuclei per cell.

Parasite: An organism that obtains its sustenance from another living organism, the host.

Pathogen: An entity that can incite disease; the term *pathogen* usually refers to a living organism. *Pathogenesis* is the chain of events or processes that lead to disease. *Pathogenicity* is the capacity of an organism to cause disease (see *Virulence*). There are *specialized pathogens* (infect metabolically active tissue of special hosts) and *opportunistic pathogens* (infect senile or stressed tissues of many plant hosts).

Pathotype and pathovar: Synonymous with form species. Pathovar and pathotype usually are used for bacteria, whereas f. sp. generally is used for fungi.

Phloem: Food (that is, carbohydrate)-conducting tissue in plants. See *Xylem*.

Phytoalexin: A compound that inhibits growth of microbes, formed by plant cells after contact with a pathogen. Phytoalexins probably result from any disturbance and are part of the wound-healing process.

Phytoplasma: A bacterium or bacteriumlike organism that lacks a rigid cell wall.

Plasmid: A circular DNA fragment found in some fungal and bacterial cells. The plasmid is self-replicating but extrachromosomal hereditary material, generally not required for survival of the organism.

Race (pathogen): A genetically stable group that infects a given set of plant cultivars or genotypes. Also used for a distinct mating group within a species.

Rickettsiae: Parasitic or symbiotic microbes similar to bacteria but generally incapable of living without a host. Many are difficult or impossible to grow in culture. The term *rickettsiae-like* is often encountered.

Resistance: The ability of a host plant to exclude or overcome (to varying degrees) the growth or effects of a pathogen. See *Immunity*.

RNA: Ribonucleic acid, a compound that serves as a template in protein synthesis. The various arrangements in RNA structure are determined by the parental DNA molecule. RNA also is the genetic material in many plant viruses. There are dsRNAs (double stranded, as in viruses), mRNAs (messenger RNAs), and so forth.

Rust: A disease of plants with a rusty appearance, caused by fungi belonging to the Uredinales.

Saprophyte: A living organism that obtains sustenance from dead organic material.

Smut: A plant disease caused by fungi in the order Ustilaginales, characterized by masses of dark, powdery spores.

Spiroplasma: Pleomorphic, wall-less but bacterialike microbes that live in phloem of host plants. Spiroplasmas often are helical and may be a form of phyotoplasma.

Sporangium: A fungal fruiting structure in which spores are produced. The sporangium may function as a single spore under some conditions (as with some *Phytophthora* species). Spores are produced on or in a *sporangiophore*.

Spore: A reproductive unit, with one or more cells, analogous to seed of higher plants.

Suscept: Any plant species or genotype that can be attacked by a given pathogen. *Susceptibility* is inability to resist a pathogen.

Teleomorph: The sexual form of pleomorphic fungi. See *Anamorph*.

Teliospore: A resting or overwintering spore of rust fungi, from which a basidium and basidiospores are produced. The *telium* is the structure that contains teliospores.

Tms cytoplasm: In maize, Texas male-sterile cytoplasm. A Tms parent was used for production of maize hybrids, to ensure that all pollen in a field came from a second parent. Tms plants are sensitive to T-toxin involved in a blight.

Toxin: In plant pathology, a pathogen-produced compound, other than an enzyme, that causes obvious damage and is known to be involved in disease development. There are *host-specific toxins* and *nonspecific toxins*. Most toxins that mediate plant diseases have low molecular weights ($<1,000$).

Urediospore (uredospore): A "repeating" spore of rust fungi. The urediospore (dikaryotic) is infectious to the same host on which it was produced.

Vector: An organism that is able to transmit a pathogen to a host; also a DNA molecule (such as a virus or plasmid) used to introduce foreign DNA into a cell.

Viroid: A viruslike entity that lacks the protein coat of viruses. Viroids, cause of diseases such as chrysanthemum stunt and cadang-cadang of coconut, are "naked" RNA strands. Found to date only in plants.

Virulence: In plant pathology, a quantitative term to denote relative disease-inducing ability of a pathogen. Often confused with *pathogenicity,* a qualitative term. A change in host range indicates a change in pathogenicity; a change in severity of disease indicates a change in virulence. There are virulent, avirulent, and intermediate levels.

Xylem: The water-conducting tissue of plants.

Zoospore: A spore that swims by action of flagella.

References

Abad, Z. G., J. A. Abad, and C. Ochoa. 1988. Historical and scientific evidence that supports the modern theory of Peruvian Andes as center of origin of *Phytophthora infestans*. In *Phytophthora infestans 150* (ed. by L. J. Dowley, E. Bannon, L. R. Cooke, T. Keane, and E. O'Sullivan). Publication of the European Association for Potato Research, pp. 239–45. Dublin: Boole Press, Ltd.

Adachi, Y., and T. Tsuge. 1994. Coinfection by different isolates of *Alternaria alternata* in single black spot lesions of Japanese pear leaves. *Phytopathology* 84:447–51.

Agrios, G. N. 1988. *Plant Pathology,* ed. 3. San Diego: Academic Press.

Ainsworth, G. C. 1981. *Introduction to the History of Plant Pathology.* London: Cambridge Univ. Press.

Akimitsu, K., L. P. Hart, J. D. Walton, and R. Hollingsworth. 1992. Covalent binding sites of victorin in oat leaf tissues detected by anti-victorin polyclonal antibodies. *Plant Physiol.* 98:121–6.

Akimitsu, K., T. Shibata, H. Otani, H. Tabira, M. Kodama, and K. Kohmoto. 1994. Light suppresses the action of the host-specific ACR-toxin on citrus leaves. *J. Fac. Agric.,* Tottori Univ., 30:17–27.

Albersheim, P., T. M. Jones, and P. D. English. 1969. Biochemistry of the cell wall in relation to infective processes. *Annu. Rev. Phytopathol.* 7:171–94.

Alexopoulos, C. J., C. W. Mims, and M. Blackwell. 1996. *Introductory Mycology,* ed. 4. New York: John Wiley & Sons.

Allen, T. C. 1957. Strains of the barley yellow dwarf virus. *Phytopathology* 47:481–90.

Anagnostakis, S. L. 1983. Conversion to curative morphology in *Endothia parasitica* and its restriction by vegetative compatibility. *Mycologia* 75:777–80.

1987. Chestnut blight: the classical problem of an introduced pathogen. *Mycologia* 79:23–37.

1988 *Cryphonectria parasitica,* cause of chestnut blight. *Advances in Plant Pathology* 6:123–36 (*Genetics of Plant Pathogenic Fungi,* ed. by G. S. Sidhu). London: Academic Press.

Anagnostakis, S. L., and R. A. Jaynes. 1973. Chestnut blight control: use of hypovirulent cultures. *Plant Dis. Rep.* 57:225–6.

Anderson, H. W. 1956. *Diseases of Fruit Crops*. New York: McGraw-Hill Book Co.

Anderson, T. R., and R. I. Buzzell. 1992. Diversity and frequency of races of *Phytophthora megasperma* f. sp. *glycinea* in soybean fields in Essex County, Ontario. *Plant Dis.* 76:587–9.

Athow, K. L., and F. A. Laviolette. 1982. *Rps 6*, a major gene for resistance to *Phytophthora megasperma* f. sp. *glycinea* in soybean. *Phytopathology* 72:1564–7.

Athow, K. L., F. A. Laviolette, E. H. Mueller, and J. R. Wilcox. 1980. A new major gene for resistance to *Phytophthora megasperma* f. sp. *sojae* in soybean. *Phytopathology* 70:977–80.

Athow, K. L., F. A. Laviolette, and J. R. Wilcox. 1979. Genetics of resistance to physiologic races of *Phytophthora megasperma* var. *sojae* in the soybean cultivar Tracy. *Phytopathology* 69:641–2.

Aylor, D. E. 1986. A framework for examining inter-regional aerial transport of fungal spores. *Agric. For. Meterol.* 38:263–88.

Bailey, J. A., and B. J. Deveral, eds. 1983. *The Dynamics of Host Defense*. Sydney: Academic Press.

Bailey, J. A., and J. W. Mansfield. 1982. *Phytoalexins*. Glasgow: Blackie & Son.

Baker, K. F., A. W. Dimock, and L. H. Davis. 1961. Cause and prevention of the rapid spread of the *Ascochyta* disease of chrysanthemum. *Phytopathology* 51:96–101.

Ballance, G. M., L. Lamari, and C. C. Bernier. 1989. Purification and characterization of a host-selective necrosis toxin from *Pyrenophora tritici-repentis*. *Physiol. Mol. Plant Pathol.* 35:203–13.

Baltenberger, D. E., H. W. Ohm, and J. E. Foster. 1987. Reactions of oat, barley, and wheat to infection with barley yellow dwarf isolates. *Crop Sci.* 27:195–8.

Baranyovits, F. 1973. The increasing problems of aphids in agriculture and horticulture. *Outlook Agric.* 7:102–8.

Bar-Joseph, M. 1978. Gross protection incompleteness: a possible cause for natural spread of citrus tristeza virus after a prolonged lag period in Israel. *Phytopathology* 68:1110–11.

Bar-Joseph, M., and G. Loebenstein. 1973. Effects of strain, source plant, and temperature on the transmissibility of citrus tristeza virus by the melon aphid. *Phytopathology* 63:716–20.

Bar-Joseph, M., S. M. Garnsey, D. Gonsalves, M. Moscovitz, D. E. Purcifull, M. F. Clark, and G. Loebenstein. 1979. The use of enzyme-linked immunosorbent assay for detection of citrus tristeza virus. *Phytopathology* 69:190–4.

Bar-Joseph, M., R. Marcus, and R. F. Lee. 1989. The continuous challenge of citrus tristeza virus control. *Annu. Rev. Phytopathol.* 27:291–316.

Barnes, E. H. 1964. Changing plant disease losses in a changing agriculture. *Phytopathology* 54:1314–1319.

Barras, F., F. van Gijsegem, and A. R. Chatterjee. 1994. Extracellular enzymes and pathogenesis of soft-rot *Erwinia*. *Annu. Rev. Phytopathol.* 32:201–34.

Bateman, D. F., and H. G. Basham. 1976. Degradation of plant cell walls and membranes by microbial enzymes. *Encyclopedia of Plant Physiology*, n.s.

4:316–55 (*Physiological Plant Pathology,* ed. by R. Heitefuss and P. A. Williams). Berlin: Springer-Verlag.

Bauer, D. W., and S. V. Beer. 1991. Further characterization of an *hrp* gene cluster in *Erwinia amylovora. Mol. Plant-Microbe Interact.* 4:493–9.

Bawden, J., P. H. Gregory, and C. G. Johnson. 1971. Possible wind transport of coffee rust across the Atlantic ocean. *Nature* 229:500–1.

Bayer, M. H. 1982. Genetic tumors: physiological aspects of tumor formation in interspecies hybrids. In *Molecular Biology of Plant Tumors* (ed. by G. Kahl and J. S. Schell), pp. 33–67. New York: Academic Press.

Bazzigher, G. 1981. Selection of blight-resistant chestnut trees in Switzerland. *Europ. J. For. Pathol.* 11:199–207.

Beachey, R. N., S. Loesch-Fries, and N. E. Tumer. 1990. Coat protein-mediated resistance against virus infection. *Annu. Rev. Phytopathol.* 28:451–74.

Bean, W. J. 1921. *Trees and Shrubs Hardy in the British Isles,* Vol. 2, ed. 3. London: J. Murray.

Becker, S., S. K. Mulinge, and J. Krantz. 1975. Evidence that urediospores of *Hemileia vasatrix* are windborne. *Phytopathol. Z.* 82:359–60.

Beckman, C. H. 1987. *The Nature of Wilt Diseases of Plants.* St. Paul, MN: APS Press.

1990. Host responses to the pathogen. In *Fusarium Wilt of Banana* (ed. by R. C. Ploetz), pp. 93–105. St. Paul, MN: APS Press.

Beckman, C. H., S. Halmos, and M. E. Mace. 1962. The interaction of host, pathogen and soil temperature in relation to susceptibility to *Fusarium* wilt of banana. *Phytopathology* 52:134–40.

Bednarski, M. A., S. Izawa, and R. P. Scheffer. 1977. Reversible effects of toxin from *Helminthosporium maydis* race T on oxidative phosphorylation by mitochondria from maize. *Plant Physiol.* 59:540–5.

Beer, S. V. and J. L. Norelli. 1976. Streptomycin-resistant *Erwinia amylovora* not found in western New York pear and apple orchards. *Plant Dis. Rep.* 60:624–6.

Behnke, H. D., V. Schaper, and E. Seëmuller. 1980. Association of mycoplasmalike organisms with pear decline symptoms in the Federal Republic of Germany. *Phytopathol. Z.* 97:89–93.

Benedict, W. V. 1981. *History of White Pine Blister Rust Control – A Personal Account.* U.S.D.A. Forest Service, FS-355.

Bennett, C. W,. and A. S. Costa. 1949. Tristeza disease of citrus. *J. Agric. Res.* 78:207–37.

Bennett, C. W., and C. Munck. 1946. Yellow wilt of sugar beet in Argentina. *J. Agric. Res.* 73:45–64.

Bennett, C. W., F. J. Hills, K. R. Ehrenfeld, B. J. Valenzuela, and K. C. Klein. 1967. Yellow wilt disease of sugar beet. *J. Amer. Soc. Sugar Beet Tech.* 14:480–510.

Berg, C. H. 1970. Plant quarantine measures against South American leaf blight. *FAO Plant Prot. Bul.* 18:1–7.

Biraghi, A. 1950. La distribuzione del cancro del castagno in Italia. *L'Italia Forestale e Montana* 5:18–21.

Bishop, C. D., and R. M. Cooper. 1983. An ultrastructural study of root invasion in three vascular wilt diseases. *Physiol. Plant Pathol.* 22:15–27.

Black, L. M. 1982. Wound tumor disease. In *Molecular Biology of Plant Tumors* (ed. by G. Kahl and J. S. Schell), pp. 69–105. New York: Academic Press.

Boccara, M., A. Diolez, M. Rouve, and A. Kotoujanske. 1988. The role of individual pectase lyases of *Erwinia chrysanthemi* strain 3937 in pathogenicity on Saintpaulia plants. *Physiol. Mol. Plant Pathol.* 33:95–104.

Bonnen, A. M., and R. Hammerschmidt. 1989. Role of cutinolytic enzymes in infection of cucumber by *Colletotrichum lagenarium*. *Physiol Mol. Plant Pathol.* 35:475–81.

Borlaug, N. E. 1972. A cereal breeder and ex-forester's evaluation of the progress and problems involved in breeding rust resistant forest trees (moderator's summary). Washington DC: U.S.D.A. Misc. Publ. 1221:612–42.

Bosland, P. W. 1988. *Fusarium oxysporum*, a pathogen of many plant species. *Advances in Plant Pathology* 6:281–9 (*Genetics of Plant Pathogenic Fungi*, ed. by G. S. Sidhu). London: Academic Press.

Bosque-Perez, N. A., and I. W. Buddenhagen. 1990. Studies on epidemiology of virus disease of chickpea in California. *Plant Dis.* 74:372–8.

Boucher, C. A., C. L. Gough, and M. Arlat. 1992. Molecular genetics of pathogenicity determinants of *Pseudomonas solanacearum*, with special emphasis on *hrp* genes. *Annu. Rev. Phytopathol.* 30:443–61.

Bourke, A. 1991. Potato blight in Europe in 1845: the scientific controversy. In *Phytophthora* (ed. by J. A. Lucas et al.), pp. 12–24. London: Cambridge Univ. Press.

Bowyer, P., B. R. Clark, P. Lunness, M. J. Daniels, and A. E. Osbourne. 1995. Host range of a plant pathogenic fungus determined by a saponin detoxifying enzyme. *Science* 267:371–4.

Boyce, J. S. 1926. Observations on white pine blister rust in Great Britain and Denmark. *J. Forestry* 24:893–6.

Brakke, M. K., and W. F. Rochow. 1974. Ribonucleic acid of barley yellow dwarf virus. *Virology* 61:240–8.

Brandes, E. W. 1919. Banana wilt. *Phytopathology* 9:339–89.

Brasier, C. M. 1988. *Ophistoma ulmi*, cause of Dutch elm disease. *Advances in Plant Pathology* 6:207–23 (*Genetics of Plant Pathogenic Fungi*, ed. by G. S. Sidhu). London: Academic Press.

1991. *Ophiostoma novo-ulmi*, sp. nov., causative agent of current Dutch elm disease pandemics. *Mycopathologia* 115:151–61.

1992. Evolutionary biology of Phytophthora. *Annu. Rev. Phytopathol.* 30:153–200 (2 parts).

Braun, A. C. 1947. Thermal studies on the factors responsible for tumor initiation in crown gall. *Amer. J. Bot.* 34:234–40.

1951. Recovery of crown gall tumor cells. *Cancer Res.* 11:839–44.

1953. Bacterial and host factors concerned in determining tumor morphology in crown gall. *Bot. Gaz.* 114:363–71.

1955. A study on the mode of action of the wildfire toxin. *Phytopathology* 45:659–64.

1958. A physiological basis for autonomous growth of the crown gall tumor cell. *Proc. Natl. Acad. Sci. U.S.A.* 44:344–9.

1959. A demonstration of the recovery of the crown gall tumor cell with the use of complex tumors of single-cell origin. *Proc. Natl. Acad. Sci. U.S.A.* 45:932–8.

1978. Plant tumors. *Biochim. Biophys. Acta* 516:167–91.

1980. Genetic and biochemical studies on the suppression of and a recovery from the tumorous state in higher plants. *In Vitro* 16:38–48.

1982. A history of the crown gall problem. In *Molecular Biology of Plant Tumors* (ed. by G. Kahl and J. S. Schell), pp. 155–210. New York: Academic Press.

Braun, A. C., and H. N. Wood. 1962. On the activation of certain essential biosynthertic systems in cells of *Vinca rosea. Proc. Natl. Acad. Sci. U.S.A.* 48:1776–82.

Braun, C. J., J. N. Siedow, and C. S. Levings III. 1990. Fungal toxins bind to the URF 13 protein in maize mitochondria and *Escherichia coli. Plant Cell* 2:153–61.

Braun, E. J. 1990. Colonization of resistant and susceptible maize plants by *Erwinia stewartii* strains differing in exopolysaccharide production. *Physiol. Mol. Plant Pathol.* 36:363–79.

Bronson, C. R. 1991. The genetics of phytotoxin production by plant pathogenic fungi. *Experientia* 47:771–6.

Bronson, C. R., and R. P. Scheffer. 1977. Heat and aging-induced tolerance of sorghum and oat tissues to host-selective toxins. *Phytopathology* 67:1232–8.

Brooks, A. D., A. Collmer, and S. Hutcheson. 1989. Molecular cloning of an exopectate lyase gene from *Erwinia chrysanthemi* and characterization of the gene product. *Phytopathology* 79:1211.

Brosch, G., R. Ransom, T. Lechner, J. D.Walton, and P. Loidl. 1995. Inhibition of maize histone acetylases by HC-toxin, the host-selective toxin of *Cochliobolus carbonum. Plant Cell* 7:1941–50.

Brown, J. K., S. D. Wyatt, and D. Hazelwood. 1984. Irrigated corn as a source of barley yellow dwarf virus and vectors in eastern Washington. *Phytopathology* 74:46–9.

Browning, J. A. 1974. Relevance of knowledge about natural ecosystems to development of pest management programs for agro-ecosystems. *Proc. Amer. Phytopathol. Soc.* 1:191–9.

Bruck, R. I., G. V. Gooding, Jr., and C. E. Main. 1982. Evidence for resistance to metalaxyl in isolates of *Peronospora hyoscyami. Plant Dis.* 66:44–5.

Bruck, R. I., W. P. Robarge, and A. McDaniel. 1989. Forest decline in the boreal montane ecosystems of the southern Appalachian Mountains. *Water, Air, and Soil Pollution* 48:161–80.

Bruehl, G. W. 1961. *Barley Yellow Dwarf.* St. Paul, MN: APS Press, Monograph no.1.

Buddenhagen, I. W. 1990. Banana breeding and *Fusarium* wilt. In *Fusarium Wilt of Banana* (ed. by R. C. Ploetz), pp. 107–13. St. Paul, MN: APS Press.

Buddenhagen, I., and A. Kelman. 1964. Biological and physiological aspects of bacterial wilt caused by *Pseudomonas solanacearum. Annu. Rev. Phytopathol.* 2:203–30.

Bulit, J., and R. Lafon. 1978. Powdery mildew of the vine. In *The Powdery Mildews* (ed. by D. M. Spencer), pp. 525–48. New York: Academic Press.

Burdon, J. J. 1993. The structure of pathogen populations in natural plant communities. *Annu. Rev. Phytopathol.* 31:305–23.

Burnet, M., and D. O. White. 1972. *Natural History of Infectious Diseases,* ed.4; London: Cambridge Univ. Press.

Burr, T. J., J. L. Norelli, B. Katz, W. F. Wilcox, and S. A. Haying. 1988. Streptomycin resistance of *Pseudomonas syringae* pv. *papulans* in apple orchards and its association with a conjugative plasmid. *Phytopathology* 78:410–13.

Burr, T. J., J. L. Norelli, C. L. Reid, L. K. Capron, L. S. Nelson, H. S. Aldwinkle, and W. F. Wilcox. 1993. Streptomycin-resistant bacteria associated with fire blight infections. *Plant Dis.* 77:63–6.

Calder, M., and P. Bernhardt. 1983. *The Biology of Mistletoes.* New York: Academic Press.

Campbell, C. L., ed. 1981. *The Fisher-Smith Controversy: Are There Bacterial Diseases of Plants?* Phytopathol. Classics, no.13. St. Paul, MN: APS Press.

Campbell, C. L. 1983. Erwin Frink Smith – pioneer plant pathologist. *Annu. Rev. Phytopathol.* 21:21–7.

Campbell, R. 1989. *Biological Control of Microbial Plant Pathogens.* London: Cambridge Univ. Press.

Cantone, F. A., and L. A. Dunkle. 1990. Resistance in susceptible maize to *Helminthosporium carbonum* race 1 induced by prior inoculation with race 2. *Phytopathology* 80:1221–4.

Carefoot, G. L., and E. R. Sprott. 1967. *Famine on the Wind.* Chicago: Rand McNally & Co.

Chatterjee, A. K. 1972. Transfer among *Erwinia* spp. and other enterobacteria of antibiotic resistance carried on R factors. *J. Bact.* 112:576–84.

Chilton, M.-D., M. H. Drummond, D. J. Merlo, D. Sciaky, A. L. Montoya, M. P. Gordon, and E. W. Nester. 1977. Stable incorporation of plasmid DNA into higher plant cells: the molecular basis of crown gall tumorigenesis. *Cell* 11:263–71.

Chilton, M.-D., D. A. Tepler, A. Petit, C. David, F. Casse-Delbart, and J. Tempe. 1982. *Agrobacterium rhizogenes* inserts T-DNA into the genome of the host plant cells. *Nature* 295:432–4.

Chiou, C.-S., and A. L. Jones. 1991. The analysis of plasmid-mediated streptomycin resistance in *Erwinia amylovora. Phytopathology* 81:710–14

  1993. Nucleotide sequence analysis of a transposon (Tn 5393) carrying streptomycin resistance genes in *Erwinia amylovora* and other gram-negative bacteria. *J. Bact.* 175:732–40.

Chou, C. K. S. 1989. Perspectives of disease threat in large-scale *Pinus radiata* monoculture. *Phytopathology* 79:1142.

Christensen, C. M. 1951. *The Molds and Man.* Minneapolis, MN: Univ. Minn. Press.

  1984. *E. C. Stakman, Statesman of Science.* St. Paul, MN: APS Press.

Ciuffetti, L. M., O. C. Yoder, and B. G. Turgeon. 1992. A microbiological assay for host-specific fungal polyketide toxins. *Fungal Genet. Newsl.* 39:18–19.

Clayton, E. E., and J. A. Stevenson. 1943. *Peronospora tabacina* Adam, the organism causing blue mold (downy mildew) disease of tobacco. *Phytopathology* 33:101–13.

Cobb, F. W., Jr., B. Uhrenholdt, and R. F. Krohn. 1969. Epidemiology of *Dothistroma pini* needle blight on *Pinus radiata*. *Phytopathology* 59:1021–2.

Cobb, W. C., and J. S. Niederhauser. 1959. The late blight of potatoes. *Sci. Amer.* (May 1959).

Coceano, P. G., and S. Peressini. 1989. Colonization of maize by aphid vectors of barley yellow dwarf virus. *Ann. Appl. Biol.* 114:443–7.

Cochran, L. C., and L. M. Hutchins. 1976. Phony. In *Virus Diseases and Noninfectious Disorders of Stone Fruits in North America*. Washington, DC: U.S.D.A., A.R.S., Agric. Handbook no. 437.

Coley-Smith, J. R., K. Verhoeff, and W. R. Jarvis, eds. 1980. *The Biology of Botrytis*. London: Academic Press.

Collier, R. 1968. *The River That God Forgot: The Story of the Amazon Rubber Boom*. New York: Dutton Publ.

Collmer, A., and N. T. Keen. 1986. The role of pectic enzymes in plant pathogenesis. *Annu. Rev. Phytopathol.* 24:383–409.

Comai, L., and T. Kosuge. 1980. Involvement of plasmid deoxyribonucleic acid in indoleacetic acid synthesis in *Pseudomonas savastanoi*. *J. Bact.* 143:950–7.

Comai, L., G. Surico, and T. Kosuge. 1982. Relation of plasmid DNA to indoleacetic acid production in different strains of *Pseudomonas syringae* pv. *savastanoi*. *J. Gen. Microbiol.* 128:2157–63.

Commack, R. H. 1961. *Puccinia polysora:* a review of some factors affecting the epiphytotic in West Africa. *Rep. 6th Commonwealth Mycol. Conf. 1960*, pp. 134–8.

Comstock, J. C., and R. P. Scheffer. 1973. Role of host-selective toxin in colonization of corn leaves by *Helminthosporium carbonum*. *Phytopathology* 63:24–9.

Contreras, M. R., and C. W. Boothroyd. 1975. Histological reactions and effects on position of epidermal nuclei in susceptible and resistant corn inoculated with *Helminthosporium maydis* race T. *Phytopathology* 65:1075–8.

Cook, R. J., and K. F. Baker. 1983. *The Nature and Practice of Biological Control of Plant Pathogens*. St. Paul, MN: APS Press.

Cooksey, D. A. 1990. Genetics of bactericide resistance in plant pathogenic bacteria. *Annu. Rev. Phytopathol.* 28:201–19.

Costa, A. S. 1974. Whitefly-transmitted plant diseases. *Annu. Rev. Phytopathol.* 14:429–49.

Cox, A. E., and E. C. Large. 1960. *Potato Blight Epidemics Throughout the World*. Washington, DC: U.S.D.A. Agric. Handbook no.174.

Cronan, C. S., and C. L. Schofield. 1979. Aluminum leaching response to acid precipitation: effects on high-elevation watersheds in the northeast. *Science* 204:304–5.

Crosby, A. W. 1986. *Ecological Imperialism: The Biological Expansion of Europe, 900–1900*. London: Cambridge Univ. Press.

Curl, E. A., and H. A. Weaver. 1958. Diseases of forage crops under sprinkler irrigation in the southeast. *Plant Dis. Rep.* 42:637–44.

D'Arcy, C. J., and P. A. Burnett. 1995. *Barley Yellow Dwarf: 40 Years of Progress*. St. Paul, MN: APS Press.

D'Arcy, C. J., A. D. Hewings, and C. E. Eastman. 1992. Reliable detection of barley yellow dwarf viruses in field samples by monoclonal antibodies. *Plant Dis.* 76:273–6.

Davey, M. R., E. C. Cocking, J. Freeman, N. Pearce, and I. Tudor. 1980. Transformation of petunia protoplasts by isolated *Agrobacterium* plasmids. *Plant Sci. Let.* 18:307–13.

Davidse, L. C. 1986. Benzimidazole fungicides: mechanism of action and biological impact. *Annu. Rev. Phytopathol.* 24:43–65.

Davis, J. M. 1987. Modeling the long-range transport of plant pathogens in the atmosphere. *Annu. Rev. Phytopathol.* 25:169–88.

Davis, M. J., A. H. Purcell, and S. V. Thomson. 1978. Pierce's disease of grapevines: isolation of the causal bacterium. *Science* 199:75–77.

Day, P. R., J. A. Dodds, J. E. Elliston, R. A. Jaynes, and S. L. Anagnostakis. 1977. Double-stranded RNA in *Endothia parasitica. Phytopathology* 67:1393–6.

Deahl, K. L., et al. 1991. Occurrence of the A2 mating type of *Phytophthora infestans* in potato fields in the United States and Canada. *Amer. Potato J.* 68:717–26.

Dekker, J. 1976. Acquired resistance to fungicides. *Annu. Rev. Phytopathol.* 14:405–28.

DeLaubenfels, D. J. 1959. Parasitic conifer found in New Caledonia. *Science* 130:97.

Delon, R., and P. Schiltz. 1989. Spread and control of blue mold in Europe, North Africa, and the Middle East. In *Blue Mold of Tobacco* (ed. by W. E. McKeen), pp.19–42. St. Paul, MN: APS Press.

Delp, C. J. 1980. Coping with resistance to plant disease control agents. *Plant Dis.* 64:652–7.

Delp, C. J., ed. 1988. *Fungicide Resistance in North America.* St. Paul, MN: APS Press.

Deverall, B. J., and J. M. Daly. 1964. Metabolism of indoleacetic acid in rust diseases. II. Metabolites of carboxyl-labeled indoleacetic acid in tissue. *Plant Physiol.* 39:1–9.

Dewey, R. E., J. N. Siedow, D. H. Timothy, and C. S. Leavings. III. 1988. A 13–kilodalton maize mitochondrial protein in *E. coli* confers sensitivity to *Bipolaris maydis* toxin. *Science* 239:293–5.

Dewey, R. E., D. H. Timothy, and C. S. Leavings III. 1987. A mitochondrial protein associated with male sterility in the T-cytoplasm of maize. *Proc. Natl. Acad. Sci. U.S.A.* 84:5374–8.

DeWit, P. J. G. M. 1992. Molecular characterization of gene-for-gene systems in plant–fungus interactions and the application of avirulence genes in control of plant pathogens. *Annu. Rev. Phytopathol.* 30:391–418.

Dickman M. B., and S. S. Patil. 1986. Cutinase deficient mutants of *Colletotrichum gloeosporiodes* are nonpathogenic to papaya fruit. *Physiol. Mol. Plant Pathol.* 28:235–42.

Dietz, A. 1978. The use of ionizing radiation to develop a blight resistant American chestnut, *Castanea dentata,* through induced mutations. In *Proc. Amer. Chestnut Symposium* (ed. by W. L. MacDonald et al.). Morgantown, WV: W. Va. Univ. Press.

Dimock, A. W. 1962. Obtaining pathogen-free stock by cultured cutting techniques. *Phytopathology* 52:1239–41.

Dimond, A. E. 1971. Birth and progress in the study of pathogenesis. In *Morphological and Biochemical Events in Plant-Parasite Interaction* (ed. by S. Akai and S. Ouchi). Tokyo: The Phytopathological Society of Japan.

Dimond, A. E., J. W. Huberger, and J. G. Horsfall. 1943. A water soluble protectant fungicide with tenacity. *Phytopathology* 33:1095–7.

Dinoor, A., and N. Eshed. 1984. The role and importance of pathogens in natural plant communities. *Annu. Rev. Phytopathol.* 22:443–66.

Dinus, R. J. 1974. Knowledge about natural ecosystems as a guide to disease control in managed forests. *Proc. Amer. Phytopathol. Soc.* 1:184–90.

Dodds, J. A., R. L. Jordon, C. N. Roistacher, and T. Jarupat. 1987. Diversity of citrus tristeza virus isolates indicated by dsRNA analysis. *Intervirology* 27:177–88.

D'Oliveira, B. 1954. Selection of coffee types resistant to *Hemileiae* leaf rust. *Coffee Tea Industry* 81:112–20.

Domsch, K. H., W. Gams, and T. H. Anderson. 1980. *Compendium of Soil Fungi* (2 vol). New York: Academic Press.

Doudrick, R. L., W. L. Nance, C. D. Nelson, G. A. Snow, and R. C. Hamelin. 1993. Detection of DNA polymorphisms in a single urediospore-derived culture of *Cronartium quercuum* f. sp. *fusiforme*. *Phytopathology* 83:388–92.

Dow, J. M., D. E. Milligan, L. Jamieson, C. E. Barber, and M. J. Daniels. 1989. Molecular cloning of a polygalacturonate lyase gene from *Xanthomonas campestris* pv. *campestris* and role of the gene product in pathogenicity. *Physiol. Mol. Plant Pathol.* 35:113–20.

Driver, C. H. 1963. Effect of certain chemical treatments on colonization of slash pine stumps by *Fomes annosus*. *Plant Dis. Rep.* 47:569–71.

Dropkin, V. H. 1989. *Introduction to Plant Nematology*. Somerset, NJ: John Wiley & Sons.

Duffus, J. E. 1965. Beet pseudo-yellows virus, transmitted by the greenhouse whitefly (*Trialeurodes vapararriorum*). *Phytopathology* 55:450–3.

1971. Role of weeds in the incidence of virus diseases. *Annu. Rev. Phytopathol.* 9:319–40.

Dunkle, L. D. 1979. Heterogeneous reaction of shattercane to *Periconia circinata* and its host-specific toxin. *Phytopathology* 69:260–2.

1981. Global occurrence of sorghums sensitive to toxin produced by *Periconia circinata*. *Phytopathology* 71:871.

Durbin, R. D. 1991. Bacterial phytotoxins: mechanisms of action. *Experientia* 47:776–83.

Durka, W., E. D. Schulze, G. Gebauer, and S. Voerkelius. 1994. Effects of forest decline on uptake and leaching of deposited nitrate determined from ^{15}N and [18]O measurements. *Nature* 372:765–7.

Durrands, P. K., and R. M. Cooper. 1988. The role of pectinases in vascular wilt disease as determined by defined mutants of *Verticillium albo-atrum*. *Physiol. Mol. Plant Pathol.* 32:363–71.

Dwinell, L. D. 1971. Interaction of *Cronortium fusiforme* and *Cronartium quercuum* with *Quercus velutina*. *Phytopathology* 61:1055–8.

1974. Susceptibility of southern oaks to *Cronartium fusiforme* and *C. quercuum*. *Phytopathology* 64:400–3.

Dwinell, L. D., and H. R. Powers, Jr. 1974. Potential for southern fusiform rust on western pines and oaks. *Plant Dis. Rep.* 58:497–500.

Eager, C., and M. B. Adams. 1992. *Ecology and Decline of Red Spruce in the Eastern United States.* Berlin: Springer-Verlag.

Eldrige, R. H., et al. 1980. Susceptibility of five provenances of ponderosa pine to *Dothistroma* blight. *Plant Dis.* 64:400–1.

Elias, K. S., and R. W. Schneider. 1992. Genetic diversity within and among races and vegetative compatibility groups of *Fusarium oxysporum* f. sp. *lycopersici. Phytopathology* 82:1421–7.

Ellingboe, A. H. 1976. Genetics of host–parasite interactions. *Encyclopedia of Plant Physiology* n. ser. 4:761–8 (*Physiological Plant Pathology*, ed. by R. Heitefuss and P. H. Williams). Berlin: Springer-Verlag.

1981. Changing concepts in host-pathogen genetics. *Annu. Rev. Phytopathol.* 19:125–43.

1992. Segregation of avirulence/virulence on three rice cultivars in 16 crosses of *Magnaporthe grisea. Phytopathology* 82:597–601.

Ellingboe, A. H., B.-c. Wu, and W. Robertson. 1990. Inheritance of avirulence/virulence in a cross of two isolates of *Magnaporthe grisea* pathogenic to rice. *Phytopathology* 80:108–11.

Ellis, M. B. 1971. *Dematiaceous Hyphomycetes.* Kew, Surrey, U.K.: Commonwealth Mycol. Inst.

Elliston, J. E. 1985. Further evidence for two cytoplasmic hypovirulence agents in a strain of *Endothia parasitica* from western Michigan. *Phytopathology* 75:1405–13.

Elton, C. S. 1925. The dispersal of insects to Spitzbergen. *Trans. Entomol. Soc. London* 1925:289–91.

Enebak, S. A., W. L. MacDonald, and B. I. Hillman. 1994. Effect of dsRNA associated with isolates of *Cryphonectria parasitica* from the central Appalachians and their relatedness to other dsRNAs from North America and Europe. *Phytopathology* 84:528–34.

Erwin, D. C. 1983. Variability within and among species of *Phytophthora*. In *Phytophthora: Its Biology, Taxonomy, Ecology and Pathology* (ed. by D. C. Erwin, S. Bartnicki-Garcia, and P. H. Tsao). St. Paul, MN: APS Press.

Fargette, D., R. M. Lister, and E. L. Hood. 1982. Grasses as a reservoir of barley yellow dwarf virus in Indiana. *Plant Dis.* 66:1041–5.

Farquet, C., and D. Farquet. 1990. African cassava mosaic virus; etiology, epidemiology, and control. *Plant Dis.* 74:404–11.

Fattouh, F. A., P. P. Weng, E. E. Kawata, D. L. Barbara, B. A. Larkins, and R. M. Lister. 1990. Luteovirus relationships assessed by cDNA clones from barley yellow dwarf virus. *Phytopathology* 80:913–20.

Fawcett, H. S., and J. M. Wallace. 1946. Evidence of the virus nature of citrus quick decline. *Calif. Citrograph* 32:50,88–9.

Filajdić, N., and T. B. Sutton. 1991. Identification and distribution of *Alternaria mali* on apples in North Carolina and susceptibility of different varieties of apples to *Alternaria* blotch. *Plant Dis.* 75:1045–8.

Filer, T. H., F. I. McCracken, G. A. Mohn, and W. K. Randall. 1971. Septoria canker on nursery stock of *Populus deltoides*. *Plant Dis. Rep.* 55:460–3.

Filler, E. C. 1933. Blister rust damage to northern white pine at Waterford, Vermont. *J. Agr. Res.* 47:297–313.

Fischer, E. 1918. Neueres über die Rostkrankheiten der forstlich wichtigsten Nadelhölzer der Schweiz. *Schweiz. Ztschr. Forstiv Jahrg.* 69:113–20.

Flor, H. H. 1947. Inheritance of reaction to rust in flax. *J. Agric. Res.* 74:241–62.

Fravel, D. R. 1988. Role of antibiotics in the biocontrol of plant diseases. *Annu. Rev. Phytopathol.* 26:75–91.

Fry, W. E. 1982. *Principles of Plant Disease Management*. Orlando FL: Academic Press.

Fry, W. E., S. B. Goodwin, J. M. Matuszak, L. J. Spielman, M. G. Milgroom, and A. Drenth. 1992. Population genetics and intercontinental migrations of *Phytophthora infestans*. *Annu. Rev. Phytopathol.* 30:107–29.

Fry, W. E., et al. 1993. Historical and recent migrations of *Phytophthora infestans:* chronology, pathways, and implications. *Plant Dis.* 77:653–61.

Fulbright, D. W. 1984. Effect of eliminating dsRNA in hypovirulent *Endothia parasitica*. *Phytopathology* 74:722–4.

Fulbright, D. W., et al. 1983. Chestnut blight and recovering American chestnut trees in Michigan. *Can. J. Bot.* 61:3164–71.

Fulton, J. P., R. C. Gergerich, and H. A. Scott. 1987. Beetle transmission of plant viruses. *Annu. Rev. Phytopathol.* 25:111–23.

Fulton, R. H., ed. 1984. Coffee rust in the Americas. St. Paul, MN: APS Press.

Fulton, R. W. 1980. Tobacco blackfire disease in Wisconsin. *Plant Dis.* 64:100.
1986. Practices and precautions in the use of cross-protection for plant virus disease control. *Annu. Rev. Phytopathol.* 24:67–81.

Galet, P. 1979. *A Practical Ampelography: Grapevine Identification* (English translation by L.T. Morton). Ithaca and London: Cornell Univ. Press (Comstock Publ. Assoc.).

Gallegly, M. E., and J. Galindo. 1958. Mating types and oospores of *Phytophthora infestans* in nature in Mexico. *Phytopathology* 48:274–7.

Gardner, J. M., Y. Kono, and J. L. Chandler. 1986. Bioassay and host-selectivity of *Alternaria citri* toxins affecting rough lemon and mandarins. *Physiol. Mol. Plant Pathol.* 29:293–304.

Gardner, J. M., I. S. Mansour, and R. P. Scheffer. 1972. Effects of the host-specific toxin of *Periconia circinata* on some properties of sorghum plasma membranes. *Physiol. Plant Pathol.* 2:197–206.

Gardner, M. W., and W. B. Hewitt. 1974. *Pierce's Disease of the Grapevine: The Anaheim Disease and the California Vine Disease*. Berkeley and Davis, Ca: Univ. CA. Special Publication.

Garnsey, S. M., D. Gonsalves, and D. E. Purcifull. 1977. Mechanical transmission of citrus tristeza virus. *Phytopathology* 67:965–8.

Garrett, S. D. 1956. *Biology of Root-Infecting Fungi*. London: Cambridge Univ. Press.

Garrod, S. W., C. P. Paul, and D. W. Fulbright. 1985. The dissemination of virulent and hypovirulent forms of a marked strain of *Endothia parasitica* in Michigan. *Phytopathology* 75:533–8.

Gäumann, E. 1950. *Principles of Plant Infection* (English translation by W. B. Brierley). New York: Hafner Publ. Co.

Giatgong, P., and R. A. Frederiksen. 1969. Pathogenic variability and cytology of monoconidial subcultures of *Piricularia oryzae*. *Phytopathology* 59:1152–7.

Gibbs, J. N. 1978. Intercontinental epidemiology of Dutch Elm disease. *Annu. Rev. Phytopathol.* 16:287–307.

Gibbs, J. N., D. R. Houston, and E. B. Smalley. 1979. Aggressive and non-aggressive strains of *Ceratocystis ulmi* in North America. *Phytopathology* 69:1215–19.

Gibson, I.A.S. 1972. *Dothistroma* blight of *Pinus radiata*. *Annu. Rev. Phytopathol.* 10:51–72.

Gilchrist, D. G., and J. J. Harada. 1989. Mode and physiological consequences of AAL-toxin interaction with *asc* locus in tomato. In *Phytotoxins and Plant Pathogenesis* (ed. by A. Graniti, R. D. Durbin, and A. Ballio), pp. 113–21. Berlin: Springer-Verlag.

Gildow, F. E., M. E. Ballinger, and W. F. Rochow. 1983. Identification of double-stranded RNAs associated with barley yellow dwarf virus infections in oats. *Phytopathology* 73:1570–2.

Gilmer, R. M., and E. C. Blodgett. 1976. X-disease. In *Virus Diseases and Noninfectious Disorders of Stone Fruits in North America*, pp. 145–55. Washington, DC: U.S.D.A. Agric. Handbook no. 437.

Glazebrook, J., and F. M. Ausubel. 1994. Isolation of phytoalexin-deficient mutants of *Arabidopsis thaliana* and characterization of their interactions with bacterial pathogens. *Proc. Natl. Acad. Sci. U.S.A.* 91:8955–9.

Godbold, D. L., E. Fritz, and A. Hutterman. 1988. Aluminum toxicity and forest decline. *Proc. Natl. Acad. Sci. U.S.A.* 85:3888–92.

Goddard, R. E., R. A. Schmidt, and F. Vande Linde. 1975. Effect of differential selection pressure on fusiform rust resistance in phenotypic selections of slash pine. *Phytopathology* 65:336–8.

Godoy, G., J. R. Steadman, M. B. Dickman, and R. Dam. 1990. Use of mutants to demonstrate the role of oxalic acid in pathogenicity of *Sclerotinia sclerotiorum* on *Phaseolus vulgaris*. *Physiol. Mol. Plant Pathol.* 37:179–91.

Goheen, A. C., G. Nyland, and S. K. Lowe. 1973. Association of rickettsialike organism with Pierce's disease of grapevines and alfalfa dwarf and heat therapy of the disease in grapevines. *Phytopathology* 63:341–5.

Goheen, D. 1993. Importing logs: a risky business. *Plant Dis.* 77:852.

Gonsalves, D., and S. M. Garnsey. 1989. Cross-protection techniques for control of plant virus diseases in the tropics. *Plant Dis.* 73:592–7.

Gooding, G. V., Jr., and H. R. Powers, Jr. 1965. Serological comparison of *Cronartium fusiforme, C. quercuum*, and *C. ribicola* by immunodiffusion tests. *Phytopathology* 55:670–4.

Goodwin, S. B., B. A. Cohen, K. L. Deahl, and W. E. Fry. 1994. Migration from northern Mexico as the probable cause of recent genetic changes in populations of *Phytophthora infestans* in the United States and Canada. *Phytopathology* 84:553–8.

Goodwin, S. B., et al. 1992. Clonal diversity and genetic differentiation of *Phytophthora infestans* populations in northern and central Mexico. *Phytopathology* 82:955–61.

Goodwin S. B., L. S. Sujkowski, A. T. Dyer, B. A. Fry, and W. E. Fry. 1995. Direct detection of gene flows and probable sexual reproduction of *Phytophthora infestans* in northern North America. *Phytopathology* 85:473–9.

Grafton, K. F., J. M. Poehlman, O. P. Sehgal, and D. T. Sechler. 1982. Tall fescue as a natural host and aphid vectors of barley yellow dwarf virus in Missouri. *Plant Dis.* 66:318–20.

Graniti, A. 1993. Siridium blight of cypress – another ecological disaster? *Plant Dis.* 77:544.

Grant, T. J., L. J. Klotz, and J. M. Wallace. 1953. The tristeza disease of citrus. In *Plant Diseases,* pp. 730–4. Washington, DC: U.S.D.A. Yearbook.

Graves, A. H. 1950. Relative blight resistance in species and hybrids of *Castanea*. *Phytopathology* 40:1125–31.

Gregory, P. H. 1973. *The Microbiology of the Atmosphere,* ed. 2. United Kingdom: Leonard Hill Books.

Grennfelt, P., C. Bengtson, and L. Skarby. 1980. An estimation of the atmospheric input of acidifying substances to a forest ecosystem. In *Effects of Acid Precipitation on Terrestrial Ecosystems* (ed. by T. C. Hutchinson and M. Havas), pp. 29–40. New York: Plenum Press.

Grente, J. 1965. Les formes hypovirulentes d' *Endothia parasitica* et les espoirs de lutte contre le chancre du chataignier. *C. R. Seances Acad. Agric. Fr.* 51:1033–7.

Grogan, R. G., and R. N. Campbell. 1966. Fungi as vectors and hosts of viruses. *Annu. Rev. Phytopathol.* 4:29–52.

Gross, D. C. 1991. Molecular and genetic analysis of toxin production by pathovars of *Pseudomonas syringae. Annu. Rev. Phytopathol.* 29:247–78.

Guerri, J., P. Moreno, and R. F. Lee. 1990. Identification of citrus tristeza virus by peptide maps of virion coat protein. *Phytopathology* 80:692–8.

Halloin, J. M., J. C. Comstock, C. A. Martinson, and C. L. Tipton. 1973. Leakage from corn tissues induced by *Helminthosporium maydis* race T toxin. *Phytopathology* 63:640–2.

Hamelin, R. C., L. Shain, R. S. Ferris, and B. A. Thielges. 1993. Quantification of disease progress and defoliation in the poplar leaf rust-eastern cottonwood pathosystem. *Phytopathology* 83:140–4.

Hamid, A. H., J. E. Ayers, and R. R. Hill. 1982. The inheritance of resistance in corn to *Cochliobolus carbonum* race 3. *Phytopathology* 72:1173–7.

Hamilton, L. M., and E. C. Stakman. 1967. Time of stem rust appearance on wheat in the western Mississippi basin in relation to the development of epidemics from 1921 to 1962. *Phytopathology* 57:609–14.

Hamilton, R. H., and M. Z. Fall. 1971. The loss of tumor-initiating ability in *Agrobacterium tumefaciens* by incubation at high temperature. *Experientia* 27:229–30.

Hammerschmidt, R., and J. Kuc, eds. 1995. *Induced Resistance to Disease in Plants.* Dordrecht: Kluwer Acad. Publ.

Hampson, M. C. 1992. Some thoughts on demography of the great potato famine. *Plant Dis.* 76:1284–6.

Hansen, E. M. 1987. Speciation in *Phytophthora:* evidence from the *Phytophthora megasperma* complex. In *Evolutionary Biology of the Fungi* (ed. by A.D.M.

Rayner, C. M. Brasier, and D. Moore), pp. 325–37. London: Cambridge Univ. Press.

Harlan, J. R. 1976. Disease as a factor in plant evolution. *Annu. Rev. Phytopathol.* 14:31–51.

1977. Sources of genetic defense. In *The Genetic Basis of Epidemics in Agriculture* (ed. by P. Day). *Ann. New York Acad. Sci.* 287:345–56.

Harper, J. L. 1977. *Population Biology of Plants.* London: Academic Press.

Harris, K. F. 1981. Arthropod and nematode vectors of plant viruses. *Annu. Rev. Phytopathol.* 19:391–426.

Harris, K. F., and K. Maramorosch. 1980. *Vectors of Plant Pathogens.* New York: Academic Press.

Hawksworth, F. G., and D. Wiens. 1972. *Biology and Classification of Dwarf Mistletoes* (Arceuthobium). Washington, DC: U.S.D.A. Handbook no. 401.

Hayami, C., H. Otani, S. Nishimura, and K. Kohmoto. 1982. Induced resistance in pear leaves by spore germination fluids of nonpathogens to *Alternaria alternata,* Japanese pear pathotype and suppression of the induction by AK-toxin. *J. Fac. Agric. Tottori Univ.* 17:9–18.

Hayashi, N., and S. Nishimura. 1989. Comparison of the parasitic fitness of AK- and AF-toxin producers of *Alternaria alternata.* In *Host-Specific Toxins: Recognition and Specificity Factors in Plant Disease* (ed. by K. Kohmoto and R.D. Durbin). Tottori University.

Hayashi, N., K. Tanabe, T. Tsuge, S. Nishimura, K. Kohmoto, and H. Otani. 1990. Determination of host-selective toxin production during spore germination of *Alternaria alternata* by high-performance liquid chromatography. *Phytopathology* 80:1088–91.

Hayward, A. C. 1991. Biology and epidemiology of bacterial wilt caused by *Pseudomonas solanacearum. Annu. Rev. Phytopathol.* 29:65–87.

He, L. Y., L. Sequeira, and A. Kelman. 1983. Characteristics of strains of *Pseudomonas solanacearum* from China. *Plant Dis.* 67:1357–61.

Heath, M. C. 1987. Evolution of parasitism in the fungi. In *Evolutionary Biology of the Fungi* (ed. by A.D.M. Raynes, C. M. Brasier, and D. Moore), pp. 149–59. London: Cambridge Univ. Press.

Hedgcock, G. G., E. Bethel, and N. R. Hunt. 1918. Piñon blister rust. *J. Agric. Res.* 14:411–24.

Hedlin, L. O. 1994. Stable isotopes, unstable forest. *Nature* 372:725–6.

Heiniger, U., and D. Rigling. 1994. Biological control of chestnut blight in Europe. *Annu. Rev. Phytopathol.* 32:581–99.

Hepting, G. H. 1971. *Diseases of Forest and Shade Trees of the United States.* Washington, DC: U.S.D.A. For. Ser. Agric. Handbook no. 386.

Hewitt, W. B. 1958. The probable home of Pierce's disease virus. *Plant Dis. Rep.* 42:211–15.

1970. Pierce's disease of *Vitis* species. In *Virus Diseases of Small Fruits and Grapevines* (ed. by N. W. Frazier), pp. 196–200. Berkeley: Univ. Calif. Press.

Hewitt, W. B., B. R. Houston, N. W. Frazier, and J. H. Freitag. 1946. Leafhopper transmission of the virus causing Pierce's disease of grape and dwarf of alfalfa. *Phytopathology* 36:117–28.

Hibino, H., G. H. Kaloostian, and H. Schneider. 1971. Mycoplasma-like bodies in the pear psylla vector of pear decline. *Virology* 43:34–40.

Hildebrand, E. M., and E. F. Phillips. 1936. The honeybee and the beehive in relation to fire blight. *J. Agric. Res.* 52:789–810.

Hill, B. L., and A. H. Purcell. 1995. Acquisition and retention of *Xyllela fastidiosa* by an efficient vector, *Graphocephala atropunctata. Phytopathology* 85:209–12.

Hodges, C. S. 1969. Modes of infection and spread of *Fomes annosus. Annu. Rev. Phytopathol.* 7:247–66.

Hodson, A. C., and E. F. Cook. 1960. Long-range aerial transport of the harlequin and the greenbug into Minnesota. *J. Econ. Entomol.* 53:604–8.

Hoitink, H.A.J., and P. C. Fahy. 1986. Basis for control of soilborne plant pathogens with composts. *Annu. Rev. Phytopathol.* 24:187–209.

Hollings, M. 1965. Disease control through virus-free stock. *Annu. Rev. Phytopathol.* 3:367–96.

Hollis, C. A., and R. A. Schmidt. 1977. Site factors related to fusiform rust incidence in north Florida slash pine plantations. *For. Sci.* 23:69–77.

Holmes, F. W. 1990. The Dutch elm disease in Europe arose earlier than was thought. *J. Arboric.* 16:281–8.

Holmes, F. W., and H. M. Heybroek. 1990. *Dutch Elm Disease: The Early Papers.* St. Paul, MN: APS Press.

Holmes, P. R. 1988. Mobility of apterous grain aphids *Sitobion avenae* within wheat fields. *Entomol. Exp. Appl.* 46:275–9.

Holsters, M., J. P. Hernalsteens, M. van Montague, and J. Schell. 1982. Ti plasmids of *Agrobacterium tumefaciens:* the nature of the TIP. In *Molecular Biology of Plant Tumors* (ed. by G. Kahl and J. S. Schell), pp. 269–98. New York: Academic Press.

Hooker, A. L. 1974. Cytoplasmic susceptibility in plant disease. *Annu. Rev. Phytopathol.* 12:167–79.

Hooker, A. L., D. R. Smith, S. M. Lim, and J. B. Beckett. 1970. Reaction of corn seedlings with male-sterile cytoplasm to *Helminthosporium maydis. Plant Dis. Rep.* 54:708–12.

Hooker, W. J., ed. 1981. *Compendium of Potato Diseases.* St. Paul, MN: APS Press.

Hooykaas, P.J.J., and A.G.M. Beijersbergen. 1994. The virulence system in *Agrobacterium tumefaciens. Annu. Rev. Phytopathol.* 32:157–79.

Hooykaas, P.J.J., and R. A. Schilperoort. 1992. *Agrobacterium* and plant genetic engineering. *Plant Mol. Biol.* 19:15–38.

Hopkins, D. L. 1985. Physiological and pathological characteristics of virulent and avirulent strains of the bacterium that causes Pierce's disease of grapevine. *Phytopathology* 75:713–17.

Hopkins, D. L., and H. H. Mollenhauer. 1973. Rickettsia-like bacterium associated with Pierce's disease of grapes. *Science* 179:298–300.

Hornby, D. 1983. Suppressive soils. *Annu. Rev. Phytopathol.* 21:65–85.

Houston, D. R. 1991. Changes in nonaggressive and aggressive subgroups of *Ophiostoma ulmi* within two populations of American elm in New England. *Plant Dis.* 75:720–2.

Howell, C. R. 1976. Use of enzyme-deficient mutants of *Verticillium dahliae* to access the importance of pectolytic enzymes in symptom expression of *Verticillium* wilt of cotton. *Physiol. Plant Pathol.* 9:279–83.

Howell, H. E. 1975. Correlation of virulence with secretion of three wall-degrading enzymes in isolates of *Sclerotinia fructigena* obtained after mutagen treatments. *J. Gen. Microbiol.* 90:32–40.

Hoy, J. W., and M. D. Grisham. 1988. Spread and increase of sugarcane smut in Louisiana. *Phytopathology* 78:1371–6.

Huang, J., S. H. Lee, C. Lin, R. Medici, E. Hack, and A. M. Myers. 1990. Expression in yeast of the T-URF protein from Texas male-sterile maize mitochondria confers sensitivity to methomyl and to Texas cytoplasm-specific fungal toxins. *EMBO Jour.* 9:339–47.

Hudson, H. J. 1971. The development of the saprophytic fungal flora as leaves senesce and fall. In *Ecology of Leaf Surface Microorganisms* (ed. by T. F. Preece and C. H. Dickinson), pp. 447–55. London: Academic Press.

Ingram, D. S., and P. H. Williams. eds. 1991. *Advances in Plant Pathology,* Vol. 7. *Phytophthora infestans.* New York: Academic Press.

Irwin, M. E., and J. M. Thresh. 1990. Epidemiology of barley yellow dwarf: a study in ecological complexity. *Annu. Rev. Phytopathol.* 28:393–424.

Itoi, S., T. Mishima, S. Arase, and M. Nozu. 1983. Mating behavior of Japanese isolates of *Pyricularia oryzae. Phytopathology* 73:155–8.

Jaynes, R. A. 1964. Interspecific crosses in the genus *Castanea. Silvae Genet.* 13:146–54.

Jaynes, R. A., S. L. Anagnostakis, and N. K. Van Alfen. 1976. Chestnut research and biological control of the chestnut blight fungus. In *Perspectives in Forest Entomology* (ed. by J. F. Anderson and H. Kaya), pp. 61–70. New York: Academic Press.

Jaynes, R. A., and J. E. Elliston. 1980. Pathogenicity and canker control by mixtures of hypovirulent strains of *Endothia parasitica* in American chestnut. *Phytopathology* 70:453–6.

1982. Hypovirulent isolates of *Endothia parasitica* associated with large American chestnut trees. *Plant Dis.* 66:769–72.

Jedlinski, H., W. F. Rochow, and C. M. Brown. 1977. Tolerance to barley yellow dwarf virus in oats. *Phytopathology* 67:1408–11.

Jennings, P. R. and A. J. Ullstrup. 1957. A histological study of three *Helminthosporium* leaf blights of corn. *Phytopathology* 47:707–14.

Jensen, D. D., W. H. Griggs, C. Q. Gonzales, and H. Schneider. 1964. Pear decline virus transmission by pear psylla. *Phytopathology* 54:1346–51.

Jewell, F. F., and D. C. Speirs. 1976. Histopathology of one- and two-year-old resisted infections by *Cronartium fusiforme* in slash pine. *Phytopathology* 66:741–8.

Johal, G. S., and S. P. Briggs. 1992. Reductase activity encoded by the *HM1* disease resistance gene in maize. *Science* 258:985–7.

Johnson, A. H. 1992. Role of abiotic stress in the decline of red spruce in high elevation forests of the eastern United States. *Annu. Rev. Phytopathol.* 30:349–67.

Johnson, D. R., and J. A. Percich. 1992. Wild rice domestication, fungal brown spot disease, and the future of commercial production in Minnesota. *Plant Dis.* 76:1193–8.

Johnson, G. I. 1989. *Peronospora hyoscyami* deBary: taxonomic history, strains, and host range. In *Blue Mold of Tobacco* (ed. by W. E. McKeen), pp. 1–18. St. Paul, MN: APS Press.

Johnson, K. B., V. O. Stockwell, D. M. Burgett, D. Sugar, and J. E. Loper. 1993. Dispersal of *Erwinia amylovora* and *Pseudomonas fluorescens* by honey bees from hives to apple and pear blossoms. *Phytopathology* 83:478–84.

Jones, A. L. 1992. Evaluation of the computer model MARYBLYT for predicting fire blight blossom infection on apple in Michigan. *Plant Dis.* 76:344–7.

Jones, A. L., J. L. Norelli, and G. R. Ehret. 1991. Detection of streptomycin-resistant *Pseudomonas syringae* pv. *papulans* in Michigan apple orchards. *Plant Dis.* 75:529–31.

Joubert, J. J., and F.H.J. Rijkenberg. 1971. Parasitic green algae. *Annu. Rev. Phytopathol.* 9:45–64.

Kahl, G., and J. S. Schell. (eds.). 1982. *Molecular Biology of Plant Tumors.* New York: Academic Press.

Kahn, R. P. 1991. Exclusion as a disease control strategy. *Annu. Rev. Phytopathol.* 29:219–46.

Karper, R. E., and J. R. Quinby. 1946. The history and evolution of milo in the United States. *J. Amer. Soc. Agron.* 38:441–53.

Kasai, T., H. Otani, K. Kohmoto, and S. Nishimura. 1975. Target tissues of *A. kikuchiana* toxin and their characteristic responses. *J. Fac. Agric.*, Tottori Univ., 10:6–14.

Kelman, A. 1953. The bacterial wilt caused by *Pseudomonas solanacearum. N.C. Agric. Exp. Sta. Tech. Bull.* no. 99.

  1995. Contributions of plant pathology to the biological sciences and industry. *Annu. Rev. Phytopathol.* 33:1–21.

Kerr, A. 1980. Biological control of crown gall through production of Agrocin 84. *Plant Dis.* 64:24–30.

Kingsolver, C. H., J. S. Melching, and K. R. Bromfield. 1983. The threat of exotic plant pathogens to agriculture in the United States. *Plant Dis.* 67:595–600.

Kistler, H. C., and E. A. Momol. 1990. Molecular genetics of plant pathogenic *Fusarium oxysporum.* In *Fusarium Wilt of Banana* (ed. by R. C. Ploetz), pp. 49–54. St. Paul, MN: APS Press.

Kiyosawa, S. 1977. Some examples of pest and disease epidemics in Japan and their causes. In *The Genetic Basis of Epidemics in Agriculture* (ed. by P. Day). *New York Acad. Sci.* 287:35–44.

Klebahn, H. 1889. Bemerkung uber den Weymouthskieferrost. *Abhandlungen herausgegeben vom Naturwissenschaftlichen Vereine zu Bremen* 10:427–8.

Klein, R. M. 1979. *The Green World: An Introduction to Plants and People.* New York: Harper & Row.

Kline, D. M., and R. R. Nelson. 1963. Pathogenicity of isolates of *Cochliobolus sativus* from cultivated and wild gramineous hosts from the western hemisphere to species of Gramineae. *Plant Dis. Rep.* 47:890–4.

Klinkowski, M. 1962. The European pandemics of *Peronospora tabacina* Adam, the causal agent of blue mould of tobacco. *Biol. Zentralbl.* 81:75–89.

Ko, W.-h. 1982. Biological control of *Phytophthora* root rot of papaya with virgin soil. *Plant Dis.* 66:446–8.

1988. Hormonal heterothallism and homothallism in *Phytophthora. Annu. Rev. Phytopathol.* 26:57–73.

1994. An alternative possible origin of the A2 mating type of *Phytophthora infestans* outside Mexico. *Phytopathology* 84:1224–7.

Kohmoto, K., S. Nishimura, and H. Otani. 1982. Action sites for AM-toxins produced by the apple pathotypes of *Alternaria alternata*. In *Plant Infection: The Physiological and Biochemical Basis* (ed. by Y. Asada et al.), pp. 254–63. Tokyo: Japan Sci. Soc. Press; Berlin: Springer-Verlag.

Kohmoto, K., and H. Otani. 1991. Host recognition by toxigenic plant pathogens. *Experientia* 47:755–64.

Kohmoto, K., R. P. Scheffer, and J. O. Whiteside. 1979. Host-selective toxins from *Alternaria citri. Phytopathology* 69:667–71.

Kohmoto, K., V. S. Verma, S. Nishimura, M. Tagami, and R. P. Scheffer. 1982. New outbreak of *Alternaria* stem canker of tomato in Japan and production of host-selective toxins by the causal fungus. *J. Fac. Agric.,* Tottori Univ., 17:1–8.

Kohmoto, K., et al. 1993. Isolation and biological activities of two host-specific toxins from the tangerine pathotype of *Alternaria alternata. Phytopathology* 83:495–502.

Kolmer, J. A., and A. H. Ellingboe. 1988. Genetic relationships between fertility and pathogenicity and virulence to rice in *Magnaporthe grisea. Can. J. Bot.* 66:891–7.

Kono, Y., and J. M. Daly. 1979. Characterization of the host-specific pathotoxin produced *by Helminthosporium maydis* race T affecting corn with Texas male sterile cytoplasm. *Biorganic Chem.* 8:391–7.

Krikorian, A. D. 1990. Baseline tissue and cell culture studies for use in banana improvement schemes. In *Fusarium Wilt of Banana* (ed. by R. C. Ploetz), pp. 127–33. St. Paul, MN: APS Press.

Kuhlman, E. G. 1990. Frequency of single-gall isolates of *Cronartium quercuum* f. sp. *fusiforme* with virulence toward three resistant loblolly pine families. *Phytopathology* 80:614–17.

Kuhlman, E. G., H. Bhattacharyya, B. L. Nash, M.L. Double, and W. L. MacDonald. 1984. Identifying hypovirulent isolates of *cryphonectria parasitica* with broad conversion capacity. *Phytopathology* 74:676–82.

Kuhlman, E. G., and W. D. Pepper. 1994. Temperature effects on basidiospore germination and infection of slash pine seedlings by *Cronartium quercuum* f. sp. *fusiforme. Phytopathology* 84:735–9.

Kuijt, J. 1969. *The Biology of Parasitic Flowering Plants.* Berkeley: Univ. Calif. Press.

1977. Haustoria of phaneroganmic parasites. *Annu. Rev. Phytopathol.* 17:91–118.

Kuo, M.-s., and R. P. Scheffer. 1964. Evaluation of fusaric acid as a factor in development of *Fusarium* wilt. *Phytopathology* 54:1041–4.

1969. Factors affecting activity of *Helminthosporium carbonum* toxin on corn plants. *Phytopathology* 59:1779–82.

1970. Comparative effects of host-specific toxins and *Helminthosporium* infections on respiration and carboxylation by host tissue. *Phytopathology* 60:1391–4.

Kuo, M.-s., O. C. Yoder, and R. P. Scheffer. 1970. Comparative specificity of the toxins from *Helminthosporium carbonum* and *H. victoriae*. *Phytopathology* 60:365–8.

Kushalappa, A. C., and A. B. Eskes. 1989. Advances in coffee rust research. *Annu. Rev. Phytopathol.* 27:503–31.

Laby, R. J., and S. V. Beer. 1992. Hybridization and functional complementation of the *hrp* gene cluster from *Erwinia amylovora* strain Ea321 with DNA of other bacteria. *Mol. Plant-Microbe Interact.* 5:412–19.

Lachmund, H. G. 1934. Damage to *Pinus monticola* by *Cronartium ribicola* at Garibaldi, British Columbia. *J. Agric. Res.* 49:239–49.

Lafon, R., and J. Bulit. 1981. Downy mildew of the vine. In *The Downy Mildews* (ed. by D. M. Spencer), pp. 601–14. London: Academic Press.

Landis, W. R., and J. H. Hart. 1972. Physiological changes in pathogen-free tissue of *Ulmus americana* induced by *Ceratocystis ulmi*. *Phytopathology* 62:909–13.

Langdon, K. R. 1965. Relative resistance or susceptibility of several clones of *Hevea brasiliensis* and *H. brasiliensis* × *H. benthamiana* to two races of *Dothidella ulei*. *Plant Dis. Rep.* 49:12–14.

Large, E. C. 1940. *The Advance of the Fungi*. London: Jonathan Cape Publ.

Layton, A. C., and D. N. Kuhn. 1988a. Heterokaryon formation by protoplast fusion of drug-resistant mutants in *Phytophthora megasperma* f. sp. *glycinea*. *Exptl. Mycol.* 12:180–94.

 1988b. The virulence of interracial heterokaryons of *Phytophthora megasperma* f. sp. *glycinea*. *Phytopathology* 78:961–6.

 1990. *In planta* formation of heterokaryons of *Phytophthora megasperma* f. sp. *glycinea*. *Phytopathology* 80:602–6.

Leath, S., and W. L. Petersen. 1986. Differences in resistance between maize hybrids with or without the *Ht* gene when infected with *Exserohilum turcicum* race 2. *Phytopathology* 76:257–60.

Lee, R. F., L. A. Calvert, J. Nagel, and J. D. Hubbard. 1988. Citrus tristeza virus: characterization of coat proteins. *Phytopathology* 78:1221–6.

Lee-Lovick, G. 1978. Smut of sugarcane – *Ustilago scitaminea*. *Rev. Plant Pathol.* 57:181–8.

Leonard, K. J. 1969. Genetic equilibria in host-pathogen systems. *Phytopathology* 59:1858–63.

 1973. Association of mating type and virulence in *Helminthosporium maydis*, and observations on the origin of the race T population in the United States. *Phytopathology* 63:112–15.

 1977. Races of *Bipolaris maydis* in the southeastern U.S. from 1974 to 1976. *Plant Dis. Rep.* 61:914–15.

Leslie, J. F. 1990. Genetic exchange within sexual and asexual populations of the genus *Fusarium*. In *Fusarium Wilt of Banana* (ed. by R. C. Ploetz), pp. 37–48. St. Paul, MN: APS Press.

Lesney, M. S., R. S. Livingston, and R. P. Scheffer. 1982. Effects of toxin from *Helminthosporium sacchari* on nongreen tissues and a reexamination of toxin binding. *Phytopathology* 72:844–9.

Leukel, R. W. 1948. *Periconia circinata* and its relation to milo disease. *J. Agric. Res.* 77:201–22.

Leung, H., and M. Taga. 1988. *Magnaporthe grisea (Pyricularia* species), the blast fungus. *Advances in Plant Pathology* 6:175–88 *(Genetics of Plant Pathogenic Fungi,* ed. by G. S. Sidhu).

Lewis, J. A., and G. C. Papavizas. 1991. Biocontrol of plant disease: the approach for tomorrow. *Crop Prot.* 10:95–105.

Liao, C.-h., H.-y, Hung, and A. K. Chatterjee. 1988. An extracellular pectate lyase is the pathogenicity factor of the soft-rotting bacterium *Pseudomonas viridiflava. Mol. Plant Microbe Interact.* 1:199–206.

Lippincott, B. B., M. H. Whatley, and J. B. Lippincott. 1977. Tumor induction by *Agrobacterium* involves attachment of the bacterium to a site on the host plant cell wall. *Plant Physiol.* 59:388–90.

Lipps, P. E. 1983. Survival of *Colletotrichum graminicola* in infested corn residues in Ohio. *Plant Dis.* 67:102–4.

Littlefield, L. J. 1981. *Biology of the Plant Rusts.* Ames, IA: Iowa State Univ. Press.

Liu, H.-y, and J. E. Duffus. 1990. Beet pseudo-yellows virus: purification and serology. *Phytopathology* 80:866–9.

Livingston, R. S., and R. P. Scheffer. 1984a. Toxic and protective effects of analogs of *Helminthosporium sacchari* toxin on sugarcane tissues. *Physiol. Plant Pathol.* 24:133–44.

1984b. Selective toxins and analogs produced by *Helminthosporium sacchari:* production, characterization and biological activity. *Plant Physiol.* 76:96–102.

Long, M., N. T. Keen, O. K. Ribeiro, J. V. Leary, D. C. Erwin, and G. A. Zentmyer. 1975. *Phytophthora megasperma* var. *sojae:* development of wild-type strains for genetic research. *Phytopathology* 65:592–7.

Long, M., and N. T. Keen. 1977a. Evidence for heterokaryosis in *Phytophthora megasperma* var. *sojae. Phytopathology* 67:670–4.

1977b. Genetic evidence for diploidy in *Phytophthora megasperma* var. *sojae. Phytopathology* 67:675–7.

Lucas, G. B. 1980. The war against blue mold. *Science* 210:147–53.

Luley, C. J., and H. S. McNabb Jr. 1989. Ascospore production, release, germination, and infection by *Mycosphaerella populorum. Phytopathology* 79:1013–18.

MacDonald, W. L., and D. W. Fulbright. 1991. Biological control of chestnut blight: use and limitations of transmissible hypovirulence. *Plant Dis.* 75:656–61.

MacHardy, W. E., and C. H. Beckman. 1973. Water relations in American elm infected with *Ceratocystis ulmi. Phytopathology* 63:98–103.

MacKenzie, J. J., and M. T. El-Ashry. 1990. *Air Pollution's Toll on Forests and Crops.* New Haven, CT: Yale Univ. Press.

Macko, V., W. Acklin, C. Hildenbrand, F. Weibel, and D. Arigoni. 1983. Structure of three isomeric host-specific toxins from *Helminthosporium sacchari. Experientia* 39:343–7.

Macko, V., M. B. Stimmel, T. J. Wolpert, L. D. Dunkle, W. Acklin, R. Bänteli, B. Jaum, and P. Arigoni. 1992. Structure of the host-specific toxins produced by the fungal pathogen *Periconia circinata. Proc. Nat. Acad. Sci. USA* 89:9574–8.

Macko, V., T. J. Wolpert, W. Acklin, B. Jaum, J. Seibl, J. Meili, and D. Arigoni. 1985. Characterization of victorin C, the major host-selective toxin from *Cochliobolus victoriae:* structure of degradation products. *Experientia* 41:1366–70.

Maekawa, N., M. Yamamoto, S. Nishimura, K. Kohmoto, M. Kuwata, and Y. Watanabe. 1984. Studies on host-specific AF-toxins produced by *Alternaria alternata* strawberry pathotype causing *Alternaria* black spot of strawberry. 1. Production of host-specific toxins and their biological activities. *Ann. Phytopathol. Soc. Japan* 50:600–9.

Maeno, S.-i, K. Kohmoto, H. Otani, and S. Nishimura. 1984. Different sensitivities among apple and pear cultivars to AM-toxin produced by *Alternaria alternata* apple pathotype. *J. Fac. Agric.*, Tottori Univ., 19:8–19.

Mahanti, N., et al. 1993. Elevated mitochondrial alternative oxidase activity in dsRNA-free, hypovirulent isolates of *Cryphonectria parasitica. Physiol. Mol. Plant Pathol.* 42:455–63.

Mahmood, T., et al. 1993. Barley yellow dwarf viruses in wheat, endophyte-infected and endophyte-free tall fescue, and other hosts in Arkansas. *Plant Dis.* 77:225–8.

Main, C. E., and J. M. Davis. 1989. Epidemiology and biometerology of tobacco blue mold. In *Blue Mold of Tobacco* (ed. by W. E. McKeen), pp. 201–15. St. Paul, MN: APS Press.

Mamiya, Y. 1983. Pathology of the pine wilt disease caused by *Bursaphelenchus xylophilus. Annu. Rev. Phytopathol.* 21:201–20.

Mangin, M. L. 1899. Sur le Piétin ou maladie du pied du Blé. *Bul. Soc. Mycol. de France* 15:210–39.

Manion, P. D. 1981. *Tree Disease Concepts.* Englewood Cliffs, NJ.: Prentice Hall, Inc.

Maramorosch, K., and S. P. Raychaudhuri. 1988. *Mycoplasma Diseases of Crops.* New York: Springer-Verlag.

Martin, R. R., P. K. Keese, M. J. Young, P. M. Waterhouse, and W. L. Gerlach. 1990. Evolution and molecular biology of luteoviruses. *Annu. Rev. Phytopathol.* 28:341–63.

Martinez, A. L., and J. M. Wallace. 1964. Studies on the transmission of the virus components of citrus seedling yellows by *Aphis gossypii. Plant Dis. Rep.* 48:131–3.

Martinez, J. A., et al. 1975. Presenca de esporos de *Hemileia vasatrix* agente causal da ferrugem do cafeeiro, em differentes altitude nas principais áreas caffeeiras dos Estados de Sao Paulo e Paraná (Brazil). *Biologico* 41:77–88.

Mather, K. 1973. *Genetical Structure of Populations.* London: Chapman & Hall.

Mathys, G., and E. A. Baker. 1980. An appraisal of the effectiveness of quarantines. *Annu. Rev. Phytopathol.* 18:85–101.

Matthews, P. 1981. Breeding for resistance to downy mildews. In *The Downy Mildews* (ed. by D. M. Spencer), pp. 256–87. London: Academic Press.

Matthews, R.E.F. 1992. *Fundamentals of Plant Virology.* Orlando, FL: Academic Press.

Matthysse, A. G., K. V. Holmes, and R.H.G. Gurlitz. 1981. Elaboration of cellulose fibrils by *Agrobacterium tumefaciens* during attachment to carrot cells. *J. Bacteriol.* 145:583–95.

Mayr, H. 1900. Naturwissenschaftliche und forstliche Studien im nordwestlichen Russland. *Allgemeine Forst. und Jagd.-Zeitung* 76:122.

McCann, T. P. 1976. *An American Company: The Tragedy of United Fruit.* New York: Crown Publ.

McCarroll, D. R., and E. Thor. 1985. Pectolytic, cellulytic, and proteolytic activities expressed by cultures of *Endothia parasitica* and inhibition of these activities by components extracted from Chinese and American chestnut inner bark. *Physiol. Plant Pathol.* 26:367–78.

McClean, A.P.D., and J. E. Vanderplank. 1955. The role of seedling yellows and stem pitting in tristeza of citrus. *Phytopathology* 45:222–4.

McClellan, W. D. 1953. Rust and other disorders of snapdragon. In *Plant Diseases,* pp. 568–72. Washington, DC: *U.S.D.A. Yearbook of Agriculture.*

McDermot, J. M., and B. A. McDonald. 1993. Gene flow in plant pathosystems. *Annu. Rev. Phytopathol.* 31:353–75.

McGrath, H., and P. R. Miller. 1958. Blue mold of tobacco. *Plant Dis. Rep.,* supplement 250:1–35.

McIntyre, J., J. A. Dodds, G. S. Walton, and G. H. Lacy. 1978. Declining pear trees in Connecticut: symptoms, distribution, symptom remission by oxytetracycline, and associated mycoplasmalike organisms. *Plant Dis. Rep.* 62:503–7.

McKeen, W. E., ed. 1989. *Blue Mold of Tobacco.* St. Paul, MN: APS Press.

McManus, P. S., and A. L. Jones. 1994. Epidemiology and genetic analysis of streptomycin-resistant *Erwinia amylovora* from Michigan and evaluation of oxytetracycline for control. *Phytopathology* 84:627–33.

Medler, J. T., and P. W. Smith. 1960. Greenbug dispersal and distribution of barley yellow dwarf virus in Wisconsin. *J. Econ. Entomol.* 53:473–4.

Meeley, R. B., and J. D. Walton. 1991. Enzymatic detoxification of HC-toxin, the host-selective cyclic peptide from *Cochliobolus carbonum. Plant Physiol.* 97:1080–6.

Meeley, R. B., G. S. Johal, S. P. Briggs, and J. D. Walton. 1992. A biochemical phenotype for a disease resistance gene of maize. *Plant Cell* 4:71–7.

Meins, F., Jr. 1982. Habituation of cultured plant cells. In *Molecular Biology of Plant Tumors* (ed. by G. Kahl and J. S. Schell), pp. 3–31. New York: Academic Press.

  1990. Habituation: heritable variation in the requirement of cultured plant cells for hormones. *Annu. Rev. Genetics* 23:395–408.

Menage, A., and G. Morel. 1964. Sur la présence d'octopine dans les tissues de crown gall. *Comp. Rend. Acad. Sci.* (Paris) 259:4795–6.

Mercer, P. C., R.K.S. Wood, and A. D. Greenwood. 1975. Ultrastructure of the parasitism of *Phaseolus vulgaris* by *Colletotriohum lindemuthianum. Physiol. Plant Pathol.* 5:203–14.

Meredith, D. S. 1970. Banana leaf spot disease (Sigatoka) caused by *Mycoshaerella musicola* Leach. (Paper no. 11). Kew, Surrey, U.K.: Commonwealth Mycol. Inst.

Mew, T. W. 1987. Current status and future prospects of research on bacterial blight of rice. *Annu. Rev. Phytopathol.* 25:359–82.

Mew, T. W., A. M. Alvarez, J. E. Leach, and J. Swings. 1993. Focus on bacterial blight of rice. *Plant Dis.* 77:5–12.

Miao, V.P.W. 1990. Using karotype variability to investigate origins and relatedness of isolates of *Fusarium oxysporum* f. sp. *cubense*. In *Fusarium Wilt of Banana* (ed. by R. C. Ploetz), pp. 55–62. St. Paul, MN: APS Press.

Miao, V.P.W., D. E. Matthews, and H. D. Van Etten. 1991. Identification and chromosomal locations of a family of cytochrome P-450 genes for pisatin detoxification in the fungus *Nectria haematococca*. *Mol. Gen. Genet.* 226:214–23.

Millardet, P.M.A. 1885. *The Discovery of Bordeaux Mixture* (transl. by F. J. Schneiderhan). *Phytopath. Classics* no. 3. St. Paul, MN: Amer. Phytopathol. Soc.

Miller, J. W. 1966. Differential clones of *Hevea* for identifying races of *Dothidella ulei*. *Plant Dis. Rep.* 50:187–90.

Miller, P. R., and J. R. McBride. 1975. Effects of air pollutants on forests. In *Responses of Plants to Air Pollution* (ed. by J. B. Mudd and T. T. Kozlowski), pp. 195–235. New York: Academic Press.

Miller, R. J., and D. E. Koeppe. 1971. Southern corn leaf blight: susceptible and resistant mitochondria. *Science* 173:67–9.

Miller, T., E. B. Cowling, H. R. Powers, Jr., and T. E. Blalock. 1976. Types of resistance and compatibility in slash pine seedlings infected by *Cronartium fusiforme*. *Phytopathology* 66:1229–35.

Mitchell, A. G., and C. M. Brasier. 1994. Contrasting structure of European and North American population of *Ophiostoma ulmi*. *Mycol. Res.* 98:576–82.

Mitchell, R. E. 1984 The relevance of non-host-specific toxins in the expression of virulence by pathogens. *Annu. Rev. Phytopathol.* 22:215–45.

1991. Structure: bacterial. In *Toxins in Plant Disease* (ed. by R. D. Durbin), pp. 259–93. New York: Academic Press.

Mohnen, V. A. 1988. The challenge of acid rain. *Sci. Amer.* 259:30–8.

Moir, W. S. 1924. White pine blister rust in western Europe. Washington, DC: U.S.D.A. Dept. Bull. 1186.

Momol, M. T., M. Timur, and W. Zeller. 1992. Identification and spread of *Erwinia amylovora* on pear in Turkey. *Plant Dis.* 76:1114–16.

Monaco, L. C. 1977. Consequences of the introduction of coffee rust into Brazil. In *The Genetic Basis of Epidemics in Agriculture* (ed. by P. Day). *New York Acad. Sci.* 287:57–71.

Montoya, A. L., M.-D. Chilton, M. P. Gordon, M. P. Scaky, and E. W. Nester. 1977. Octopine and nopaline metabolism in *Agrobacterium tumefaciens* and crown gall tumor cells: role of plasmid genes. *J. Bact.* 129:101–7.

Moreno, P., J. Guerri, and N. Muñoz. 1990. Identification of Spanish strains of citrus tristeza virus by analysis of double-stranded RNA. *Phytopathology* 80:477–82.

Müller-Dombois, D. 1992. A natural dieback theory, cohort senescence as an alternative to the decline disease theory. In *Forest Decline Concepts* (ed. by P. D. Manion and D. Lachance), pp. 26–37. St. Paul, MN: APS Press.

Mulcahy, D. L., and R. Bernatzky. 1992. Speeding restoration of the American chestnut by using genetic markers in a backcrossing program: an homage to Dr. Charles Burnham. *J. Amer. Chestnut Foundation* 7:33–6.

Murach, D., and B. Ulrich. 1988. Destabilization of forest ecosystems by acid deposition. *GeoJournal* 172:253–60.

Murai, N., F. Skoog, M. E. Doyle, and R. S. Hanson. 1980. Relationships between cytokinin production, presence of plasmids, and fasciation caused by strains of *Corynebacterium fascians*. *Proc. Nat. Acad. Sci. U.S.A.* 77:619–23.

Musselman, L. J. 1980. The biology of *Striga, Orobanche,* and other root-parasitic weeds. *Annu. Rev. Phytopathol.* 18:463–89.

Nagarajan, S., and D. V. Singh. 1990. Long-distance dispersal of rust pathogens. *Annu. Rev. Phytopathol.* 28:139–53.

Nakajima, H., and R. P. Scheffer. 1987. Interconversions of aglycone and host-selective toxin from *Helminthosporium sacchari*. *Phytochemistry* 26:1607–11.

Nakashima, T., T. Ueno, and H. Fukami. 1982. Structure elucidation of AK-toxins, host-specific phytotoxic metabolites produced by *Alternaria kikuchiana Tanaka*. *Tetrahed. Lett.* 23:4469–72.

Namiki, F., M. Yamamoto, S. Nishimura, S. Nakatsuka, T. Goto, K. Kohmoto, and H. Otani. 1986. Protective effect of AF-toxin II on AF-toxin I-induced toxic action and fungal infection. *Ann. Phytopathol. Soc. Japan* 52:428–36.

Nelson, R. R. 1960. Evolution of sexuality and pathogenicity. I. Interspecific crosses in the genus *Helminthosporium*. *Phytopathology* 50:375–7.

1961. Evidence of gene pools for pathogenicity in species of *Helminthosporium*. *Phytopathology* 51:736–7.

1970. Genes for pathogenicity in *Cochliobolus carbonum*. *Phytopathology* 60:1335–7.

1978. Genetics of horizontal resistance to plant diseases. *Annu. Rev. Phytopathol.* 16:359–78.

Nelson, R. R., and D. M. Kline. 1969. Genes for pathogenicity in *Cochliobolus heterostrophus*. *Can. J. Botany* 47:1311–14.

Nelson, R. R., R. P. Scheffer, and R. B. Pringle. 1963. Genetic control of toxin production in *Helminthosporium victoriae*. *Phytopathology* 53:385–7.

Nesmith, W. C. 1984. The North American blue mold warning system. *Plant Dis.* 68:933–6.

Nester, E. W., M.-D. Chilton, M. Drummond, D. Merlo, A. Montoya, D. Scialky, and M. P. Gordon. 1977. Search for bacterial DNA in crown gall tumors. In *Recombinant Molecules: Impact on Science and Society* (ed. by R. F. Beers, Jr. and E. G. Bassett), pp. 179–88. New York: Raven Press.

Newcombe, G., and G. A. Chastagner. 1993. A leaf rust epidemic of hybrid poplar along the lower Columbia River caused by *Melampsora medusae*. *Plant Dis.* 77:528–31.

Newhouse, J. R., W. L. MacDonald, and H. C. Hoch. 1990. Virus-like particles in hyphae and conidia of European hypovirulent (dsRNA-containing) strains of *Cryphonectria parasitica*. *Can. J. Botany* 68:90–101.

Nicholson, R. L. 1992. *Colletotrichum graminicola* and the anthracnose disease of maize and sorghum. In *Colletotrichum: Biology, Pathology and Control* (ed. by J. A. Bailey and M. J. Jeger), pp. 186–202. Wallingford, U.K.: C.A.B. International.

Niederhauser, J. S. 1991. *Phytophthora infestans:* the Mexican connection. In *Phytophthora* (ed. by J. A. Lucas et al.), pp. 25–45. London: Cambridge Univ. Press.

1993. International cooperation in potato research and development. *Annu. Rev. Phytopathol.* 31:1–21.

Niederhauser, J. S., J. Cervantes, and L. Servin. 1954. Late blight in Mexico and its implications. *Phytopathology* 44:406–8.

Nishimura, S., and K. Kohmoto. 1983. Host-specific toxins and chemical structures from *Alternaria* species. *Annu. Rev. Phytopathol.* 21:87–116.

Nishimura, S., K. Kohmoto, and H. Udagawa. 1976. Tolerance to polyoxin in *Alternaria kikuchiana* Tanaka, causing black spot disease of Japanese pears. *Rev. Plant Prot. Res.* 9:47–57.

Nishimura, S., M. Sugihara, K. Kohmoto, and H. Otani. 1978. Two different phases in pathogenicity of the *Alternaria* pathogen causing black spot disease of Japanese pear. *J. Fac. Agric.,* Tottori Univ., 13:1–10.

Norelli, J. L., T. J. Burr, A. M. Lo Cicero, M. T. Gilbert, and B. H. Katz. 1991. Homologous streptomycin-resistant gene present among diverse gram-negative bacteria in New York apple orchards. *Appl. Environ. Microbiol.* 57:486–91.

Nuss, D. L., and Y. Koltin. 1990. Significance of dsRNA genetic elements in plant pathogenic fungi. *Annu. Rev. Phytopathol.* 28:37–58.

Nutsugah, S. K., K. Kohmoto, H. Otani, M. Kodama, and R. R. Sunkeswari. 1994. Production of a host-specific toxin by germinating spores of *Alternaria tenuissima* causing leaf spot of pigeon pea. *J. Phytopathol.* 140:19–30.

Nutsugah, S. K., P. Park, H. Otani, M. Kodama, and K. Kohmoto. 1993. Ultrastructural changes in pigeon pea cells caused by a host-specific toxin from *Alternaria tenuissima*. *Ann. Phytopathol. Soc. Japan* 59:407–15.

Nyland, G., and A. C. Goheen. 1969. Heat therapy of virus diseases of perennial plants. *Annu. Rev. Phytopathol.* 7:331–54.

Orlob, G. B. 1971. History of plant pathology in the middle ages. *Annu. Rev. Phytopathol* 9:7–20.

1973. Ancient and medieval plant pathology. *Planzenschutz-Nachrichten Bayer* 26:65–294.

Ostofsky, W. D., T. Rumpf, D. Struble, and R. Bradbury. 1988. Incidence of white pine blister rust in Maine after 70 years of a *Ribes* eradication program. *Plant Dis.* 72:967–70.

Ostry, M. E., P. M. Pijut, and D. D. Skilling. 1991. Screening larch *in vitro* for resistance to *Mycosphaerella laricina*. *Plant Dis.* 75:1222–24.

Oswald, J. W., and B. R. Houston. 1953a. The yellow dwarf virus disease of cereal crops. *Phytopathology* 43:128–36.

1953b. Host range and epiphytology of the cereal yellow dwarf disease. *Phytopathology* 43:309–13.

Otani, H., and K. Kohmoto. 1992. Host-specific toxins of *Alternaria* species. In *Alternaria: Biology, Plant Diseases and Metabolites* (ed. by J. Chelkowski and A. Visconti), pp. 123–56. Amsterdam: Elsevier Press.

Otani, H., K. Kohmoto, S. Nishimura, T. Nakashima, T. Ueno, and H. Fukami. 1985. Biological activities of AK-toxins I and II, host-specific toxins from *Alternaria alternata* Japanese pear pathotype. *Ann. Phytopathol. Soc. Japan* 51:285–93.

Otani, H., S. Nishimura, and K. Kohmoto. 1974. Nature of specific susceptibility to *Alternaria kikuchiana* in Nijisseiki pears. III. Chemical and thermal protection against effects of host-specific toxin. *Ann. Phytopathol. Soc. Japan* 40:59–66.

Ou, S. H. 1972. Blast. In *Rice Diseases* (ed. by S. H. Ou). Kew, Surrey, U.K.: Commonwealth Mycol. Inst.

1980. Pathogen variability and host resistance in rice blast disease. *Annu. Rev. Phytopathol.* 18:167–87.

Ou, S. H., and M. R. Ayad. 1968. Pathogenic races of *Piricularia oryzae* originating from single lesion and monoconidial cultures. *Phytopathology* 58:179–82.

Paddock, W. C. 1953. Histological study of suscept–pathogen relationships between *Helminthosporium victoriae* and seedling oat leaves. *Cornell Agric. Exp. Sta. Mem.* 315.

Padmanabhan, S. Y. 1973. The great Bengal famine. *Annu. Rev. Phytopathol.* 11:11–26.

Panaccione, D. G., J. S. Scott-Craig, J. A. Pocard, and J. D. Walton. 1992. A cyclic peptide synthetase gene required for pathogenicity of the fungus *Cochliobolus carbonum* on maize. *Proc. Nat. Acad. Sci. U.S.A.* 89:6590–4.

Panopoulos, N. J., and R. C. Peet. 1985. The molecular genetics of plant pathogenic bacteria and their plasmids. *Annu. Rev. Phytopathol.* 23:381–419.

Paul, C. P., and D. W. Fulbright. 1988. Double-stranded RNA molecules from Michigan hypovirulent isolates of *Endothia parasitica* vary in size and sequence homology. *Phytopathology* 78:751–5.

Pavari, A. 1949. Chestnut blight in Europe. *Unasylva* 3:8–13.

Payne, G. A., H. E. Duncan, and C. R. Adkins. 1987. Influence of tillage on the development of gray leaf spot and the number of airborne conidia of *Cercospora zea-maydis. Plant Dis.* 71:329–32.

Pearson, R. C., and A. C. Goheen, eds. 1988. *Compendium of Grape Diseases.* St. Paul, MN: APS Press.

Pegg, G. F. 1976a. The involvement of ethylene in plant pathogenesis. In *Physiological Plant Pathology* (ed. by R. Heitefuss and P. H. Williams), pp. 582–91, *Encyclopedia of Plant Physiology,* n.s., vol. 4. Berlin: Springer-Verlag.

1976b. Endogenous gibberellins in healthy and diseased plants. In *Physiological Plant Pathology* (ed. by R. Heitefuss and P. H. Williams), pp. 592–606, *Encyclopedia of Plant Physiology,* n.s., vol. 4. Berlin: Springer-Verlag.

Peterson, G. W. 1967. *Dothistroma* needle blight of Austrian and ponderosa pines: epidemiology and control. *Phytopathology* 57:437–41.

1973. Infection of Austrian and ponderosa pines by *Dothistroma pini* in eastern Nebraska. *Phytopathology* 63:1060–3.

1984. Resistance to *Dothistroma pini* within geographic seed sources of *Pinus ponderosa. Phytopathology* 74:956–60.

Peterson, G. W., and R. A. Read. 1971. Resistance to *Dothistroma pini* within geographic sources of *Pinus nigra. Phytopathology* 61:149–50.

Peterson, R. S., and F. F. Jewell. 1968. Status of American stem rusts of pine. *Annu. Rev. Phytopathol.* 6:23–40.

Petit, A., S. Delhaye, J. Tempe, and G. Morel. 1970. Recherches sur les guanidines des tissus de crown gall. Mise en évidence d'une relation biochimique spécifique entre les sources d'*Agrobacterium tumefaciens* et les tumeurs qu'elles induisent. *Physiol. Veg.* 8:205–13.

Ploetz, R. C. 1990. Population biology of *Fusarium oxysporum* f. sp. *cubense*. In *Fusarium Wilt of Banana* (ed. by R. C. Ploetz), pp. 63–76. St. Paul, MN: APS Press.

Ploetz, R. C., and J. C. Correll. 1988. Vegetative compatibility among races of *Fusarium oxysporum* f. sp. *cubense*. *Plant Dis*. 72:325–8.

Ploetz, R. C., J. Herbert, K. Sebasigari, J. H. Hernandez, K. G. Pegg, J. A. Ventura, and L. S. Mayato. 1990. Importance of *Fusarium* wilt in different banana-growing regions. In *Fusarium Wilt of Banana* (ed. by R. C. Ploetz), pp. 9–26. St. Paul, MN: APS Press.

Ploetz, R. C., and E. S. Shepard. 1988. Fusarial wilt of banana in south Florida. *Plant Dis*. 72:994.

Populer, C. 1981. Epidemiology of downy mildews. In *The Downy Mildews* (ed. by D. M. Spencer), pp. 57–105. London: Academic Press.

Powers, H. R., Jr. 1975. Relative susceptibility of five southern pines to *Cronartium fusiforme*. *Plant Dis. Rep*. 59:312–14.

   1980. Pathogenic variation among single aeciospore isolates of *Cronartium quercuum* f. sp. *fusiforme*. *For. Sci*. 26:280–2.

Powers, H. R., Jr., and H. J. Duncan. 1976. Increasing fusiform rust resistance by intraspecific hybridization. *For. Sci*. 22:267–8.

Powers, H. R., Jr., D. Lin, and M. Hubbes. 1989. Interspecific and intraspecific differentiation within the genus *Cronartium* by isozyme and protein pattern analysis. *Plant Dis*. 73:691–4.

Powers, H. R., Jr., and F. R. Matthews. 1979. Interactions between virulent isolates of *Cronartium quercuum* f. sp. *fusiforme* and loblolly pine families of varying resistance. *Phytopathology* 69:720–2.

   1980. Comparison of six geographic sources of loblolly pine for fusiform rust resistance. *Phytopathology* 70:1141–3.

Powers, H. R., Jr., F. R. Matthews, and L. D. Dwinell. 1977. Evaluation of pathogenic variability of *Cronartium fusiforme* on loblolly pine in the southern USA. *Phytopathology* 67:1403–7.

Powers, H. R., Jr., R. A. Schmidt, and G. A. Snow. 1981. Current status and management of fusiform rust on southern pines. *Annu. Rev. Phytopathol*. 19:353–71.

Powers, H. R., Jr., G. A. Snow, D. Lin, and M. Hubbes. 1991. Isozyme analysis as an indication of synonymy of the causal agents of gall rust on sand and Virginia pine. *Plant Dis*. 75:1225–7.

Powers, H. R., Jr., and A. F. Verrall. 1962. A closer look at *Fomes annosus*. *For. Farmer* 21:8–9, 16–17.

Prakash, C. S., and B. A. Thielges. 1987. Pathogenic variation in *Melampsora medusae* leaf rust of poplars. *Euphytica* 36:563–70.

Pringle, R. B. 1976. Comparative biochemistry of the phytopathogenic fungus *Helminthosporium* XVI. The production of victoxinine by *H. sativum* and *H. victoriae*. *Can. J. Biochem*. 54:783–7.

Pringle, R. B., and R. P. Scheffer. 1963. Purification of the selective toxin of *Periconia circinata*. *Phytopathology* 53:785–7.

   1964. Host-specific plant toxins. *Annu. Rev. Phytopathol*. 2:133–56.

Psallidas, P. G., ed. 1993. Sixth International Workshop on Fire Blight. Int. Soc. Hort. Sci., *Acta Horticulturae*, no. 338.

Puhalla, J. E. 1985. Classification of strains of *Fusarium oxysporum* on the basis of vegetative compatibility. *Can. J. Botany* 63:179–83.

Purcell, A. H. 1982. Insect vector relationships with procaryotic plant pathogens. *Annu. Rev. Phytopathol.* 20:397–417.

Purdy, L. H., S. V. Krupa, and J. L. Dean. 1985. Introduction of sugar cane rust into the Americas and its spread to Florida. *Plant Dis.* 69:689–93.

Quinby, J. R., and R. E. Karper. 1949. The effect of milo disease on grain and forage yields of sorghum. *Agron. J.* 41:118–22.

Raccah, B., G. Loebenstein, and S. Singer. 1980. Aphid-transmissibility variants of tristeza virus in infected citrus trees. *Phytopathology* 70:89–93.

Raddi, P., and H. R. Powers, Jr. 1982. Relative susceptibility of several European species of pine to fusiform rust. *Eur. J. For. Pathol.* 12:442–7.

Raloff, J. 1995. When nitrate reigns: air pollution can damage forests. *Science News* 147:90–1.

Rao, B. S. 1973. Potential threat of South American leaf blight to the plantation rubber industry in southeast Asia and the Pacific region. *FAO Plant Prot. Bull.* 21:107–13.

Raychaudhuri, S. P., and J. P. Verma. 1977. Therapy by heat, radiation, and meristem culture. In *Plant Disease: An Advanced Treatise,* vol. 1 (ed. by J. G. Horsfall and E. B. Cowling), pp. 177–89. New York: Academic Press.

Rayner, E. W. 1960. Rust disease of coffee. 2. Spread of the disease. *World Crops* 12:222–4.

Ream, W. 1989. *Agrobacterium tumefaciens* and interkingdom genetic exchange. *Annu. Rev. Phytopathol.* 27:583–618.

Reid, J. L., and A. Collmer. 1988. Construction and characterization of an *Erwinia chrysanthemi* mutant with directed deletions in all of the pectate lyase structural genes. *Mol. Plant Microbe Interact.* 1:32–8.

Reil, W. O., W. J. Moller, and S. V. Thomson. 1979. An historical analysis of fire blight epidemics in the central valley of California. *Plant Dis. Rep.* 63:545–9.

Renfro, B. L., and S. Shankara-Bhat. 1981. Role of wild hosts in downy mildew diseases. In *The Downy Mildews* (ed. by D. M. Spencer), pp.107–19. London: Academic Press.

Richards, K. E., and T. Tamada. 1992. Mapping functions on the multipartite genome of beet necrotic yellow vein virus. *Annu. Rev. Phytopathol.* 30:291–313.

Richens, R. H. 1983. *Elms.* London: Cambridge Univ. Press.

Ritchie, D. F., and E. J. Klos. 1977. Isolation and partial characterization of *Erwinia amylovora* bacteriophage from aerial parts of apple trees. *Phytopathology* 67:101–4.

Roane, M. K., G. J. Griffin, and J. R. Elkins. 1986. *Chestnut Blight, Other* Endothia *Diseases, and the Genus* Endothia. APS Monograph Series. St. Paul, MN: APS Press.

Robertsen, B. 1984. An alkaline extracellular protease produced by *Cladosporium cucumerinum* and its possible importance in the development of scab disease of cucumber seedlings. *Physiol. Plant Pathol.* 24:83–92.

Rocha-Peña, M. A., R. F. Lee, R. Lastra, C. L. Niblett, F. M. Ochoa-Corona, S. M. Garnsey, and R. K. Yokomi. 1995. Citrus tristeza virus and its aphid vector *Toxoptera citricola:* threats to citrus production in the Caribbean and Central and North America. *Plant Dis.* 79:437–45.

Rochow, W. R. 1970. Barley yellow dwarf virus: phenotypic mixing and vector specificity. *Science* 167:875–8.

Rodrigues, C. J., Jr., A. J. Betancourt, and L. Rifo. 1975. Races of the pathogen and resistance to coffee rust. *Annu. Rev. Phytopathol.* 13:49–70.

Roelfs, A. P. 1982. Effects of barberry eradication on stem rust in the United States. *Plant Dis.* 66:177–81.

1985. Epidemiology in North America. In *The Cereal Rusts,* vol. 1 (ed. by W. R. Bushnell and A. P. Roelfs), pp. 403–34. Orlando, FL: Academic Press.

Roelfs, A. P., and J. V. Groth. 1980. A comparison of virulence phenotypes in wheat stem rust populations reproducing sexually and asexually. *Phytopathology* 70:855–62.

1988. *Puccinia graminis* f. sp. *tritici,* black stem rust of *Triticum* spp. *Advances in Plant Pathology* (ed. by D. S. Ingram and P. H. Williams), 6:345–61. London: Academic Press.

Roelfs, A. P., D. L. Long, and J. J. Roberts. 1993. Races of *Puccinia graminis* in the United States during 1992. *Plant Dis.* 77:1122–5.

Roistacher, C. N., R. L. Blue, E. M. Nauer, and E. C. Calavan. 1974. Suppression of tristeza virus symptoms in Mexican lime seedlings grown at warm temperatures. *Plant Dis. Rep.* 58:757–60.

Roosje, G. S. 1979. The struggle against fire blight in Europe. *Proc. Symp.,* IX Int. Congress of Plant Prot., pp. 329–32. Washington, DC.

Rosner, A., R. F. Lee, and M. Bar-Joseph. 1986. Differential hybridization with cloned cDNA sequences for detecting a specific isolate of citrus tristeza virus. *Phytopathology* 76:820–4.

Ross, E. W. 1975. *Fomes annosus* in eastern North America. In *Biology and Control of Soil-Borne Plant Pathogens* (ed. by G. W. Bruehl), pp. 107–10. St. Paul, MN: APS Press.

Rowan, S. J. 1970. Fusiform rust gall formation and cytokinin of loblolly pine. *Phytopathology* 60:1225–6.

Rowell, J. B., and R. W. Romig. 1966. Detection of urediospores of wheat rusts in spring rains. *Phytopathology* 56:807–11.

Rufty, R. C. 1989. Genetics of host resistance to tobacco blue mold. In *Blue Mold of Tobacco* (ed. by W. E. McKeen), pp. 141–64. St. Paul, MN: APS Press.

Russell, J. A. 1942. Fordlandia and Balterra rubber plantations on the Tapajoz River, Brazil. *Econ. Geogr.* 18:125–43.

Rutherford, F. S., E.W.B. Ward, and R. I. Buzzell. 1985. Variation in virulence in successive single-zoospore propagations of *Phytophthora megasperma* f. sp. *glycinea. Phytopathology* 75:371–4.

Ryley, M. J., and N. R. Obst. 1992. Race-specific resistance in soybean cv. Davis to *Phytophthora megasperma* f. sp. *glycinea. Plant Dis.* 76:665–8.

Salaman, R. N. 1949. *The History and Social Influence of the Potato.* London: Cambridge Univ. Press.

Samaddar, K. R., and R. P. Scheffer. 1968. Effects of the specific toxin in *Helminthosporium victoriae* on host cell membranes. *Plant Physiol.* 43:21–8.

1971. Early effects of *Helminthosporium victoriae* toxin on plasma membranes and counteraction by chemical treatments. *Physiol. Plant Pathol.* 1: 319–28.

Santiago, J. C. 1968. Differentiating biotypes within physiologic races and its importance in epidemiological studies. *Proc. Cereal Rusts Conference 1965,* Oeiras, Portugal.

Sawamura, K., and H. Yanase. 1963. On *Alternaria* sp., causal organism of *Alternaria* blotch of apple and Japanese name of the disease. Bull. Hort. Res. Sta. Japan, Series C., no.1, pp. 71–94.

Scheffer, R. P. 1976. Host-specific toxins in relation to pathogenesis and disease resistance. In *Physiological Plant Pathology* (ed. by R. Heitefuss and P. H. Williams), pp. 247–69, *Encyclopedia of Plant Physiology,* n.s. vol. 4. Berlin: Springer-Verlag.

1983. Toxins as chemical determinants of plant disease. In *Toxins and Plant Pathogenesis* (ed. by J. M. Daly and B. J. Deverall), pp. 1–40. Sydney: Academic Press.

1989a. Ecological consequences of toxin production by *Cochliobolus* and related fungi. In *Phytototoxins and Plant Pathogenesis* (ed. by A. Graniti, R. D. Durbin, and A. Ballio), pp. 285–300. Berlin: Springer-Verlag.

1989b. Host-specific toxins in phytopathology: origins and evolution of the concept. In *Host-Specific Toxins: Recognition and Specificity Factors in Plant Disease* (ed. by K. Kohmoto and R. D. Durbin), pp. 1–17. Tottori University, Japan.

1991. Role of toxins in evolution and ecology of plant pathogenic fungi. *Experientia* 47:804–11.

1992. Ecological and evolutionary roles of toxins from *Alternaria* species pathogenic to plants. In *Alternaria: Biology, Plant Diseases and Metabolites* (ed. by J. Chelkowski and A. Visconti), pp. 101–22. Amsterdam: Elsevier Press.

Scheffer, R. P., and S. P. Briggs. 1981. A perspective of toxin studies in plant pathology. In *Toxins in Plant Disease* (ed. by R. D. Durbin), pp. 1–20. New York: Academic Press.

Scheffer, R. P., and R. S. Livingston. 1980. Sensitivity of sugarcane clones to toxin from *Helminthosporium sacchari,* as determined by electrolyte leakage. *Phytopathology* 70:400–4.

1984. Host-selective toxins and their role in plant diseases. *Science* 223:17–21.

Scheffer, R. P., and R. R. Nelson. 1967. Geographical distribution and prevalence of *Helminthosporium victoriae. Plant Dis. Rep.* 51:110–11.

Scheffer, R. P., R. R. Nelson, and A. J. Ullstrup. 1967. Inheritance of toxin production and pathogenicity in *Cochliobolus carbonum* and *Cochliobolus victoriae. Phytopathology* 57:1288–91.

Scheffer, R. P., and R. B. Pringle. 1961. A selective toxin produced by *Periconia circinata. Nature* 191:912–13.

Scheffer, R. P., and A. J. Ullstrup. 1965. A host-specific toxic metabolite from *Helminthosporium carbonum. Phytopathology* 55:1037–8.

Scheffer, R. P., and O. C. Yoder. 1972. Host-specific toxins and selective toxicity. In *Phytotoxins in Plant Diseases* (ed. by R.K.S. Wood, A. Ballio, and A. Graniti), pp. 251–72. London: Academic Press.

Schell, M. A., D. P. Roberts, and T. P. Denny. 1988. Analysis of *Pseudomonas solanacearum* polygalacturonase encoded by *pglA* and its involvement in phytopathogenicity. *J. Bact.* 170:4501–8.

Schertz, K. F., and Y. P. Tai. 1969. Inheritance of reaction of *Sorghum bicolor* (L.) Moench to toxin produced by *Periconia circinata* (Mang.) Sacc. *Crop Sci.* 9:621–4.

Schieber, E. 1972. Economic impact of coffee rust in Latin America. *Annu. Rev. Phytopathol.* 10:491–510.

1975. Present status of coffee rust in South America. *Annu. Rev. Phytopathol.* 13:375–82.

Schieber, E., and G. A. Zentmyer. 1984. Coffee rust in the Western Hemisphere. *Plant Dis.* 68:89–93.

Schiltz, P. 1981. Downy mildew of tobacco. In *The Downy Mildews* (ed. by D. M. Spencer), pp. 577–99. London: Academic Press.

Schipper, A. L., and D. H. Dawson. 1974. Poplar leaf rust – a problem in maximum wood fiber production. *Plant Dis. Rep.* 58:721–2.

Schmidt, R. A., R. C. Holley, M. C. Klapproth, and T. Miller. 1986. Temporal and spatial patterns of Fusiform rust epidemics in young plantations of susceptible and resistant slash and loblolly pines. *Plant Dis.* 70:661–6.

Schmitthenner, A. F. 1985. Problems and progress in control of *Phytophthora* root rot of soybean. *Plant Dis.* 69:362–8.

1988. *Phytophthora* rot of soybean. In *Soybean Diseases of the North Central Region* (ed. by T. D. Wyllie and D. H. Scott), pp. 71–80. St. Paul, MN: APS Press.

Schmitthenner, A. F., M. Hobe, and R. G. Bhat. 1994. *Phytophthora sojae* races in Ohio over a 16 year interval. *Plant Dis.* 78:269–76.

Schneider, H. 1970. Graft transmission and host range of the pear decline causal agent. *Phytopathology* 60:204–7.

Schroth, M. N., S. V. Thomson, D. C. Hilderbrand, and W. J. Mollor. 1974. Epidemiology and control of fire blight. *Annu. Rev. Phytopathol.* 12:389–412.

Schuh, W. 1990. The influence of tillage systems on incidence and spatial pattern of tan spot of wheat. *Phytopathology* 80:804–7.

Schumann, G. L. 1991. *Plant Diseases: Their Biology and Social Impact.* St. Paul, MN: APS Press.

Scott-Craig, J. S., D. G. Panaccione, F. Cervone, and J. D. Walton. 1990. Endopolygalacturonase is not required for pathogenicity of *Cochliobolus carbonum* on maize. *Plant Cell* 2:1191–1200.

Scott-Craig, J. S., D. G. Panaccione, J. A. Pocard, and J. D. Walton. 1993. The cyclic peptide synthetase catalyzing HC-toxin production in the filamentous fungus *Cochliobolus carbonum* is encoded by a 15.7-kilobase open reading frame. *J. Biol. Chem.* 267:26044–9.

Seemüller, E. A., and S. V. Beer. 1976. Absence of cell wall polysaccharide degradation by *Erwinia amylovora*. *Phytopathology* 66:433–6.

Semancik, J. S., ed. 1987. *Viroids and Viroid-like Pathogens.* Boca Raton, FL: CRC Press.

Sequeira, L. 1973. Hormone metabolism in diseased plants. *Annu. Rev. Plant Physiol.* 24:353–80.

Shain, L., and R. A. Franich. 1981. Induction of *Dothistroma* blight symptoms with dothistromin. *Physiol. Plant Pathol.* 19:49–55.

Shepard, J. F., and L. E. Chaflin. 1975. Critical analyses of the principles of seed potato certification. *Annu. Rev. Phytopathol.* 13:271–93.

Shortle, W. C., and K. T. Smith. 1988. Aluminum-induced calcium deficiency syndrome in declining red spruce. *Science* 240:1017–18.

Sigee, D. C. 1993. *Bacterial Plant Pathology*. London and New York: Cambridge Univ. Press.

Siler, D. J., and D. G. Gilchrist. 1983. Properties of host-specific toxins produced by *Alternaria alternata* f. sp. *lycopersici* in culture and in tomato plants. *Physiol. Plant Pathol.* 23:265–74.

Simmonds, N. W. 1966. *Bananas,* ed. 2. London: Longmans.

Simmonds, N. W., and K. Shepherd. 1955. The taxonomy and origin of the cultivated bananas. *J. Linnean Soc.,* London, 55:302–12.

Sinclair, J. B., and P. A. Backman, eds. 1989. *Compendium of Soybean Diseases,* ed. 3. St. Paul, MN: APS Press.

Sinclair, W. A., H. H. Lyon, and W. T. Johnson. 1987. *Diseases of Trees and Shrubs*. Ithaca, NY: Cornell Univ. Press (Comstock Publ. Co.).

Sinclair, W. A., D. S. Welch, K. G. Parker, and L. J. Tyler. 1974. Selection of American elms for resistance to *Ceratocystis ulmi*. *Plant Dis. Rep.* 58:784–8.

Sinkar, V. P., F. F. White, I. J. Furner, M. Abrahmsen, F. Pythoud, and M. P. Gordon. 1988. Reversion of aberrant plants transformed with *Agrobacterium rhizogenes* in association with the transcriptional inactivation of the T_L-DNA genes. *Plant Physiol.* 86:584–90.

Skelley, J. M., and J. L. Innes. 1994. Waldsterben in the forests of central Europe and eastern North America: fantasy or reality? *Plant Dis.* 78:1021–32.

Skipp, R. A., and B. J. Deverall. 1972. Relationship between fungal growth and host changes visible by light microscopy during infection of bean hypocotyls (*Phaseolus vulgaris*) susceptible and resistant to physiological races of *Collectotrichum lindemuthianum*. *Physiol. Plant Pathol.* 2:357–74.

Smalley, E. B., and R. P. Guries. 1993. Breeding elms for resistance to Dutch elm disease. *Annu. Rev. Phytopathol.* 31:325–52.

Smidt, M., and T. Kosuge. 1978. The role of indole-3-acetic acid accumulation by alpha methyl tryptophan-resistant mutants of *Pseudomonas savastanoi* in gall formation on oleanders. *Physiol. Plant Pathol.* 13:203–13.

Smith, D. R., A. L. Hooker, and S. M. Lim. 1970. Physiologic races of *Helminthosporium maydis*. *Plant Dis. Rep.* 54:819–22.

Smith, R., and J. C. Walker. 1930. A cytological study of cabbage plants in strains susceptible or resistant to yellows. *J. Agric. Res.* 41:17–35.

Snell, W. H. 1928. Forest damage and the white pine blister rust. *J. For.* 29:68–78.

Snow, G. A., R. J. Dinus, and A. G. Kais. 1975. Variation in pathogenicity of diverse sources of *Cronartium fusiforme* on selected slash pine families. *Phytopathology* 65:170–5.

Snow, G. A., R. J. Dinus, and C. H. Walkinshaw. 1976. Increase in virulence of *Cronartium fusiforme* on resistant slash pine. *Phytopathology* 66:511–13.

Sobiczewski, P., C.-s. Chiou, and A. L. Jones. 1991. Streptomycin-resistant epiphytic bacteria with homologous DNA for streptomycin resistance in Michigan apple orchards. *Plant Dis.* 75:1110–3.

Spaulding, P. 1911. The blister rust of white pine. U.S.D.A., Bur. Plant Ind., Bull. no. 206.

1922. Investigations of the white pine blister rust. U.S.D.A. Prof. Paper, Bull. no. 957.

1929. White pine blister rust: a comparison of European with North American conditions. U.S.D.A. Tech. Bull. no. 87.

Spencer, D. M., ed. 1981. *The Downy Mildews*. London: Academic Press.

Spiegel, S., E. A. Frison, and R. H. Converse. 1993. Recent developments in therapy and virus-detection procedures for international movement of clonal plant germ plasm. *Plant Dis.* 77:1176–80.

Spielman, L. J., A. Drenth, L. C. Davidse, L. S. Sujkowski, W. K. Gu, P. W. Tooley, and W. E. Fry. 1991. A second world-wide migration and population displacement of *Phytophthora infestans*. *Plant Pathol.* 40:422–30.

Sprague, R. 1950. *Diseases of Cereals and Grasses in North America*. New York: Ronald Press.

Sreeramulu, T., and B.P.R. Vittal. 1972. Spore dispersal of the sugarcane smut (*Ustilago scitaminea*). *Trans. Brit. Mycol. Soc.* 58:301–2.

Stahl, D. J., and W. Schäfer. 1992. Cutinase is not required for fungal pathogenicity on pea. *Plant Cell* 4:621–9.

Stakman, E. C. 1934. Relation of barberry to the origin and persistence of physiologic forms of *Puccinia graminis*. *J. Agric. Res.* 48:953–69.

Stakman, E. C., and G. Harrar. 1957. *Principles of Plant Pathology*. New York: Ronald Press.

Stakman, E. C., et al. 1944. Identification of physiologic races of *Puccinia graminis tritici*. *U.S.D.A. Bur. Ent. and Plant Quar. Bull.* E-617.

Staub, T. 1991. Fungicide resistance: practical experience with antiresistance strategies and the role of integrated use. *Annu. Rev. Phytopathol.* 29:421–42.

Stedman, J. R., and L. Sequeira. 1970. Abscissic acid in tobacco plants: tentative identification and its relation to stunting induced by *Pseudomonas solanacearum*. *Plant Physiol.* 45:691–7.

Stermer B. A., R. P. Scheffer, and J. H. Hart. 1984. Isolation of toxins of *Hypoxylon mammatum* and demonstration of some toxic effects on selected clones of *Populus tremuloides*. *Phytopathology* 74:654–8.

Stewart, F. C. 1906. An outbreak of the European currant rust (*Cronartium ribicola*). NY State Agric. Exp. Sta. Tech. Bull. no. 2.

Stewart, W. W. 1971. Isolation and proof of structure of wildfire toxin. *Nature* 229:174–8.

Stichlen, M. B., and J. Sherald, eds. 1993. *Dutch Elm Disease: Cellular and Molecular Approaches*. New York: Springer-Verlag.

Stipes, R. J., and R. J. Campana, eds. 1981. *Compendium of Elm Diseases*. St. Paul, MN: APS Press.

Stoessl, A. 1983. Secondary plant metabolites in preinfectional and postinfectional resistance. In *The Dynamics of Host Defense* (ed. by J. A. Bailey and B. J. Deverall), pp. 71–122. Sydney: Academic Press.

Stover, R. H. 1962a. Intercontinental spread of banana leaf spot (*Mycosphaerella musicola*). *Trop. Agric. Trinidad* 39:327–38.

1962b. Fusarial Wilt (Panama Disease) of Bananas and Other *Musa* Species. Phytopathol. Paper no. 4. Kew, Surrey, U.K.: Commonwealth Mycol. Inst.

1972. *Banana, Plantain, and Abaca Diseases*. Kew, Surrey, U.K.: Commonwealth Mycol. Inst.

1977. Distribution and probable origin of *Mycosphaerella fijiensis* in southeast Asia. *Trop. Agric. Trinidad* 55:65–8.

1980. Sigatoka leaf spots of bananas and plantains. *Plant Dis.* 64:750–6.

1986. Disease management strategies and the survival of the banana industry. *Annu. Rev. Phytopathol.* 24:83–91.

1990. *Fusarium* wilt of banana: some history and current status of the disease. In *Fusarium Wilt of Banana* (ed. by R. C. Ploetz), pp. 1–7. St. Paul, MN: APS Press.

Su, H., S. Hwang, and W.-h. Ko. 1986. *Fusarium* wilt of Cavendish bananas in Taiwan. *Plant Dis.* 70:814–18.

Sumner, D. R., B. Doupnik, Jr., and M. G. Boosalis. 1981. Effects of reduced tillage and multiple cropping on plant diseases. *Annu. Rev. Phytopathol.* 19:167–87.

Sun, E. J., H. J. Su, and W.-h. Ko. 1978. Identification of *Fusarium oxysporum* f. sp. *cubense* race 4 from soil or host tissue by cultural characteristics. *Phytopathology* 68:1672–3.

Surico, G., N. S. Iacobellis, and A. Sisto. 1985. Studies on the role of indole-3-acetic acid and cytokinins in the formation of knots on olive and oleander plants by *Pseudomonas syringae* pv. *savastanoi*. *Physiol. Plant Pathol.* 26:309–20.

Sutherland, M. L., and G. F. Pegg. 1992. The basis of host recognition in *Fusarium oxysporum* f. sp. *lycopersici*. *Physiol. Mol. Plant Pathol.* 40:423–36.

Sutton, T. B., and A. L. Jones. 1975. Monitoring *Erwinia amylovora* populations on apple in relation to disease incidence. *Phytopathology* 65:1009–12.

Tainter, F. H., and R. L. Anderson. 1993. Twenty-six new pine hosts of fusiform rust. *Plant Dis.* 77:17–20.

Takai, S. 1989. Host-specific factors in Dutch elm disease. In *Host-Specific Toxins: Recognition and Specificity Factors in Plant Disease* (ed. by K. Kohmoto and R. D. Durbin), pp. 75–96. Tottori, Japan: Tottori Univ.

Takai, S., W. C. Richards, and K. J. Stevenson. 1983. Evidence for the involvement of cerato-ulmin, the *Ceratocystis ulmi* toxin, in the development of Dutch elm disease. *Physiol. Plant Pathol.* 23:275–80.

Tanaka, S. 1933. Studies on black spot disease of the Japanese pears (*Pirus serotina* Rehd.). Mem. Coll. Agr., Kyoto Imp. Univ., 28:1–31.

ten Houten, J. C. 1974. Plant pathology: changing agricultural methods and human society. *Annu. Rev. Phytopathol.* 12:1–11.

Thielges, B. A., and J. C. Adams. 1975. Genetic variation and variability of *Melampsora* rust resistance in eastern cottonwood. *For. Sci.* 21:278–82.

Thielges, B. A., C. S. Praskash, A. Sabdono, R. C. Hamelin, and R. J. Rousseau. 1988. Attaining durable disease resistance in forest plantations: implications from studies in the cottonwood-leaf rust pathosystem. In *Proc. 10th North Amer. For. Biol. Conf.*, pp. 152–7. Vancouver: Univ. B. Columbia Press.

Thimann, K. V., and T. Sachs. 1966. The role of cytokinin in the fasciation disease caused by *Corynebacterium fascians*. *Amer. J. Bot.* 53:731–9.

Thomas, T. M., and A. L. Jones. 1992. Severity of fire blight on apple cultivars and strains in Michigan. *Plant Dis.* 76:1049–52.

Thompson, G. E. 1941. Leaf-spot diseases of poplars caused by *Septoria musiva* and *S. populicola*. *Phytopathology* 31:241–54.

Thresh, J. M. 1982. Cropping practices and virus spread. *Annu. Rev. Phytopathol.* 20:193–218.

Thurston, H. D. 1973. Threatening plant diseases. *Annu. Rev. Phytopathol.* 11:27–52.

———. 1984. *Tropical Plant Diseases*. St. Paul, MN: APS Press.

Tollenaar, D. 1959. Rubber growing in Brazil in view of the difficulties caused by South American leaf blight *(Dothidella ulei)*. *Netherlands J. Agric. Sci.* 7:173–89.

Tomlinson, G. H., II, ed. 1990. *Effects of Acid Deposition on the Forests of Europe and North America*. Boca Raton, FL: CRC Press.

Tomlinson, J. A., A. L. Carter, W. T. Dale, and C. J. Simpson. 1970. Weed plants as sources of cucumber mosaic virus. *Ann. Appl. Biol.* 66:11–16.

Tooley, P. W., C. R. Grau, and M. C. Stough. 1986. Microplot comparison of rate-reducing and race-specific resistance to *Phytophthora megasperma* f. sp. *glycinea* in soybean. *Phytopathology* 76:554–7.

Tsuchizaki, T. A., A. Sasaki, and Y. Saito. 1978. Purification of citrus tristeza virus from diseased citrus fruits and the detection of the virus in citrus tissues by fluorescent antibody techniques. *Phytopathology* 68:139–42.

Turchetti, T. 1982. Hypovirulence in chestnut blight (*Endothia parasitica*) and some practical aspects in Italy. *Eur. J. For. Pathol.* 12:414–17.

Turgeon, R., H. N. Wood, and A. C. Braun. 1976. Studies on the recovery of crown gall tumor cells. *Proc. Nat. Acad. Sci. U.S.A.* 73:3562–4.

Udagawa, H., T. Kohguchi, H. Otani, K. Kohmoto, and S. Nishimura. 1983. A decade of transition of polyoxin-tolerant strains of *Alternaria alternata* Japanese pear pathotype in the field ecosystem. *J. Fac. Agric.*, Tottori Univ., 18:9–17.

Ueno, T., T. Nakashima, Y. Hayashi, and H. Fukami. 1975a. Structures of AM-toxin I and II, host specific phytotoxic metabolites produced by *Alternaria mali*. *Agric. Biol. Chem.* 39:1115–22.

———. 1975b. Isolation and structure of AM-toxin III, a host specific phytotoxic metabolite produced by *Alternaria mali*. *Agric. Biol. Chem.* 39:2081–2.

Ullstrup, A. J. 1941. Two physiologic races of *Helminthosporium maydis* in the corn belt. *Phytopathology* 31:508–21.

———. 1972. The impacts of the southern corn leaf blight epidemics of 1970–1971. *Annu. Rev. Phytopathol.* 10:37–50.

Ulrich, B. 1985. Interaction of indirect and direct effects of air pollutants in forests. In *Air Pollution and Plants* (ed. by C. Trojanowski), pp. 149–80. Weinheim: VCH Publishers.

———. 1986. Natural and anthropogenic components of soil acidification. *Z. Pflanzenernaehr. Bodenk.* 149:702–17.

Ulrich, B., and J. Pankrath, eds. 1983. Effects of accumulation of air pollutants in the forest ecosystem. Dordrecht, Boston, and London: Reidel Publ.

Vakili, N. G. 1965. *Fusarium* wilt resistance in seedlings and mature plants of *Musa* species. *Phytopathology* 55:135–40.

Valent, B., and F. G. Chumley. 1991. Molecular genetic analysis of the rice blast fungus. *Annu. Rev. Phytopathol.* 29:443–67.

Valent, B., et al. 1994. Molecular characterization of avirulence genes from the rice blast fungus. In *Host Specific Toxin: Biosynthesis, Receptor, and Molecular Biology* (ed. by K. Kohmoto and O. C. Yoder), p. 275. Tottori, Japan: Tottori Univ.

Valleau, W. D. 1947. Can tobacco plant beds in Kentucky and Tennessee be infected by *Peronospora tabacina* blown in from Texas? *Plant Dis. Rep.* 31:480–2.

Valsangiacomo, C., and C. Gessler. 1988. Role of cuticular membrane in ontogenic and V$_f$-resistance of apple leaves against *Venturia inaequalis*. *Phytopathology* 78:1066–9.

Van Alfen, N. K., R. A. Jaynes, S. L. Anagnostakis, and P. R. Day. 1975. Chestnut blight: biological control by transmissable hypovirulence in *Endothia parasitica*. Science 189:890–1.

Van Arsdel, E. P., A. J. Riker, T. F. Kouba, V. E. Suomi, and R. A. Bryson. 1961. The climatic distribution of blister rust on white pine in Wisconsin. U.S.D.A. For. Ser., Lake States For. Exp. Sta. Paper no. 87.

Van der Plank, J. E. 1963. *Plant Diseases: Epidemics and Control*. New York: Academic Press.

1984. *Disease Resistance in Plants,* ed. 2. Orlando, FL: Academic Press.

Van der Zwet, T. 1968. Recent spread and present distribution of fire blight in the world. *Plant Dis. Rep.* 52:698–702.

Van der Zwet, T., A. R. Biggs, R. Hefelbower, and G. W. Lightner. 1994. Evaluation of the MARYBLYT computer model for predicting blossom blight on apple in West Virginia and Maryland. *Plant Dis.* 78:225–30.

Van der Zwet, T., and H. L. Keil. 1979. *Fire Blight: A Bacterial Disease of Rosaceous Plants*. U.S.D.A. Agric. Handbook no. 510. Washington, DC.

Van Dijkman, A. 1972. Natural resistance of tomato plants to *Cladosporium fulvum:* a biochemical study (Ph.D. thesis). Utrecht: Netherlands State Univ.

Van Etten, H. D., D. E. Matthews, and P. S. Matthews. 1989. Phytoalexin detoxification: importance for pathogenicity and practical implications. *Annu. Rev. Phytopathol.* 27:143–64.

Van Kan, J.A.L., M.H.A. Joosten, G.F.J.M. Van den Ackerveken, and P.J.G.M. de Wit. 1994. Molecular characterization of avirulence determinants of the tomato pathogen *Cladosporium fulvum*. In *Host-Specific Toxin-Biosynthesis, Receptor and Molecular Biology* (ed. by K. Kohmoto and O. C. Yoder). Tottori Univ., Japan.

Van Montagu, M., et al. 1980. The interaction of *Agrobacterium* Ti-plasmid DNA and plant cells. *Proc. Roy. Soc. London. B.* 210:351–65.

Van Vuuren, S. P., R. P. Collins, and J. V. da Graca. 1993. Evaluation of citrus tristeza virus isolates for cross protection of grapefruit in South Africa. *Plant Dis.* 77:24–9.

Vavilov, N. I. 1926. *Studies on the Origin of Cultivated Plants*. Leningrad: Inst. Appl. Bot. Plant Breed., Leningrad.

Vela, C., M. Cambra, E. Cortés, P. Moreno, J. G. Miguet, e. de San Román, and A. Sanz. 1986. Production and characterization of monoclonal antibodies specific for citrus tristeza virus and their use for diagnosis. *J. Gen. Virol.* 67:91–6.

Viennot-Bourgin, G. 1981. History and importance of downy mildews. In *The Downy Mildews* (ed. by D. M. Spencer), pp. 1–15. New York: Academic Press.

Vonallmen, J. M., W. H. Rottmann, B. G. Gengenbach, A. J. Harvey, and D. M. Lonsdale. 1991. Transfer of methomyl and HmT-toxin sensitivity from T-cytoplasm maize to tobacco. *Mol. Gen. Genet.* 229:405–12.

Von Schmeling, B., and M. Kulka. 1966. Systemic fungicide activity of 1,4-oxathin derivatives. *Science* 152:659–60.

Waggoner, P. E., and G. S. Taylor. 1955. Tobacco blue mold epiphytotics in the field. *Plant Dis. Rep.* 39:79–85.

Waite, B. H. 1963. Wilt of *Heliconia* spp. caused by *Fusarium oxysporum* f. *cubense* race 3. *Trop. Agric.* 40:299–305.

Waite, B. H., and V. C. Dunlop. 1953. Preliminary host range studies with *Fusarium oxysporum* f. sp. *cubense. Plant Dis. Rep.* 37:79–80.

Walker, J. C. 1953. Disease resistance in the vegetable crops II. *Bot. Rev.* 19:606–43.

  1959. Progress and problems in controlling plant diseases by host resistance. In *Plant Pathology: Problems and Progress* (ed. by C. S. Holton et al.), pp. 32–41. Madison: Univ. Wis. Press.

  1969. *Plant Pathology,* ed. 3. New York: McGraw-Hill.

Walker, J. C., and M. A. Stahmann. 1955. Chemical nature of disease resistance in plants. *Annu. Rev. Plant Physiol.* 6:351–66.

Walkinshaw, C. H. 1978. Cell necrosis and fungus content in fusiform rust-infected loblolly, longleaf, and slash pine seedlings. *Phytopathology* 68:1705–10.

Walkinshaw, C. H., and C. F. Bey. 1981. Reaction of field-resistant slash pines to selected isolates of *Cronartium quercuum* f. sp. *fusiforme. Phytopathology* 71:1090–2.

Walkinshaw, C. H., and T. A. Roland. 1990. Incidence and histology of stem-girdling galls caused by fusiform rust. *Phytopathology* 80:251–5.

Wallace, J. M., P.C.J. Oberholzer, and J.D.J. Hofmeyer. 1956. Distribution of viruses of tristeza and other diseases of citrus in propagative material. *Plant Dis. Rep.* 40:3–10.

Wallin, J. R., and D. V. Loonan. 1971. Low-level jet winds, aphid vectors, local weather and barley yellow dwarf virus outbreaks. *Phytopathology* 61:1068–70.

Walton, J. D. 1987. Two enzymes involved in biosynthesis of the host-selective phytotoxin HC-toxin. *Proc. Nat. Acad. Sci. U.S.A.* 84:8444–7.

Walton, J. D., E. D. Earle, O. C. Yoder, and R. M. Spanswick. 1979. Reduction of adenosine triphosphate levels in susceptible maize mesophyll protoplasts by *Helminthosporium maydis* race T toxin. *Plant Physiol.* 63:806–10.

Walton, J. D., and F. R. Holden. 1988. Properties of two enzymes involved in the biosynthesis of the fungal pathogenicity factor HC-toxin. *Mol. Plant Microb. Interact.* 1:128–34.

Walton, J. D., and D. G. Panaccione. 1993. Host-selective toxins and disease specificity: perspectives and progress. *Annu. Rev. Phytopathol.* 31:275–303.

Ward, H. M. 1882. Researches on the life history of *Hemileia vasatrix. J. Linn. Soc.* 19:299–335.

Warren, H. L., and R. L. Nicholson. 1975. Kernel infection, seedling blight, and wilt of maize caused by *Colletotrichum graminicola*. *Phytopathology* 65:620–3.

Warren, H. L., and P. L. Shepherd. 1976. Relationship of *Colletotrichum graminicola* to foliar and kernel infection. *Plant Dis. Rep.* 60:1084–6.

Waterworth, H. 1993. Processing foreign germ plasm at the national germ plasm quarantine center. *Plant Dis.* 77:854–60.

Waterworth, H., and G. H. White. 1982. Plant introductions and quarantine: the need for both. *Plant Dis.* 66:87–90.

Watson, B., et al. 1975. Plasmid required for virulence of *Agrobacterium tumefaciens*. *J. Bact.* 123:255–64.

Webster, R. K. 1974. *Introduction to Fungi,* ed. 2. London: Cambridge Univ. Press.

Wei, Z.-m., and S. V. Beer. 1993. *Hrp 1* of *Erwinia amylovorae* functions in secretion of harpin and is a member of a new protein family. *J. Bact.* 175:7958–67.

Wei, Z.-m., R. J. Laby, C. H. Zumoff, D. W. Bauer, S. Y. He, A. Collmer, and S. V. Beer. 1992. Harpin, elicitor of the hypersensitive response produced by the pathogen *Erwinia amylovora*. *Science* 257:85–8.

Wellman, F. L. 1952. Peligro de introducción de la *Hemileia* del café a las Américas. *Turialba* 2:47–50.

   1961. *Coffee: Botany, Cultivation, and Utilization*. New York: Interscience Publ.

   1970. Coffee yellow leaf rust: world history: minimizing losses in tropical America. *Proc. Reunión Técnica sobre las Royas del Cafeto*. San Jose, Costa Rica: Inst. Inter-Amer. Cien Agric. Org. Estad. Am.

Wellman, F. L., and E. Echandi. 1981. The coffee rust situation in Latin America in 1980. *Phytopathology* 71:968–71.

Wells, O. O., and P. C. Wakeley. 1966. Geographic variation in survival, growth, and fusiform-rust infection of planted loblolly pines. *For. Sci. Monogr. 11* (USA).

Weltzien, H. C. 1981. Geographical distribution of downy mildews. In *The Downy Mildews* (ed. by D. M. Spencer), pp. 31–43. New York: Academic Press.

Welz, H. G., and K. J. Leonard. 1993. Phenotypic variation and parasitic fitness of races of *Cochliobolus carbonum* on corn in North Carolina. *Phytopathology* 83:593–601.

White, F. F., G. Ghidossi, M. P. Gordon, and E. W. Nester. 1982. Tumor induction by *Agrobacterium rhizogenes* involves the transfer of plasmid DNA to the plant genome. *Proc. Nat. Acad. Sci. U.S.A.* 79:3193–7.

White, F. F., and E. W. Nester. 1980a. Hairy root: plasmid encodes virulence traits in *Agrobacterium rhizogenes*. *J. Bact.* 141:1134–41.

   1980b. Relationships of plasmids responsible for hairy root and crown gall tumorigenicity. *J. Bact.* 144:710–20.

Wiese, M. V., and J. E. DeVay. 1970. Growth regulator changes in cotton associated with defoliation caused by *Verticillium albo-atrum*. *Plant Physiol.* 45:304–9.

Willis, J. W., J. K. Engwall, and A. K. Chatterjee. 1987. Cloning of genes for *Erwinia carotovora* subsp. *carotovora* pectolytic enzymes and further characterization of polygalacturonases. *Phytopathology* 77:1199–1205.

Wingfield, M. J., ed. 1987. *Pathogenicity of the Pine Wood Nematode*. St. Paul, MN: APS Press.

Wise, R. P., A. E. Fliss, D. R. Pring, and B. G. Gengenbach. 1987. *Urf 13-T* of T-cytoplasm maize mitochondria encodes a 13-kd polypeptide. *Plant Mol. Biol.* 9:121–6.

Wise, R. P., D. R. Pring, and B. G. Gengenbach. 1987. Mutation to male fertility and toxin insensitivity in Texas (T) cytoplasm maize is associated with a frameshift in a mitochondrial open reading frame. *Proc. Nat. Acad. Sci. U.S.A.* 84:2858–62.

Wolf, F. A. 1947. Tobacco downy mildew, endemic to Texas and Mexico. *Phytopathology* 37:721–9.

Wolpert, T. J., and V. Macko. 1991. Immunological comparison of the *in vitro* and *in vivo* labeled victorin binding protein from susceptible oats. *Plant Physiol.* 95:917–20.

Wolpert, T. J., D. A. Navarre, D. L. Moore, and V. Macko. 1994. Identification of the 100 kd victorin binding protein from oats. *Plant Cell* 6:1145–55.

Wood, H. N., and A. C. Braun. 1961. Studies on the regulation of certain essential biosynthetic systems in normal and crown gall tumor cells. *Proc. Nat. Acad. Sci. U.S.A.* 47:1907–13.

Woodham-Smith, C. 1962. *The Great Hunger: Ireland 1845–1849*. New York: Harper & Row.

Xiao, J.-z., S. Nakatsuka, M. Tsuda, N. Doke, and S. Nishimura. 1990. Isolation and characterization of a factor from *Bipolaris zeicola* race 3 that induces susceptibility in rice plants. *Ann. Phytopathol. Soc. Japan* 56:605–12.

1991. Rice-specific toxins produced by *Bipolaris zeicola* race 3: evidence for role as pathogenicity factors for rice and maize plants. *Physiol. Mol. Plant Pathol.* 38:67–82.

Yamada, T. 1993. The role of auxin in plant disease development. *Annu. Rev. Phytopathol.* 31:253–73.

Yamamoto, M., F. Namiki, S. Nishimura, and K. Kohmoto. 1985. Studies on host-specific AF-toxins produced by *Alternaria alternata* strawberry pathotype. 3. Use of toxin for determining inheritance of disease reaction in strawberry cultivar Morioka-16. *Ann. Phytopathol. Soc. Japan* 51:530–35.

Yamamoto, M., S. Nishimura, K. Kohmoto, and H. Otani. 1984. Studies on host-specific AF-toxins produced by *Alternaria alternata* strawberry pathotype. 2. Role of toxins in pathogenesis. *Ann. Phytopathol. Soc. Japan* 50:610–19.

Yoder, O. C., and R. P. Scheffer. 1973a. Effects of *Helminthosporium carbonum* toxin on nitrate uptake and reduction by corn tissues. *Plant Physiol.* 52:513–17.

1973b. Effects of *Helminthosporium carbonum* toxin on absorption of solutes by corn roots. *Plant Physiol.* 52:518–23.

Yoder, O. C., and B. G. Turgeon. 1994. Molecular determinants of the plant/fungus interaction. In *Host-Specific Toxin: Biosynthesis, Receptor, and Molecular Biology* (ed. by K. Kohmoto and O. C. Yoder), pp. 23–32. Tottori Univ., Japan.

Yokomi, R. K., S. M. Garnsey, E. L. Civerolo, and D. J. Gumpf. 1989. Transmission of exotic citrus tristeza virus isolates by a Florida colony of *Aphis gossypii*. *Plant Dis.* 73:552–6.

Zadoks, J. C., and R. D. Schein. 1979. *Epidemiology and Plant Disease Management*. New York and Oxford: Oxford Univ. Press.

Zaenen, I., N. Van Larebeke, H. Teuchy, M. Van Montagu, and J. Schell. 1974. Supercoiled circular DNA in crown gall inducing *Agrobacterium* strains. *J. Mol. Biol.* 86:109–27.

Ziller, W. G. 1965. Studies on western tree rusts. VI. The aecial host ranges of *Melampsora albertensis, M. medusae,* and *M. occidentalis. Can. J. Bot.* 43:217–30.

Index

abiotic plant diseases, 9, 251–63
abscisic acid, in plant disease, 31
acid rain, vegetation damage from, 252, 254, 258, 260
ACT-toxins, 202, 205
Adirondack Mountains, forest dieback in, 262
aecium, 11, 13
AF-toxins, 201, 202, 205
agriculture, basic premises of, 1
Agrobacterium, 9
Agrobacterium radiobacter, 68
Agrobacterium rhizogenes, 29, 30
Agrobacterium rubi, 30
Agrobacterium tumefaciens, 20, 21, 23, 25, 26, 27, 30; control of, 68
Agrocin 84, from *Agrobacterium*, 23, 68
agropine, 26, 270
air pollution, role in plant diseases, 9, 251–63
AK-toxins, 196, 197, 199, 202, 205
alfalfa, dwarf disease of, 101
algae, 11; leaf necrosis from, 9
almond, 233
Alternaria, 15, 16, 47, 48, 49, 53, 54, 115, 178; control of, 62; diseases with toxins from, 196–209; epidemics of, 176; pathogenicity of, 193, 196
Alternaria alternata, 49, 50, 55, 199, 202, 203, 205, 207, 208, 209; pathotypes of, 196
Alternaria alternata f. sp. *citri*, 49, 50, 203, 204, 205
Alternaria alternata f. sp. *kikuchiana*, 50, 196, 197
Alternaria alternata f. sp. *lycopersici*, 50, 206

Alternaria alternata f. sp. *mali*, 50, 200, 201
Alternaria alternata f. sp. *tenuis*, 17, 208
Alternaria alternata f. sp. *tenuissima*, 49, 207
Alternaria blotch disease, 200–1
Alternaria brassicae, 50, 207
Alternaria eichhorniae, 208
Alternaria longipes, 50, 207
Alternaria solani, 208
alternate host: definition of, 267
alternative host, 267
AL-toxin, 206
aluminum toxicity, to plants, 258, 263
Amelanchier, 103
amino acids, novel, in crown gall tissues, 26
amino peptidase, Zn-containing, 17
amoebae, response to bacteria, 45
AM-toxin, 196, 200–1
amylase, of plant pathogens, 18
Anabaena, nitrogen fixation by, 172
Anaheim disease. *See* Pierce's disease
anamorph: definition of, 267
anastomosis: definition of, 267
Andes Mountains, potato origin in, 124
angular leafspot disease, 16
anthracnose, 233; of beans, 56, 153; of maize, 13, 232; minimum tillage encouragement of, 232
anthropogenic, definition, 267
anthropogenic factors, in plant-disease epidemics, 3–4, 42
anthropogenic reintroduction, of plant diseases, 237–50
antibiotics, 62, 63, 109
antigibberellins, commercial use of, 32
Antirrhinum majus. See snapdragon

313